An Introduction to Random Sets

An Introduction to Random Sets

Hung T. Nguyen

New Mexico State University
Las Cruces, New Mexico, U.S.A.

CRC Press
Taylor & Francis Group
Boca Raton London New York

CRC Press is an imprint of the
Taylor & Francis Group, an **informa** business

A CHAPMAN & HALL BOOK

CRC Press
Taylor & Francis Group
6000 Broken Sound Parkway NW, Suite 300
Boca Raton, FL 33487-2742

First issued in paperback 2019

© 2006 by Taylor & Francis Group, LLC
CRC Press is an imprint of Taylor & Francis Group, an Informa business

No claim to original U.S. Government works

ISBN-13: 978-1-58488-519-1 (hbk)
ISBN-13: 978-0-367-39099-0 (pbk)

Library of Congress Cataloging-in-Publication Data

Catalog record is available from the Library of Congress

Visit the Taylor & Francis Web site at
http://www.taylorandfrancis.com

and the CRC Press Web site at
http://www.crcpress.com

Contents

Preface

This text is designed for a one-semester course at the advanced undergraduate or beginning graduate level. It is also intended for use as a reference book for researchers in fields such as probability and statistics, artificial intelligence, computer science, engineering. It is a friendly but solid introduction to the topic of random sets for those who need a strong background for further study. After completing the course, the students should be able to read more specialized and advanced books on the subject as well as articles in technical and professional journals.

The material presented in this text is drawn from many sources in the literature, including our own research. The presentation of the material is *from the ground up*. The prerequisite consists simply of a good upper-level undergraduate course in probability and statistics. A summary of concepts and results in probability theory is given in the Appendix.

The theory of random sets is viewed as a natural generalization of probability and statistics on random vectors, i.e., of multivariate statistical analysis. Random set data can be also viewed as imprecise/incomplete observations which are frequent in today's technological societies. As models for set-valued observations as well as for the process underlying the gathering of perception-based information, via coarsening schemes, random sets are a new type of data. As such, new mathematical tools for statistical inference and decision making need to be developed. In the foreword to Mathéron's book on random sets [73], G. Watson expressed his vision of statistics as follows: "Modern statistics must be defined as the applications of computers and mathematics to data analysis. It must grow as new types of data are considered and as computing technology advances."

We would like to express our thanks to all participants of the statistics seminar at New Mexico State University, 2002–2004, for their discussions on statistics of random sets and especially for their insistence that the topic of random sets should be included in a first course in probability.

We thank the Department of Mathematics and Statistics of Bowling Green State University, Ohio, for providing an excellent environment during our stay as visiting distinguished Lukacs professor, spring 2002, where parts of the advanced topics on random sets in this text grew out of our lectures given there.

Thanks are especially due to Vladik Kreinovich, De-Jun Feng, Ding Feng, and Elbert Walker for reading the manuscript and offering many penetrating comments and suggestions.

Hung T. Nguyen
Las Cruces, New Mexico

About the Author

Hung T. Nguyen is Professor of Mathematical
Sciences at New Mexico State University, Las
Cruces, New Mexico, U.S.A.

Chapter 1

Generalities on Probability

This chapter is a short review of the basics of probability that are needed to discuss random sets in subsequent chapters. For more details, see the Appendix.

1.1 Survey Sampling Revisited

Gathering information for decision making is frequent and essential in human activities. Rather than being complete, information is in general uncertain in many respects. Throughout this course we will encounter different types of uncertainty, but first, let us start out with a familiar type of uncertainty, namely *randomness*. After all, the analysis of randomness will serve as a guideline for studying other types of uncertainty.

Below is a simple example of information gathering.

Suppose that we are interested in the annual income of individuals in the population of Las Cruces, say, in 2004. Suppose that, for some reasons (e.g., costs and time), we are unable to conduct a census (i.e., a complete enumeration) throughout the entire population, and hence we can only rely on the information obtained from a small part of that population, i.e., from a *sample* of that population. Suppose that the physical population of individuals is identified as a *finite* set $U = \{u_1, \ldots, u_N\}$, where N is the population size. A sample in U is a *subset* of U. Our *variable* of interest is θ, the annual income. We use $\theta(u_k)$ to denote the annual income of the individual u_k. Thus, θ is a map from U to $\mathbb{R}$, the set of real numbers. To obtain partial knowledge about θ, we are going to conduct a *sampling survey*, i.e., to select a sample A from U. Then, from the knowledge of the map θ on A, i.e., the values $\theta(u)$, $u \in A$, we wish to "guess" or estimate θ, or some function of it, e.g., the *population total*

$$\mathcal{T}(\theta) = \sum_{u \in U} \theta(u)$$

by $\sum_{u \in A} \theta(u)$.

This is *inductive logic:* making statements about the whole population from the knowledge of a part of it. Then the basic question is: *How* to make this inductive logic valid? For example, for some chosen subset A of U, how do we know that $\sum_{u \in A} \theta(u)$ will be a "good" guess for $\mathcal{T}(\theta)$? Can we specify the error in our estimation process? Questions such as these are clearly related to the quality of the data we collected, e.g., are our data, gathered on a selected sample, *representative* or *typical* for the whole population? It all boils down to "how to select a good sample?" It seems that, to eliminate the bias in the selection of samples, and to gain public acceptance (with regards to objectivity), we could select samples *at random*, that is, we could introduce a *manmade randomization* into our process of samples selection in the hope of making our intended inductive logic valid. In other words, we should create a *chance model*.

For example, to select at random a subset of U of given size n, we can create a uniform chance model with probability $\dfrac{1}{\binom{N}{n}}$, where $\binom{N}{n} = \dfrac{N!}{n!(N-n)!}$,

i.e., each subset of U of size n has the same chance of $\dfrac{1}{\binom{N}{n}}$ to be selected.

Now, we are somewhat familiar with *games of chance*, such as tossing coins, rolling dice. These gambling devices have known structures, but their *outcomes* remain uncertain before playing. Our manmade randomization in sampling survey is a game of chance, called a *probability sampling plan*. As we will see by doing so, we will obtain more than just getting a "good" data set, namely, we will be able to assess the qualities of our estimation procedure.

Observe that when selecting samples according to a probability sampling plan, we actually perform a *random experiment* whose outcomes are *subsets* of a set U. Since subsets of U are samples, we call the set of all outcomes of a random experiment its *sample space*. An outcome of a random experiment (i.e., an experiment whose outcomes cannot be predicted with certainty in advance) is obtained at random. Thus, in survey sampling, a sample is a set obtained at random. Sets obtained at random are called *random sets*.

To carry our inductive logic, we need a body of modeling concepts and techniques, i.e., a *science of statistics*. Statistics is a science for making inference from samples to populations. Starting with providing useful information for states (hence the name statistics), the framework and methodology of statistical science spread out to almost all fields of our societies.

The science of statistics consists of using probability theory to arrive at valid inductive logic. It is so mainly because inference cannot be absolutely certain, and hence we need the *language of probability* to formulate results.

1.2   Mathematical Models for Random Phenomena

Starting with *classical probability theory*, i.e., modeling and analyzing of games of chance, we quickly realize that *random objects* encountered in real-world applications are much more general. To cover all forms of possible random objects, an *abstract theory* is needed.

The following is a mathematical model for general random experiments, and, by analogy principle, for random phenomena in nature.

To motivate the abstract models for general random phenomena, let us look at a game of chance.

Consider the game of rolling a pair of dice. Suppose you are interested in betting on the *event* that the sum of two numbers shown will be 7. A little bit of analysis will reveal the following. First, we can list all possible outcomes of the game, i.e., we can specify its *sampling space*

$$\Omega = \{(i,j) : i,j = 1, 2, \ldots, 6\}.$$

Next, what you are interested in is a subset A of Ω, i.e., an *event*, namely,

$$A = \{(i,j) : i + j = 7\}.$$

More specifically, you are interested in the chance of A to occur. You might be interested in the chances of other events as well, i.e., for any $B \subseteq \Omega$, can you specify its chance, called *probability*, and denoted by $P(B)$? In other words, besides Ω and its *power set* 2^Ω (the set of all subsets of Ω), representing all possible events of the game, you need a map $P : 2^\Omega \to [0,1]$ which assigns, to each $B \subseteq U$, its probability value $P(B)$.

Note that we denote by 2^Ω its power set, by identifying each subset B of Ω with its *indicator function* $I_B : \Omega \to \{0,1\}$,

$$I_B(\omega) = \begin{cases} 1 \text{ if } \omega \in B \\ 0 \text{ if } \omega \notin B \end{cases},$$

i.e., 2^Ω is $\{\varphi : \Omega \to \{0,1\}\}$, which is also denoted as $\{0,1\}^\Omega$. We use 2 to abbreviate the set $\{0,1\}$, which has 2 elements.

Although it is trivial in our game of dice, we take 2^Ω as the domain for our probability map P, since we can assign $P(B)$ to any $B \in 2^\Omega$, namely

$$P(B) = \frac{\#(B)}{\#(\Omega)},$$

where $\#(B)$ or $|B|$ denotes the cardinality (the number of elements) of B.

From this definition of P, we see that P satisfies two basic properties, namely,

i) $P(\Omega) = 1$

ii) If A and B are disjoint, i.e., $A \cap B = \emptyset$ (the intersection is empty), then $P(A \cup B) = P(A) + P(B)$, where $\cup$ denotes *union* of sets. The is referred to as (finite) *additivity* of P.

Also, 2^Ω has the following basic properties:

a) $\Omega \in 2^\Omega$.

b) If $A \in 2^\Omega$ then its *set-complement* $A^c \in 2^\Omega$, where

$$A^c = \{\omega : \omega \in \Omega, \omega \notin A\}.$$

c) If $A, B \in 2^\Omega$ then $A \cup B \in 2^\Omega$.

Classes of subsets of Ω satisfying a), b), and c) are called *fields* or *algebras* (of subsets) of Ω.

Thus, for *finite games* (i.e., games that have finite sample spaces), the triple $(\Omega, 2^\Omega, P)$ above is a complete description of the random evolution of the game.

The situation is more subtle for infinite games or for random experiments that have an infinite number of outcomes, countable or uncountable. For example, consider the random experiment of picking at random a number in the interval $[0, 1)$. The sample space of this experiment is $\Omega = [0, 1)$. However, it is not clear how to assign probabilities for $A \subseteq \Omega$. This is so because, while it is intuitive to assign $P(A)$ as the "length" of the set A, it is not clear what is the "length" of an *arbitrary subset* A of Ω. The length of an interval $A = [a, b]$ is of course $b - a$, and the length of a subset B, which is a countable union of pairwise disjoint intervals $[a_n, b_n]$ is $\sum_{n \geq 1} (b_n - a_n)$. The question is: what is the largest class of subsets of $[0, 1)$ for which their lengths can be defined? It turns out that this class $\mathcal{A}$ is different from 2^Ω. Also, in view of the infinite countable summation above, we need to strengthen the basic properties for both the class of events $\mathcal{A}$ and the mapping P defined on $\mathcal{A}$. It should be clear from all of the above that a general mathematical model for an arbitrary random experiment is the following.

DEFINITION 1.1   *A mathematical model for a random experiment is a probability space $(\Omega, \mathcal{A}, P)$, where:*

α) Ω is a set, representing the sample space of the experiment,

β) $\mathcal{A}$ is a σ-field (representing events), i.e.:

i) $\Omega \in \mathcal{A}$

ii) If $A \in \mathcal{A}$, then $A^c \in \mathcal{A}$

iii) If $A_n \in \mathcal{A}$ for $n \geq 1$, then $\bigcup_{n \geq 1} A_n \in \mathcal{A}$.

The pair $(\Omega, \mathcal{A})$ is called a measurable space.

$\gamma)$ $P : \mathcal{A} \to [0,1]$ *is a probability measure, i.e.:*

a) $P(\Omega) = 1$

b) *If $\{A_n, n \geq 1\}$ is a sequence (finite or infinitely countable) of pairwise disjoint (i.e., $A_n \cap A_m = \emptyset$ for $n \neq m$) elements of $\mathcal{A}$, then $P(\bigcup_{n\geq 1} A_n) = \sum_{n\geq 1} P(A_n)$, a property that is referred to as σ-additivity of P.*

For example, in sampling from a finite population U with $\#(U) = N$, if we decide to select samples of given size n, then $\Omega = 2^U$, $\mathcal{A} =$ power set of 2^U, and $P : \mathcal{A} \to [0,1]$ is defined as $P(\mathbb{A}) = \sum_{A \in \mathbb{A}} P(A)$, $\mathbb{A} \subseteq 2^U$, where for $A \in 2^U$,

$$P(A) = \begin{cases} \dfrac{1}{\binom{N}{n}} & \text{if } \#(A) = n \\ 0 & \text{if } \#(A) \neq n \end{cases}.$$

1.3 Random Elements

In our previous example of games of chance, we perform the random experiment of rolling two dice, but we are interested in the sum of two numbers shown. In other words, we are interested in some *variable* associated with the experiment. A *variable* is the name of a quantity of interest, here "sum." Although we know the *range* of this variable, namely, the set of all its possible values $\mathcal{R}(X) = \{2,3,\ldots,12\}$, the concrete value taken by X depends on the outcomes of the random experiment. Thus, on the one hand, X is a *random variable*, and on the other hand, the variable X is in fact a *map* from the sample space of the experiment $\Omega = \{(i,j) : i,j = 1,2,\ldots,6\}$ to the real line $\mathbb{R}$.

When we are interested in quantities (variables) from natural random phenomena, we can view, by analogy with games of chance, variables as random variables with similar structures. Now probabilities of interest are in fact in the range $\mathbb{R}$ of a variable X, e.g., $P(X \in A)$ for $A \subseteq \mathbb{R}$, where $(X \in A)$ denotes the subset $\{\omega : X(\omega) \in A\}$ of Ω. This subset is also written as $X^{-1}(A)$. For $P(X \in A)$ to be well defined, $X^{-1}(A)$ has to be in the domain of P, i.e., $X^{-1}(A) \in \mathcal{A}$.

For random phenomena, the probability space $(\Omega, \mathcal{A}, P)$ is rather abstract, or just be taken to be, say, $\mathbb{R}$ together with some σ-field $\mathcal{B}$ on $\mathbb{R}$. The standard σ-field on $\mathbb{R}$ is the *Borel σ-field* $\mathcal{B}(\mathbb{R})$ generated by the open sets of $\mathbb{R}$, i.e., the smallest σ-field containing all open sets of $\mathbb{R}$. The probability measure on

$\mathbb{R}$, which governs the random evolution of X, is $P_X = PX^{-1} : \mathcal{B}(\mathbb{R}) \to [0,1]$, where for $A \in \mathcal{B}(\mathbb{R})$, $P_X(A) = P(X \in A) = P(X^{-1}(A))$. P_X is also referred to as the *probability law* of X. For the above to be well defined, the variable X, as a mapping from X to $\mathbb{R}$, should possess some appropriate properties.

DEFINITION 1.2   *A random variable X is a map from a set Ω (in fact, by abuse of language, defined on a probability space $(\Omega, \mathcal{A}, P)$) to $\mathbb{R}$ such that $X^{-1}(\mathcal{B}(\mathbb{R})) \subseteq \mathcal{A}$, i.e., $\forall B \in \mathcal{B}(\mathbb{R}))$, $X^{-1}(B) \in \mathcal{A}$, in other words, X is a $\mathcal{A}$-$\mathcal{B}(\mathbb{R})$-measurable map.*

The above framework can be made general to cover all possible forms of random "quantities."

DEFINITION 1.3   *Let $(\Omega, \mathcal{A}, P)$ be a probability space and $(U, \mathcal{U})$ be a measurable space. By a* random element, *we mean a map $X : \Omega \to U$, which is $\mathcal{A}$-$\mathcal{U}$-measurable. The probability law of X is the probability measure on $\mathcal{U}$ defined by $P_X = PX^{-1}$.*

Examples.

i) $U = 2^V$ for some finite set V, $\mathcal{U}$ power set of U.

Let $X : \Omega \to U$ be a map such that $\forall A \subseteq V, \{\omega : X(\omega) = A\} \in \mathcal{A}$. Then X is a *finite random set*.

ii) $U = \mathbb{R}^d, \mathcal{U} = \mathcal{B}(\mathbb{R}^d)$, the Borel σ-field of $\mathbb{R}^d$.

Let $X : \Omega \to \mathbb{R}^d$ be $\mathcal{A}$-$\mathcal{B}(\mathbb{R}^d)$-measurable. Then X is a *random vector*.

iii) Let $U = C([0,1])$, the space of continuous functions defined on $[0,1]$ with the topology generated by the metric

$$\rho(f.g) = \sup_{0 \le x \le 1} |f(x) - g(x)|,$$

$\mathcal{U}$ is the Borel σ-field on $C([0,1])$.

Let $X : \Omega \to C([0,1])$ be $\mathcal{A}$-$\mathcal{U}$-measurable. Then X is a *random function*.

1.4   Distribution Functions of Random Variables

The complete information about a random variable

$$X : (\Omega, \mathcal{A}, P) \to (\mathbb{R}, \mathcal{B}(\mathbb{R}))$$

is its law $P_X = PX^{-1}$ on $\mathcal{B}(\mathbb{R})$. From a practical point of view, it is a probability measure on the measurable space $(\mathbb{R}, \mathcal{B}(\mathbb{R}))$. Thus, specifying probability measure on $\mathcal{B}(\mathbb{R})$ is at the heart of the analysis. Now, a probability measure Q on $\mathcal{B}(\mathbb{R})$ is a set function with a huge domain $\mathcal{B}(\mathbb{R})$. How to specify it?

A probability measure Q on a *finite* (or countable) set U is determined by its values on singletons of U, i.e., $Q(\{u\})$, $u \in U$. Indeed, by additivity (or σ-additivity) of Q, we have $Q(A) = \sum_{u \in A} Q(\{u\})$, $\forall A \subseteq U$. The function $f : U \to \mathbb{R}^+$, $f(u) = Q(\{u\})$, is called the *probability density* (or *mass*) *function* associated with Q. Thus the specification of Q in this case reduces to a much simpler level, namely, from 2^U to U.

There is an equivalent way to specify Q in the discrete case. Suppose U is a *discrete set* (finite or infinitely countable), which is *a subset of* $\mathbb{R}$. The cumulative distribution function, or simply the *distribution function* associated with Q is $F : \mathbb{R} \to [0, 1]$ defined by

$$F(x) = Q((-\infty, x]).$$

Then there is a one-to-one relationship between f and F. Indeed, if f is given, then F is obtained from f via

$$F(x) = \sum_{y \leq x} f(y),$$

with convention that $f(y) = 0$ for $y \notin U$. This is so because

$$Q((-\infty, x]) = \sum_{y \leq x, y \in U} f(y) + Q((-\infty, x] \setminus U),$$

where $A \setminus B = A \cap B^c$ and $Q((-\infty, x] \setminus U) = 0$.

Conversely, if F is given, then we recover f via

$$f(x) = F(x^+) - F(x^-),$$

where $F(x^+), F(x^-)$ denote right and left limits at x, i.e.,

$$F(x^+) = \lim_{y \searrow x} F(y), \quad F(x^-) = \lim_{y \nearrow x} F(y),$$

noting that in the *discrete case*, F is a jump function (specifically, F is right continuous but not continuous on $\mathbb{R}$). The above follows from basic properties of the probability measure Q.

The above equivalence between f and F is interesting. It does not mean simply that we can use either f or F to specify Q, but rather it suggests also what to use in the *nondiscrete* case, i.e., for more general cases. For example, when F is continuous on $U = \mathbb{R}$, the function $f(x) = F(x^+) - F(x^-)$ is identically zero on $\mathbb{R}$, and this cannot be used to specify Q. In other words, there is no longer equivalence between f and F in general cases, and the distribution function F is more fundamental (than f) in *characterizing* Q (i.e., for specifying Q uniquely). Consider the random experiment of picking a number at random in the interval $U = [0,1] = \{x \in \mathbb{R} : 0 \le x < 1\}$. While it is intuitive that each point $x \in [0,1)$ is possible to be picked, and all points are "equally likely," the probability that any x is picked is zero! Indeed, let $f(x) = Q(\text{getting } x) = \alpha$, $\forall x \in [0,1)$. If $\alpha \ne 0$, then we can choose n numbers $x_1, \ldots, x_n$ in $[0,1)$ with $n > [1/\alpha]$, the integer part of $1/\alpha$ (i.e., the largest integer less than or equal to $1/\alpha$), we have that

$$Q(\{x_1, \ldots, x_n\}) = \sum_{i=1}^{n} f(x_i) = n\alpha > 1.$$

Thus α must be zero. Thus, to specify Q, we should not assign nonzero probabilities on singletons, but on more general subsets of $[0,1)$, or of $\mathbb{R}$. On the other hand, it is intuitive that uniformity means that if $A = [a,b] \subseteq [0,1)$, then $Q([a,b]) = b - a$. We can in fact assign Q on a larger class of subsets of $[0,1)$, namely, the class $\mathcal{S}$ of subsets, which are finite unions of disjoint intervals (which forms a field of subsets of $[0,1)$), by

$$Q\left(\bigcup_{i=1}^{m}[a_i, b_i]\right) = \sum_{i=1}^{m}(b_i - a_i).$$

Thus, the problem of specification (or construction) of a probability measure Q on $[0,1)$, say, is this. We specify a class $\mathcal{S}$ of subsets of $[0,1)$ (which is a field but not a σ-field) on which we define Q which is σ-additive. But the domain of Q should be a σ-field. Thus, we would like to *extend* Q further. It turns out that we can extend Q *uniquely* to a probability measure on the σ-field generated by $\mathcal{S}$ denoted $\sigma(\mathcal{S})$ (i.e., the smallest σ-field containing $\mathcal{S}$). It turns out that $\sigma(\mathcal{S})$ is the Borel σ-field of $[0,1)$.

Thus, in general, the question of interest is this. How to construct probability measures Q on $\mathcal{B}(\mathbb{R})$? That is, how to suggest models from empirical observations? It turns out that the key is in the concept of distribution functions.

Let P be a probability on $\mathcal{B}(\mathbb{R})$. The distribution function of Q is

$$F : \mathbb{R} \to [0,1], \quad F(x) = Q((-\infty, x]).$$

This function F satisfies the following basic properties:

(i) F is monotone nondecreasing, i.e., if $x \le y$ then $F(x) \le F(y)$,

(ii) $\lim\limits_{x \nearrow +\infty} F(x) = 1$, $\lim\limits_{x \searrow -\infty} F(x) = 0$,

(iii) F is right continuous on $\mathbb{R}$, i.e., for each $x \in \mathbb{R}$, $F(x) = \lim\limits_{y \searrow x} = F(x^+)$.

These properties are consequences of the properties of Q.

The upshot is this. All functions on $\mathbb{R}$ satisfying (i), (ii), and (iii) above are distribution functions of probability measures on $\mathcal{B}(\mathbb{R})$. In other words, there is a bijection between functions satisfying (i), (ii), and (iii) above and probability measures on $\mathcal{B}(\mathbb{R})$; see a text like Nguyen and Wang [90]. Thus, instead of specifying a probability measure Q on $\mathcal{B}(\mathbb{R})$, we simply need to specify a *function* F on $\mathbb{R}$ with the above properties. When F is specified, the unique probability measure Q on $\mathcal{B}(\mathbb{R})$ such that $F(x) = Q((-\infty, x])$, $\forall x \in \mathbb{R}$, is often denoted as dF.

In summary, for $X : (\Omega, \mathcal{A}, P) \rightarrow (\mathbb{R}, \mathcal{B}(\mathbb{R}))$, we have $P_X = PX^{-1} = dF$ on $\mathcal{B}(\mathbb{R})$ where

$$F : \mathbb{R} \rightarrow [0, 1], \quad F(x) = P(X \leq x),$$

i.e., the distribution function F characterizes the probability law of the random variable X.

How about the counterpart of *probability density* (mass) *function* in the discrete case?

As mentioned earlier, if F is continuous, the probability measure dF on $\mathcal{B}(\mathbb{R})$ does not charge any singleton sets, i.e., $dF(\{x\}) = 0$, $\forall x \in \mathbb{R}$. Such probability measures are said to be *nonatomic*. On the other hand, being monotone, any distribution function is *differentiable* almost everywhere (a.e.). Let $f = \dfrac{dF}{dx}$ be the a.e. derivative of F. The function f will be useful if F can be recovered from it, i.e., $\forall x \in \mathbb{R}$,

$$F(x) = \int_{-\infty}^{x} f(y)\, dy. \tag{1.1}$$

If such a situation holds, then $f(x) = \dfrac{dF}{dx}$ will play the role of *probability density function*, although $f(x) \neq P(X = x) = Q(\{x\})$. Note that, as in the discrete case, we have

(i) $f \geq 0$,

(ii) $\int_{-\infty}^{+\infty} f(y)\, dy = 1$.

But the above is a familiar situation in calculus! Indeed, the answer is in the *Fundamental Theorem of Calculus*, namely, (1.1) holds if and only F is *absolutely continuous*, i.e., $\forall \varepsilon > 0$, $\exists \delta > 0$ such that $\sum_{j=1}^{n} |F(b_j) - F(a_j)| < \varepsilon$ for any nonoverlapping open intervals (a_j, b_j), $j = 1, 2, \ldots, n$, $n \geq 1$, with $\sum_{j=1}^{n} (b_j - a_j) < \delta$.

For this reason, the random variable X is said to be of *absolutely continuous type* when its distribution function F is absolutely continuous, i.e., when F

has a probability density function f. From this, we see that *not all* random variables with range $\mathbb{R}$ have density functions, since F might be continuous but not absolutely continuous.

1.5   Distribution Functions of Random Vectors

In multivariate statistical analysis, the sample space of observations is the cartesian product space $\mathbb{R}^d$, d integer ≥ 1. Few facts about $\mathbb{R}^d$ are needed.

A partial order $\leq$ on $\mathbb{R}^d$ is defined as follows: For $x = (x_1, x_2, \ldots, x_d)$, $y = (y_1, \ldots, y_d)$ in $\mathbb{R}^d$, we write $x \leq y$ to mean $x_i \leq y_i$, $\forall i = 1, 2, \ldots, d$.

We write $y \nearrow x$ (resp. $y \searrow x$) to mean $y_i \nearrow x_i$ (resp. $y_i \searrow x_i$), for all $i = 1, 2, \ldots, d$.

When $x \leq y$, we denote by $(x, y]$ the "interval" (or d-dimensional rectangle),

$$\prod_{i=1}^{d}(x_i, y_i] = \{z = (z_1, z_2, \ldots, z_d) : x_i < z_i \leq y_i, i = 1, 2, \ldots, d\}.$$

For example, $(-\infty, x] = \prod_{i=1}^{d}(-\infty, x_i]$.

Like $\mathbb{R}$, $\mathbb{R}^d$ is a metric space (with euclidean distance). Its Borel σ-field is generated also by d-dimensional rectangles of the form $(x, y]$, $x, y \in \mathbb{R}^d$.

Similar to $\mathbb{R}$, the characterization of probability measures on $\mathcal{B}(\mathbb{R}^d)$ is done via the concept of distribution functions (the Lebesgue-Stieltjes theorem). Let $X : (\Omega, \mathcal{A}, P) \to (\mathbb{R}^d, \mathcal{B}(\mathbb{R}^d))$. The distribution function F of X is

$$F : \mathbb{R}^d \to [0, 1], \quad F(x) = P(X \leq x) = PX^{-1}((-\infty, x]).$$

From basic properties of P, F satisfies the following:

(i) $\lim_{x_j \to -\infty} F(x_1, \ldots, x_d) = 0$ for at least one j, and

$$\lim_{x_j \to +\infty} F(x_1, \ldots, x_d) = 1 \text{ for all } j = 1, 2, \ldots, d.$$

(ii) F is right continuous on $\mathbb{R}^d$, i.e., $\lim_{y \searrow x} F(y) = F(x)$, $\forall x \in \mathbb{R}^d$.

(iii) For any $(a, b] \subseteq \mathbb{R}^d$,

$$\Delta_{a,b}(F) \geq 0,$$

where $\Delta_{a,b}(F) = \sum_{x} s(x) \cdot F(x)$, the summation is over the 2^d vertices x of $(a, b]$, and $s(x)$ (in fact, $s_{a,b}(x)$) is $+1$ or -1 according to whether the number of i satisfying $x_i = a_i$ is even or odd.

Remark. For $d = 2$, $a = (a_1, a_2) \leq b = (b_1, b_2)$, the four vertices of $(a_1, b_1] \times (a_2, b_2]$ are (a_1, a_2), (a_1, b_2), (b_1, a_2), and (b_1, b_2), and

$$s(a_1, b_1) = s(b_1, b_2) = +1, \quad s(a_1, b_2) = s(b_1, a_2) = -1,$$

so that

$$\Delta_{a,b}(F) = F(b_1, b_2) - F(a_1, b_2) - F(b_1, a_2) + F(a_1, a_2) = P((a, b]).$$

The property (iii) is stronger than the monotonicity of F (i.e., $x \leq y \Rightarrow F(x) \leq F(y)$), and is referred to as *monotonicity of infinite order*.

Again, the upshot is that each function $F : \mathbb{R}^d \to [0, 1]$ satisfying the above (i), (ii), and (iii) is the distribution function of a unique probability measure Q on $\mathcal{B}(\mathbb{R}^d)$, i.e.,

$$F(x) = Q((-\infty, x]), \forall x \in \mathbb{R}^d.$$

We write $Q = dF$.

Example. Let $F_1, F_2, \dots, F_d$ be d distribution functions on $\mathbb{R}$. Then

$$F(x_1, x_2, \dots, x_d) = \prod_{i=1}^{d} F_i(x_i)$$

is a distribution function on $\mathbb{R}^d$. This F is of the form

$$F(x_1, x_2, \dots, x_d) = C(F_1(x_1), F_2(x_2), \dots, F_d(x_d)),$$

where $C : [0, 1]^d \to [0, 1]$, $C(t_1, t_2, \dots, t_d) = \prod_{i=1}^{d} t_i$. This fact is general. The above function C is a special *copula*. Copulas (or d-copulas) are distribution functions F on $\mathbb{R}^d$ whose one-dimensional marginal distributions F_i are uniform on $(0, 1)$, where

$$F_i(x_i) = F(\infty, \dots, \infty, x_i, \infty, \dots, \infty), i = 1, 2, \dots, d.$$

The concept of copulas (which is a Latin word meaning a link) appeared in probability literature since the work of Sklar in 1959 (see, e.g., [80]) related to Frechet's problem on distributions with given marginals. They are functions that join multivariate distribution functions to their marginals. Specifically,

DEFINITION 1.4   *A d-copula is a function* $C : [0,1]^d \to [0,1]$ *such that*

(i) $\forall x_j \in [0,1]$, $C(1,\ldots,1,x_j,1,\ldots,1) = x_j$.

(ii) *For any* $[a,b] = \prod_{i=1}^{d} [a_i,b_i] \subseteq [0,1]^d$, $\Delta_{a,b}(C) \geq 0$.

(iii) C *is grounded, i.e.,* $C(x_1,\ldots,x_d) = 0$ *for all* $(x_1,\ldots,x_d) \in [0,1]^d$ *such that* $x_i = 0$ *for at least one* x_i.

Examples of copulas are xy, $\min(x,y)$, $\max(0, x+y-1)$.

The *Sklar theorem* (for $d = 2$) is this. Let F be a bivariate distribution function whose marginal distribution functions are H and G. Then there exists a 2-copula C such that

$$F(x,y) = C(H(x),G(y)), \forall(x,y) \in \mathbb{R}^2.$$

Conversely, if C is a 2-copula, and H, G are one-dimensional distribution functions, then $C(H(x),G(y))$ on $\mathbb{R}^2$ is a 2-dimensional distribution function, with marginal distribution functions H and G.

1.6   Exercises

1.1 Let U be a finite population of size N. Let $1 < n < N$ be a fixed sample size. Consider the sampling plan for selecting subsets of size n from U with probability $\dfrac{1}{\binom{N}{n}}$.

Show that this sampling plan can be implemented sequentially, i.e., by a series of draws as follows. With probability $\dfrac{1}{N}$, select one element from U in the first draw; select the second element in the second draw with probability $\dfrac{1}{N-1}$ from the remaining $N-1$ elements of U, and so one, until the nth draw.

1.2 (Poincaré's equalities). Let $(\Omega, \mathcal{A}, P)$ be a probability space. Show that, for any $n \geq 1$, and $A_1, A_2, \ldots, A_n \in \mathcal{A}$,

(i) $P\left(\bigcup_{i=1}^{n} A_i\right) = \sum_{\emptyset \neq I \subseteq \{1,2,\ldots,n\}} (-1)^{\#(I)+1} P\left(\bigcap_{i \in I} A_i\right)$

(ii) $P\left(\bigcap_{i=1}^{n} A_i\right) = \sum_{\emptyset \neq I \subseteq \{1,2,\ldots,n\}} (-1)^{\#(I)+1} P\left(\bigcup_{i \in I} A_i\right)$

1.3 Let $\Omega = \{1, 2, 3, \dots\}$ and $\mathcal{A} = \{A \subseteq \Omega : A \text{ is finite or } A^c \text{ is finite}\}$.

(i) Verify that $\mathcal{A}$ is a σ-field.

(ii) Define $P : \mathcal{A} \to [0, 1]$ by

$$P(A) = \begin{cases} 1 \text{ if } A \text{ is infinite} \\ 0 \text{ if } A \text{ is finite} \end{cases}$$

Show that P is finitely additive, but not σ-additive.

1.4 Show that the Borel σ-field $\mathcal{B}(\mathbb{R})$ on the real line $\mathbb{R}$ is generated by open intervals with rational end points.

1.5 Let $f : \mathbb{R} \to \mathbb{R}^+ = [0, +\infty)$, measurable, such that $\int\limits_{-\infty}^{+\infty} f(x)dx = 1$. Define $F : \mathbb{R}^+ \to [0, 1]$ by $F(x) = \int\limits_{-\infty}^{x} f(y)dy$. Show that

(i) F is absolutely continuous on $\mathbb{R}$.

(ii) F is a distribution function.

(iii) f is the derivative of F almost everywhere.

1.6 Show that the following functions $C : [0, 1] \times [0, 1] \to [0, 1]$ are 2-copulas:

i) $C(x, y) = xy$

ii) $C(x, y) = \max(x + y - 1, 0)$

iii) $C(x, y) = \dfrac{xy}{x + y - xy}$

1.7 Let $(\Omega, \mathcal{A}, P)$ be a probability space. Show that, in general, for any $A, B \in \mathcal{A}$, $P(B^c \cup A) \geq P(A|B)$.

1.8 Let P be the product probability measure on $(\mathbb{R}^2, \mathcal{B}(\mathbb{R}^2))$ and let $F : \mathbb{R}^2 \to [0, 1]$ be defined by $F(x_1, x_2) = P((-\infty, x_1] \times (-\infty, x_2])$.

(i) Show that F is monotone nondecreasing in each of its arguments.

(ii) Show that if $(x_1, x_2) \leq (y_1, y_2)$, then

$$F(y_1, y_2) - F(y_1, x_2) - F(x_1, y_2) + F(x_1, x_2) \geq 0.$$

1.9 Let (X, Y) be a bivariate random vector, defined on $(\Omega, \mathcal{A}, P)$. Let

$$F : \mathbb{R}^2 \to [0, 1], \quad F(x, y) = P(X \le x, Y \le y)$$

and $F_X(x) = P(X \le x)$, $F_Y(y) = P(Y \le y)$. Show that

$$F_X(x) = \lim_{y \to +\infty} F(x, y), \quad \forall x \in \mathbb{R}.$$

$$F_Y(y) = \lim_{x \to +\infty} F(x, y), \quad \forall y \in \mathbb{R}.$$

1.10 Writing 0 and 2 for tails and heads, respectively, consider X_n as the outcome of the nth toss of a fair coin. Let $Y = \sum_{n \ge 1} \frac{X_n}{3^n}$.

(i) Show that Y is a random variable.

(ii) Show that the distribution function of Y is continuous but not absolutely continuous.

Chapter 2

Some Random Sets in Statistics

We provide here some situations in statistics and related fields where random sets appear naturally. This will bring out necessary concepts and techniques that need to be developed in subsequent chapters.

2.1 Probability Sampling Designs

We have mentioned in Chapter 1 that manmade randomization is necessary to make inductive logic valid in sample surveys. A probability sampling design (or plan) is nothing else than a random set S, defined on some probability space $(\Omega, \mathcal{A}, P)$, taking values as subsets of a finite set U. Such a random set is called a *finite random set*. It corresponds to a game of chance where a probability measure P_S is specified on the power set of the power set of U. Different P_S lead to different sampling plans, so that there is a need to look at distributions of random sets. We will see in Chapter 3 that the class of all sampling plans (having the same specified inclusion probabilities) can be completely determined.

A manmade chance model is specified by a function $f : 2^U \rightarrow [0,1]$ such that $\sum_{A \subseteq U} f(A) = 1$, with $f(A)$ being the probability of selecting the subset A of U. Such an f is referred to as a probability sampling design. A selected sample A is an outcome of the random experiment that is performed according to the probability sampling design f.

Thus, this random experiment is a random set S, in the sense that its outcomes are sets (subsets of U). The probability sampling design f is the probability "density" function of S, i.e.,

$$f(A) = P(S = A), \quad \forall A \subseteq U.$$

Random sets as sampling designs are similar to games of chance that are introduced in the teaching of probability before embarking on natural random phenomena. Let us pursue the context of survey sampling a little further to show that we can use survey sampling as a concrete motivating example to address the framework of statistical inference.

With the notation of the situation considered in the beginning of Chapter 1, it is natural to use the statistic $\varphi(S) = \sum\limits_{u \in S} \theta(u)$ to estimate the population total $T(\theta) = \sum\limits_{u \in U} \theta(u)$. To adjust the bias, let us compute the expected value of $\varphi(S)$. We have

$$E\varphi(S) = \sum_{A \subseteq U} \varphi(A)f(A) = \sum_{A \subseteq U} \left[\sum_{u \in A} \theta(u) \right] f(A) =$$

$$\sum_{u \in U} \theta(u) \left[\sum_{A:A \ni u} f(A) \right] = \sum_{u \in U} \theta(u)\pi(u),$$

where $\pi(u) = \sum\limits_{A:A \ni u} f(A) = P(u \in S)$.

The function $\pi : U \to [0,1]$ so defined is called the *covering function* of the random set S. The value $\pi(u)$ is the probability that the individual u will be included in the sampling that is "drawn" according to the random mechanism governed by the density f.

Thus, if we consider the Horvit-Thompson estimator $\sum\limits_{u \in S} \dfrac{\theta(u)}{\pi(u)}$, then we obtain an unbiased estimator for $T(\theta)$, provided our sampling design f is such that $\pi(u) > 0$. for any $u \in U$. With the random set S playing the role of a finite random variable, we proceed to compute other quantities of interest in the analysis of estimation performance, such as the variance of the above unbiased estimator.

An interesting computation involving the covering function π is the expected sample size of the random set S.

Let $|A|$ denote the cardinality of set A, then, formally,

$$E|S| = \sum_{A \subseteq U} |A|f(A).$$

In survey sampling, a direct calculation leads to $E|S| = \sum\limits_{u \in U} \pi(u)$. It turns out that this result is a special case of *Robbins' formula* [105]. For more general random sets see 2.2 below.

It is interesting to note that the practical and primitive situation in survey sampling sets up the general framework of statistical inference.

Let $U = \{1, 2, \ldots, N\}$ be a finite population, and $\theta_0 : U \to \Re$ be our quantity of interest, where $\Re$ could be of an arbitrary nature, such as $\mathbb{R}^+$, $\mathbb{R}^d$, $C[0,1]$, ... We are interested in, say, estimating θ_0 or some function $\tau(\theta_0)$ from the observations of θ_0 only on some subset A of U, i.e., by revealing the restriction of θ_0 on A, denoted as $\theta_0|_A$.

As explained previously, one way to carry out our intended inductive logic is to collect the data in some random fashion. The simplest approach to random selection is to introduce a manmade random mechanism, called a *probability sampling* plan (or scheme, or design). This is achieved by specifying a *probability density function* f on the power set of U, i.e.,

$$f : \mathcal{P}(U) \to [0,1], \quad \sum_{A \subseteq U} f(A) = 1,$$

with the meaning $f(A)$ being the probability for selecting the subset A of U. Of course, the choice of f is left to the "discretion" of the statisticians in the field! Seriously speaking, the choice of f is general dictated by subjective judgments and domain knowledge of the problem under study. These factors lead to the consideration of a variety of sampling plans such as simple random sampling, Poisson sampling, stratified sampling, sampling with probability proportional to size, etc.

In any case, let f be given. First, our *population parameter* θ_0 is unknown, but it is located in the *parameter space* $\Theta = \Re^U = \{\theta : U \to \Re\}$. Next, when we perform our random experiment, i.e. "draw" a subset of U according to the probability law f, we obtain, of course, a subset A of U, but we are really interested in $\theta_0|_A$. Thus, we can view that the "outcome" of our random experiment is $\theta_0|_A$, and not A per se.

The *sample space* of our random experiment (which is the collection of all its possible outcomes) is then

$$\mathcal{X} = \bigcup_{A \subseteq U} \Re^A.$$

Remark. In probabilistic survey sampling, the outcomes of our random experiment are samples, so that we call its space of all possible outcomes its *sample space* (the space of samples). Thus, in general, the set of all possible outcomes of *any* random experiment is also called its sample space!

It is clear that an observation $x \in \mathcal{X}$ is "drawn" from a probability law related to f and $\theta \in \Theta$. Specifically, for $x \in \mathcal{X}$, i.e., $x = \theta|_A$ for some $A \subseteq U$, the probability of observing such x is

$$f_\theta(x) = f_\theta(\theta|_A) = f(A).$$

Thus, let $f_\theta : \mathcal{X} \to [0,1]$ be

$$f_\theta(x) = \begin{cases} f(A) & \text{if} \quad x = \theta|_A \\ 0 & \text{otherwise,} \end{cases}$$

we see that $f_\theta(x)$ is a *probability density function* (or density) on $\mathcal{X}$ with *finite support*. Indeed, for any $\theta \in \Theta$, since U is finite, the set of restrictions $\{\theta|_A : A \subseteq U\}$ is finite, and is the support of $f_\theta(x)$. Now,

$$\sum_{x \in \mathcal{X}} f_\theta(x) = \sum_{A \subseteq U} f_\theta(\theta|_A) = \sum_{A \subseteq U} f(A) = 1.$$

Let P_θ denote the *probability measure* on $\mathcal{X}$ (or more rigorously, on some σ-field $\sigma(\mathcal{X})$) generated by $f_\theta(\cdot)$, i.e.,

$$P_\theta = \sum_{A \subseteq U} f(A)\delta_{(\theta|_A)},$$

where $\delta_{(\theta|_A)}$ is the *Dirac measure* on $\mathcal{X}$ at the point $\theta|_A$, i.e., for $B \in \sigma(\mathcal{X})$,

$$\delta_{(\theta|_A)}(B) = A(x) = \begin{cases} 1 \text{ if } \theta|_A \in B \\ 0 \text{ if } \theta|_A \notin B. \end{cases}$$

We are led to a *family of probability spaces* indexed by Θ,

$$\{(\mathcal{X}, \sigma(\mathcal{X}), P_\theta) : \theta \in \Theta\}.$$

The statistical interpretation of this *statistical model* is this. The true but unknown parameter of interest is θ_0. We make one observation $x \in \mathcal{X}$, which is "drawn" according to P_{θ_0}. The problem is how to "guess" θ_0? Note that we take in fact *one global* observation $x \in \mathcal{X}$. The *sample size* is the size of this global observation. In a simple random sampling plan, we fix in advance an integer n, so that $x = \theta|_A$, when A is realized under the sampling scheme, has n values $(\#(A) = n)$. Otherwise, the sampling size is random.

The above statistical model $\{(\mathcal{X}, \sigma(\mathcal{X}), P_\theta) : \theta \in \Theta\}$ is general. In fact, standard statistical models are of the following form:

$$\mathcal{X} = \mathbb{R}^n, \quad \sigma(\mathcal{X}) = \mathcal{B}(\mathbb{R}^n), \quad \Theta \subseteq \mathbb{R}^m,$$

and P_θ is a *product measure* on $\mathcal{B}(\mathbb{R}^n)$ of the form

$$P_\theta = \otimes_{j=1}^n dF_{j,\theta}, \qquad dF_{j,\theta} = dF_\theta, \quad j = 1, 2, \ldots, n.$$

In other words, the data is a random vector $\mathbf{X} = (X_1, X_2, \ldots, X_n)$ whose components X_j's are *independent and identically distributed* (i.i.d.) as the population X with distribution function F_θ.

Note that while the above statistical model is general, any specific form of it is dictated by the types of data gathered. For example, if the data is collected by recording the observations of some phenomena at points in time (so that the assumption of i.i.d. might not hold), we are facing the problem of *Time Series* in which the theory of *Stochastic Processes* is needed.

2.2   Confidence Regions

Consider a parametric statistical model

$$\{f(x, \theta) : x \in \mathcal{X} \subseteq \mathbb{R}^m, \theta \in \Theta \subseteq \mathbb{R}^d\}$$

and a parameter of interest $\varphi(\theta)$. Given a random sample $X_1, X_2, \ldots, X_n$ from the population, the essence of confidence region estimation is to find a random set $\mathcal{C}(X_1, X_2, \ldots, X_n)$, which contains $\varphi(\theta_0)$, θ_0 being the true parameter, with high probability. Specifically, the random set $\mathcal{C}(X_1, X_2, \ldots, X_n)$ is a confidence set for $\varphi(\theta)$ at confidence level $1 - \alpha \in (0, 1)$, if for any $\theta \in \Theta$,

$$P_\theta(\varphi(\theta) \in \mathcal{C}(X_1, X_2, \ldots, X_n)) \geq 1 - \alpha$$

where $dP_\theta = f(x, \theta)dx$.

In simple situations, the construction of the "best" confidence region for $\varphi(\theta)$ can be carried out without using the formal concept of random sets and their associated distributions. For example, let the population X be normally distributed $\mathcal{N}(\mu, \sigma^2)$. Here $\theta = (\mu, \sigma^2)$. Consider $\varphi(\theta) = \mu$. A $(1 - \alpha)\%$ confidence interval for the man μ can be obtained by using the pivotal quantity $\sqrt{n}(\overline{X}_n - \mu)/V$, where $\overline{X}_n = \dfrac{X_1 + \ldots + X_n}{n}$ and $V^2 = \dfrac{1}{n-1} \sum_{i=1}^{n} (X_i - \overline{X}_n)^2$ are sample mean and sample variance, respectively, based on a random sample $X_1, X_2, \ldots, X_n$ drawn from X. This is so because we know the distribution of $\dfrac{\sqrt{n}}{V}(\overline{X}_n - \mu)$ which is t-distribution with $n - 1$ degrees of freedom, and we can actually pivot this quantity to get $t_1 < \dfrac{\sqrt{n}}{V}(\overline{X}_n - \mu) < t_2$ if and only if $L(X_1, \ldots, X_n) < \varphi(\theta) < U(X_1, \ldots, X_n)$. Obviously, there are many points (t_1, t_2) such that $P\left(t_1 < \dfrac{\sqrt{n}}{V}(\overline{X}_n - \mu) < t_2\right) = 1 - \alpha$. The best confidence interval, at level $1 - \alpha$, can be defined as the one with the shortest length (representing the precision of the estimation of $\varphi(\theta)$). Here the length is a random variable, namely, $|S| = U - L = \dfrac{V}{\sqrt{n}}(t_2 - t_1)$, so that we could choose t_1, t_2 by minimizing the expected length $E|S|$ of the random set $S = [L, U] = \left[\overline{X}_n - \dfrac{V}{\sqrt{n}}t_2, \overline{X}_n - \dfrac{V}{\sqrt{n}}t_1\right]$.

In higher dimensions, the length of a random set S is replaced by its Lebesgue volume $\Lambda(S)$, so that we will face first the computation of $E\Lambda(S)$. For that to be possible, we need to define the concept of a random set, say, in $\mathbb{R}^d$, in such a way that $\Lambda(S)$ is a (nonnegative) random variable. Random sets encountered in most statistical problems are random *closed sets* of euclidean spaces $\mathbb{R}^d$. A theory of random closed sets on $\mathbb{R}^d$ (or more generally, on Hausdorff, locally compact, second countable spaces) is developed by Mathéron [73]: Let $\mathcal{F}$ denote the space of all closed sets of $\mathbb{R}^d$, and $\mathcal{B}(\mathcal{F})$ its Borel σ-field generated by the hit-or-miss topology on $\mathcal{F}$; see details in Chapter 5. Then a *random closed set* S is a map from Ω to $\mathcal{F}$ which is $\mathcal{A}$-$\mathcal{B}(\mathcal{F})$-measurable (i.e., S is a random element defined on a probability space $(\Omega, \mathcal{A}, P)$, with values in the measurable space $(\mathcal{F}, \mathcal{B}(\mathcal{F}))$). The probability law of S is the probability measure image $P_S = PS^{-1}$ as usual.

Let $\varphi : \mathbb{R}^d \times \mathcal{F} \to \{0, 1\}$ be

$$\varphi(x, F) = \begin{cases} 1 \text{ if } x \in F, \\ 0 \text{ if } x \notin F. \end{cases}$$

Then the restriction of the Lebesgue measure Λ on $\mathbb{R}^d$ to $\mathcal{F}$ is

$$\Lambda(F) = \int_{\mathbb{R}^d} \varphi(x, F) \Lambda(dx)$$

so that, by the Fubini theorem (see Appendix), $\Lambda : \mathcal{F} \to \overline{\mathbb{R}}$ is $\mathcal{B}(\mathcal{F})$-$\mathcal{B}(\overline{\mathbb{R}})$-measurable. Thus $\Lambda(S) = \Lambda \circ S$ is $\mathcal{A}$-$\mathcal{B}(\overline{\mathbb{R}})$-measurable, i.e., $\Lambda(S)$ is a (non-negative) random variable. On the other hand φ is measurable since

$$\varphi^{-1}(\{0\}) = \{(x, F) : x \in \mathbb{R}^d, F \in \mathcal{F}, x \notin F\} = \bigcup_{B \in \mathbb{B}} (B \times \mathcal{F}^B) \in \mathcal{F}(\overline{\mathbb{R}}) \otimes \mathcal{B}(\mathcal{F})$$

where $\mathbb{B}$ is a countable base for the topology of $\mathbb{R}^d$, and

$$\mathcal{F}^B = \{F \in \mathcal{F} : F \cap B = \emptyset\}.$$

As such, we have, by the Fubini theorem,

$$\int_{\mathbb{R}^d \times \mathcal{F}} \varphi(x, F) d(\Lambda \otimes P_S) = \int_{\mathbb{R}^d} \int_{\mathcal{F}} \varphi(x, F) dP_S(F) \Lambda(dx) =$$

$$\int_{\mathbb{R}^d} \int_{\mathcal{F}} P(x \in S) \Lambda(dx) = \int_{\mathcal{F}} \int_{\mathbb{R}^d} \varphi(x, F) \Lambda(dx) dP_S(F) =$$

$$\int_{\mathcal{F}} \Lambda(F) dP_S(F) = \int_{\mathcal{F}} \Lambda(S)(\omega) dP(\omega) = E(\Lambda(S))$$

which yields *Robbins' formula*:

$$E(\Lambda(S)) = \int_{\mathbb{R}^d} \pi(x) \Lambda(dx)$$

where $\pi)x) = P(x \in S)$ is the covering function of the random set S. Since the Fubini theorem is valid for σ-finite measures, Robbins's formula yields

$$E|S| = \sum_{u \in U} \pi(u)$$

for finite random sets S with values as subsets of a finite population U, by considering counting measure $|\cdot|$.

2.3 Robust Bayesian Statistics

Perhaps robust Bayesian statistics is the closest example for illustrating a framework in which random sets appear in a formal way. But before talking about robust Bayesian methodology, we give an example of a situation of "imprecise probabilities."

The following example is about imprecision in the sense that we can only specify partially the distribution of a random variable.

Let X be a finite random variable with values in $\{a, b, c, d\} \subseteq \mathbb{R}$, and with density f_0 partially known as

$$f_0(a) \geq 0.4, \quad f_0(b) \geq 0.2, \quad f_0(c) \geq 0.2, \quad f_0(d) \geq 0.1. \tag{2.1}$$

Let P_{f_0} denote the probability measure on $\Theta = \{a, b, c, d\}$ associated with f_0, i.e.,

$$P_{f_0}(A) = \sum_{\theta \in A} f_0(\theta), \quad A \subseteq \Theta.$$

Let $\mathcal{P}$ denote the class of all probability measures P_f on Θ where f satisfies (2.1) at the place of f_0. Then, we only know that

$$P_{f_0} \in \mathcal{P},$$

and hence $F \leq P_{f_0} \leq G$ where $F = \inf_{\mathcal{P}} P$, $G = \sup_{\mathcal{P}} P$. Note that since F and G are conjugate to each other in the sense that

$$F(A) + G(A^c) = 1 \quad \text{for} \quad A \subseteq \Theta$$

(where A^c denotes the set complement of A), we need only to consider one of them.

The set function F on 2^Θ is nonadditive and theoretically known! For example, for $A = \{b, c, d\}$, $F(A) = \inf_{\mathcal{P}} P(A) = 0.5$ by picking P in $\mathcal{P}$, which assigns masses $0.5, 0.2, 0.2, 0.1$ to a, b, c, d, respectively.

Since F is defined on the locally finite ordered set $(2^\Theta, \subseteq)$ since Θ is finite here, it has the Möbius transform $\phi : 2^\Theta \to \mathbb{R}$ given by (see Chapter 4 and [1] for background on combinatorial theory):

$$\phi(A) = \sum_{B \subseteq A} (-1)^{\#(A \backslash B)} F(B)$$

where $A \backslash B = A \cap B^c$ and $\#(A)$ denotes the cardinality of A.

Thus, the set-function ϕ on 2^Θ is also known. On the other hand, it can be shown that F satisfies the following properties (see Chapters 3 and 4):

(i) $F(\emptyset) = 0$, $F(\Theta) = 1$

(ii) For any $n \geq 2$, and $A_1, \ldots, A_n \subseteq \Theta$,

$$F\left(\bigcup_{i=1}^{n} A_i\right) \geq \sum_{\emptyset \neq I \subseteq \{1,2,\ldots,n\}} (-1)^{\#(I)+1} F\left(\bigcap_{j \in I} A_j\right).$$

These properties of F imply that ϕ is nonnegative and

$$\sum_{A \subseteq \Theta} \phi(A) = 1.$$

Thus, ϕ is a bona fide probability mass function of some random element S taking values in 2^{Θ}, namely,

$$\phi(A) = P(S = A), \quad A \subseteq \Theta.$$

The random element S is a random set on Θ with distribution ϕ or equivalently F, since by Möbius inversion, we have

$$F(A) = \sum_{B \subseteq A} \phi(B).$$

From the above analysis, the original problem concerning the population X can be transformed into a problem concerning the random S with known "density" ϕ:

$$\phi : 2^{\Theta} \to [0, 1],$$

$$\phi(\{a\}) = 0.4, \quad \phi(\{b\}) = 0.2, \quad \phi(\{c\}) = 0.2,$$

$$\phi(\{d\}) = 0.1, \quad \phi(\Theta) = 0.1 \quad \text{and}$$

$$\phi(A) = 0 \text{ for all other subsets } A \text{ of } \Theta.$$

The point is this. A problem with imprecision about the distribution of a random variable can be transformed into a precise problem concerning a random set. In other words, random sets seem to offer an appropriate framework for problems with imprecise probabilities, especially in robust Bayesian statistics.

The viewpoint of robust Bayesian inference is as follows. Consider a statistical model of the form $\{F(x|\theta), \theta \in \Theta\}$ for some random vector X, where $F(x|\theta)$ denotes the distribution function under θ, and Θ is a parameter space. The Bayesian approach starts out by assuming that there is a prior probability measure π on the measurable space $(\Theta, \sigma(\Theta))$. However, it is realistic to view both the model and the prior as approximations. Thus robust Bayesian inference refers to the analysis in which priors are replaced by sets of priors. Specifically, the imprecision in specifying priors is made explicit by considering a class of probability measure $\mathcal{P}$ on $(\Theta, \sigma(\Theta))$ containing the "true" prior π_0. Now the set of priors $\mathcal{P}$ induces lower and upper envelops on π_0:

$$L(A) = \inf_{\mathcal{P}} \pi(A), \quad U(A) = \sup_{\mathcal{P}} \pi(A), \quad \forall A \in \sigma(\Theta).$$

Then inference will be essentially based upon L, say. Note that $L(\cdot)$ is not a probability measure on $(\Theta, \sigma(\Theta))$, since it is nonadditive. In view of this fact, not only from the framework of random sets, but also from a general axiomatic consideration, there is an explosion of interests in using Choquet capacities in statistics. Statistical problems involving these nonadditive set-functions will be studied in subsequent chapters.

2.4 Probability Density Estimation

Let X be a random vector with values in $\mathbb{R}^d$, with unknown density f. The support of f is the set $S(f) = \{x \in \mathbb{R}^d : f(x) > 0\}$. If we desire to estimate $S(f)$ from a random sample $X_1, X_2, \ldots, X_n$ drawn from X, then the estimator will be a random set (a set depending on the sample). Recently, interests are shifted to estimation of the α-level sets of f, i.e., for $\alpha > 0$,

$$A_\alpha(f) = A_\alpha = \{x \in \mathbb{R}^d : f(x) \geq \alpha\}.$$

See [50, 70, 92, 100, 101, 120], in which motivations for applications are also outlined.

We outline here the problem of density estimation via the estimation of level sets. This is an alternative approach to kernel method and orthogonal functions in nonparametric density estimation when qualitative information about the density (such as its shape, geometric properties of its contour clusters) is available rather than analytic information.

The density function $f : \mathbb{R}^d \to \mathbb{R}^+$ can be written in terms of its α-level sets as

$$\forall x \in \mathbb{R}^d, \quad f(x) = \int_0^{+\infty} A_\alpha(x) d\alpha,$$

where $A_\alpha(x) = I_{A_\alpha}(x)$ denotes the *indicator function* of the set A_α. Thus, if each A_α is estimated by a random set $A_{\alpha,n}$ (measurable with respect to the sample $X_1, \ldots, X_n$), then it is natural to consider the plug-in estimator

$$f_n(x) = \int_0^{+\infty} A_{\alpha,n}(x) d\alpha$$

for $f(x)$. Note that here $A_{\alpha,n}$ is a random set of more general nature than finite random sets.

The point is this. We are led to set estimation and random sets, as set estimators, arise naturally.

It is interesting to note that the *excess mass approach* [50] to level-set estimation bears some resemblance with maximum likelihood principle in statistics. Indeed, for a fixed level α, the target parameter is the set A_α. The qualitative information about f leads to the statistical model: $A_\alpha \in \mathcal{C}$, where $\mathcal{C}$ is a specified class of subsets of $\mathbb{R}^d$, e.g., closed convex subsets, ellipsoids.

We are in the standard framework of statistical estimation theory: the parameter space is $\mathcal{C}$, so that estimators of A_α should be random sets with values in $\mathcal{C}$. Due to the nature of the target parameters A_α, it is possible to find a general principle to suggest estimators for it.

Let dF denote the probability law of X on $\mathbb{R}^d$ (i.e., the Stieltjes measure associated with f), and again, μ denotes the Lebesgue measure on $\mathbb{R}^d$.

Clearly, $(dF - \alpha\mu)(A_\alpha)$ is the "excess mass" of the set A_α at level α. Thus we can consider the signed measure $dF - \alpha\mu = \mathcal{E}_\alpha$ on $\mathcal{B}(\mathbb{R}^d)$, with $\mathcal{E}_\alpha(A)$ as the excess mass of A at level α. Writing $A = AA_\alpha \cup AA_\alpha^c$, we see that

$$\mathcal{E}_\alpha(A) \leq \mathcal{E}_\alpha(A_\alpha), \quad \forall A \in \mathcal{B}(\mathbb{R}^d),$$

i.e., the level set A_α has the largest excess mass, at level α, among all Borel sets. This suggests a way to estimate A_α using the empirical counterpart of the measure $dF - \alpha\mu$.

Let dF_n denote the empirical measure associated with the sample $X_1, X_2, \ldots, X_n$, i.e.,

$$dF_n = \frac{1}{n} \sum_{j=1}^{n} \delta_{X_j}$$

with δ_x being the Dirac measure at $x \in \mathbb{R}^d$. Then the empirical excess mass, at level α, of $A \in \mathcal{B}(\mathbb{R}^d)$ is

$$\mathcal{E}_{\alpha,n}(A) = (dF_n - \alpha\mu)(A).$$

Thus, it is natural to hope that "good" estimators $A_{n,\alpha}$ of A_α can be obtained by maximizing $\mathcal{E}_{\alpha,n}(A)$ over $A \in \mathcal{C}$. Note that this optimization problem needs special attention: the variable is neither a vector, nor a function, but a set. This type of optimization of set-functions occurs in many areas of applied mathematics such as shape optimization. Later we will discuss a variational calculus of set-functions, which seems appropriate for optimization of set-functions (Chapter 8).

As a routine in statistical practice, desirable properties of the above random set estimator $A_{\alpha,n}$ need to be assessed. In particular, for large sample statistics, consistency and limiting distribution problems need to be examined. This will be carried out by considering statistical convergence in distribution of sequences of random sets (Chapter 7).

2.5   Coarse Data Analysis

Coarse data are a typical situation where the observations are sets rather than points in a sample space. By coarse data we mean rough data or data with low quality. This happens, for example, when the available data are imprecise, say, due to imperfection of the data acquiring procedure (e.g., inaccuracy of the measuring instruments). In such cases, rather than trying to ascribe unique values to the observations, it might be more preferable to represent the outcomes of the random experiment as subsets containing the "true" observation values. Familiar examples of coarse data are missing data, censored or grouped data in biostatistics. A general framework for set-valued observations was proposed by Schreiber [113] using random sets. Specifically, let X be the random variable of interest. The observation process is modeled by a random set S on the range of X. Each unobservable outcome X_j, $j = 1, 2, \ldots, n$, is in the observed random set outcomes S_j, $j = 1, 2, \ldots, n$, which is an i.i.d. sample from S. Thus, X is an *almost sure selector* of S. The statistical inference problem about X, say, estimating the probability density functions of X, will be based upon the *random set data* S_j, $j = 1, 2, \ldots, n$. The point is this. In order to study the above estimation problem, we need a rigorous theory of random sets, especially their distributions.

The practical situation in coarse data analysis is this. Being unable to observe with accuracy the values of a random sample $X_1, X_2, \ldots, X_n$ from X, the statistician tries to locate these observations in random sets S_j, $j = 1, 2, \ldots, n$. There are of course various ways for doing so. Each way represents a coarsening of the data. The observed random sets S_j are viewed as a random sample from a coarsening S, which is a random set. A random set S is called a coarsening of X if S contains X almost surely (i.e., with probability one), i.e., X is an almost sure selector of S. Note that here X is given first, and S is a *random set model* for X. Thus, selector or coarsening depends on which is given first!

A useful model for coarsening is the CAR model [51], where CAR stands for *coarsening at random*. Here is an example. Let $U \subseteq \mathbb{R}$ be the range of X, and $\{A_1, A_2, \ldots, A_k\}$ be a (measurable) partition of U. Consider the coarsening scheme

$$S : (\Omega, \mathcal{A}, P) \to \{A_1, A_2, \ldots, A_k\}.$$

Suppose the unknown probability density function of X is of a parametric form, i.e., $f(x|\theta)$, $\theta \in \Theta$. Then the likelihood function based on the random set sample S_j, $j = 1, 2, \ldots, n$, is

$$L(\theta|S_1, S_2, \ldots, S_n) = \prod_{j=1}^{n} \int_{S_j} f(x|\theta)dx.$$

Thus, the maximum likelihood estimator of θ can be computed using only the observed $S_1, S_2, \ldots, S_n$. However, the investigation of large sample properties of the estimator requires also distributional aspects of the random set model S.

2.6   Perception-Based Information

Coarse data analysis can be viewed as a special procedure in perception-based information gathering process by humans. This type of data is used in the field of *artificial intelligence* to imitate remarkable human intelligent behavior, say, in decision making and control. The search for the source of how humans gather the information in their environments is of course useful in many applications, such as robotics. For example, being unable with the naked eye to accurately measure the distance to some referential location, humans try to extract useful information by considering some simple schemes over the range of the measurements, such as a partition of it. In other words, when we cannot exactly measure the values of some variable of interest X, we coarsen it, e.g., use some random set S such that $P(X \in S) = 1$ to extract information about X.

For example, a perception-based information of the form "Tony is young" has the following structure. The underlying variable of interest is $X = $ age(of Tony) with range $U = [0, 100]$. The linguistic label "young" does not have sharply defined boundaries, and hence can be modeled as a *fuzzy subset* of U, i.e., as a function $A : U \to [0, 1]$.

Students interested in the theory of *fuzzy sets and logics* can read a book like [89]. Since the value of X cannot be observed with accuracy, $A = $ "young" is taken as an observation value instead. This fuzzy value A is in fact one of the possible fuzzy values, not of X, but of some coarsening S of X. For example, a "fuzzy partition" of U could be "very young," "young," "middle-aged," "old," so that S is a (random) fuzzy set, i.e., a random element taking fuzzy subsets of U as values. We will *not* discuss *random fuzzy sets* in this text. Interested students can read, e.g., Li et al. [66] and Nguyen and Wu [91].

If we consider the special case of random sets as coarsening schemes in some perception-based information gathering processes, then various types of uncertainties arising in artificial intelligence can be rigorously justified. To avoid topological details at this stage, suppose the range U is a finite set. Let S be a random set, which is a coarsening of a random variable X that takes values in U, i.e., $P(X \in S) = 1$. Let $A \subseteq U$ be an event. A is said to occur if $X(\omega) \in A$. But if we cannot observe $X(\omega)$, but only $S(\omega)$ is observed, then clearly we are even uncertain about the occurrence of A. If $S(\omega) \subseteq A$, the clearly A occurs. So, from a *pessimistic viewpoint*, we *quantify* our *degree of belief* in the occurrence of A by $P(S \subseteq A)$, which is less than the actual probability that A occurs, namely $P(X \in A)$, since X is an almost sure selector of S. This fact is a starting point of the well-known Dempster-Shafer theory of evidence or *belief functions* that is popular in the field of artificial intelligence (also in some aspects of robust bayesian statistics). Interested students can read a text like [114].

On the other hand, if $S(\omega) \cap A \neq \emptyset$, then it is *possible* that A occurs. A plausible way to *quantify this possibility* is to take $P(S \cap A \neq \emptyset)$. This seems to be consistent with the common sense that possibilities are always larger than probabilities since possibilities tend to represent the *optimistic assessment* as opposed to beliefs. This is so since X is an almost sure selector of S, we clearly have $\{X \in A\} \subseteq \{S \cap A \neq \emptyset\}$ and hence $P(X \in A) \leq P(S \cap A \neq \emptyset)$.

However, as argued by Zadeh [128], unlike *probability logic*, reasoning with possibilities by humans is *semi-truth-functional*. Thus, the question is: Is there a random set S, which is a coarsening of X, such that the set-function

$$\Pi : 2^U \rightarrow [0, 1]$$

defined by $\Pi(A) = P(S \cap A \neq \emptyset)$, is semi-truth-functional, i.e., the knowledge of, say, $\Pi(A)$ and $\Pi(B)$ will determine $\Pi(A \cup B)$? As we will see in the subsequent chapters, the answer is affirmative.

2.7 Stochastic Point Processes

From an application viewpoint, random sets are stochastic models for the analysis of spatial patterns, such as for the *tumor-growth* phenomenon, where, as emphasized by clinicians, the tumor's appearance is extremely important in characterizing its growth (e.g. [17]). However, data analysis when the data are sets, rather than points in some space, is not a nice situation to work with! Of course, statisticians can employ some techniques of standard exploratory data analysis to look at the set data, but to characterize *geometrically* the pattern, other methods might be required.

Random set theory is relatively recent. Choquet presented some key ideas in 1953 [15]. Kendall (1974 [61]) and Mathéron (1975 [73]) provided the foundations. The literature on theoretical developments of the theory and applications grows significantly ever since. Although difficulties are expected, not only because of the complexity of *set-valued analysis*, but also because of the lack of tractable random set models, the advent of image-analyzing computers has provided a practical stimulus for the topic.

Random set models can be used to construct stochastic models for the pattern under study as well as to describe an appropriate scheme of randomization for the probing of a specimen.

We are going to describe the most popular random (closed) set model in applications, known as the *Boolean model*. By doing so, we point out that *point processes* are special cases of random closed sets. For background on stochastic processes, see, e.g., Bosq and Nguyen [12].

Recall the Poisson process on $\mathbb{R}^+$. In a classical representation, a homogeneous Poisson process with intensity Λ is described by a counting process

$N = (N_t, t \geq 0)$, where N_t denotes the number of "events" occurring during $[0, t]$, according to the following hypotheses:

(i) Each random variable $N_t - N_s$, $s < t$, is Poisson distributed with mean $\Lambda(t - s)$, i.e.,

$$P(N_t - N_s = j) = e^{-\Lambda(t-s)}(\Lambda(t - s))^j / j!$$

for $j \in \mathbb{N} = \{0, 1, 2, \ldots\}$.

(ii) For any $t_1 < t_2 < \ldots < t_n$, the random variables $N_{t_j} - N_{t_{j-1}}$, $j = 1, 2, \ldots, n$, are mutually independent.

If $(T_n, \ n \geq 1)$ is the collection of successive (random) "arrival times" of the process, then the Poisson process can be also described in terms of the T_n's. Observe that the T_n's form a random distribution of points in $\mathbb{R}^+$, i.e., $\{T_n, n \geq 1\}$ is a *random set* on $\mathbb{R}^+$. In fact $\{T_n, \ n \geq 1\}$ is almost surely a random *closed* set of $\mathbb{R}^+$. Indeed, for any $t < +\infty$, $EN_t = \Lambda t < +\infty$, so that $N_t < +\infty$, a.s. Thus, for $\omega \in \mathcal{W}^c$, where $\mathcal{W} \in \mathcal{A}$ with $P(\mathcal{W}) = 0$, the subset $\{T_n(\omega), \ n \geq 1\}$ of $\mathbb{R}^+$ is *locally finite* in $\mathbb{R}^+$ in the sense that each bounded subset B of $\mathbb{R}^+$ contains only a finite number of the points $T_n(\omega)$'s. This is so because we can take t so that $B \subseteq [0, t]$, and hence

$$\#(\{T_n(\omega), \ n \geq 1\} \cap B) \leq N_t(\omega) < +\infty.$$

As such, $\{T_n(\omega), \ n \geq 1\}$ is *necessarily closed.* Here is the proof! Let $a \notin \{T_n(\omega), \ n \geq 1\}$. Consider the closed "ball" centered at a with radius $\varepsilon > 0$, in $\mathbb{R}^+$, $B = \{y \in \mathbb{R}^+ : |x - y| \leq \varepsilon\}$. If $T_j(\omega), j = 1, 2, \ldots, m$, say, are the only points of $\{T_n(\omega), \ n \geq 1\}$, which are in B, then

$$B^\circ \cap \{T_j(\omega), \ j = 1, 2, \ldots, m\}^c$$

is an open neighborhood of a that does not contain any element of $\{T_n(\omega), \ n \geq 1\}$ (an hence $\{T_n(\omega), \ n \geq 1\}$ is equal to its closure).

On the other hand, if we let

$$S(\omega) = \{T_n(\omega), \ n \geq 1\}$$

then, being a closed set of $\mathbb{R}^+$, S is the *support* of the *random* measure τ on $\mathcal{B}(\mathbb{R}^+)$, where

$$\tau(A) = \sum_{T \in S} \delta_T(A)$$

where δ_t is the Dirac measure at t. Note that the *support* of a (Borel) measure τ is the smallest closed set F such that $\tau(F^c) = 0$. Here note that the support of τ is

$$supp(\tau)(\omega) = \{x \in R^+ : \tau(\omega)(\{x\}) \neq 0\}$$
$$= \{T_n(\omega), \ n \geq 1\}.$$

The measure τ associated with S is very special. It is a random *point measure*. A point measure τ is a Borel measure (i.e., defined here on $\mathcal{B}(\mathbb{R}^+)$) of the form

$$\tau(A) = \sum_{n \geq 1} \delta_{x_n}(A)$$

where $x_n \in \mathbb{R}^+$, such that $\forall K \in \mathcal{K}(\mathbb{R}^+)$, $\tau(K) < +\infty$ (this last property of τ makes τ a *Radon measure*).

Indeed, $\tau(K) = \sum_{T \in S} \delta_T(K) = \sum_{T \in S \cap K} \delta_T(K)$, which is finite since S is locally finite.

The correspondence $S \rightarrow \tau$ is a bijection. Indeed, if τ is a point measure, then clearly its support is a *locally finite random set* (see also [104]). In view of this, the Poisson process can be defined as a point measure τ. This also allows the extension of Poisson processes to arbitrary spaces, such as $\mathbb{R}^d$, as a point measure on $\mathcal{B}(\mathbb{R}^d)$. Note that we continue to call a point measure a *point process* even if there is no dynamics involved (time-dependent concepts). Poisson processes in $\mathbb{R}^d$ can serve as stochastic models for a random set of points in $\mathbb{R}^d$, e.g., positions of visible stars in a patch of the sky. For further background on point processes, see e.g., [103].

Remark. From a mathematical view point, a random set S in $\mathbb{R}^d$ can also be viewed as a *random function*, namely the indicator function of S. However, as noted e.g., in [17], the random function approach fails to capture the geometric complexities of the pattern such as in image analysis.

Now given the data as in the tumor-growth phenomenon (e.g., [17]), a random set model seems appropriate. The model is somewhat an extension of the Poisson model to the case of a random distribution of sets in the space $\mathbb{R}^d$. In the first step, "germs" are distributed according to a Poisson process in $\mathbb{R}^d$, then in a second step, these germs cause sets of points (grains) modeled as random closed sets of $\mathbb{R}^d$. The union of these grains is a random closed set in $\mathbb{R}^d$, called the *Boolean model*. For practical aspects of the Boolean model as well as statistical inference involved, we refer the reader to [73, 78]. Here, we elaborate on its mathematical structure.

We need to consider Poisson model in the space $\mathcal{F} \backslash \{\emptyset\}$ of nonempty closed sets of $\mathbb{R}^d$. This space, with relative topology induced by $\mathcal{F}$ is locally compact. A point process is a random point measure on $\mathcal{F} \backslash \{\emptyset\}$. Given a Radon measure τ on $\mathcal{F} \backslash \{\emptyset\}$, a point process N is called a Poisson process with mean measure τ if

(i) For any Borel set B of $\mathcal{F} \backslash \{\emptyset\}$, and integer k,

$$P(N(B) = k) = \begin{cases} e^{-\tau(B)} \tau(B)^k / k! & \text{if } \tau(B) < \infty \\ 0 & \text{if } \tau(B) = +\infty \end{cases}$$

(ii) For any $k \geq 1$, and $B_1, \ldots, B_k$ disjoint Borel sets, the variables $N(B_j)$, $j = 1, 2, \ldots, k$ are independent.

In the case of a Poisson process on $\mathbb{R}^d$, the locally finite random (closed) set associated is the collection of points generated by the process. When these points become random sets, its union is the counterpart of the above locally finite random set. For a Poisson process in $\mathcal{F}\setminus\{\emptyset\}$, let us consider a realization of the form $(F_n(\omega),\ n \geq 1)$, where each $F_n(\omega) \in \mathcal{F}\setminus\{\emptyset\}$. Then, again, $S(\omega) = \bigcup_{n\geq 1} F_n(\omega)$ is a closed set of $\mathbb{R}^d$. This is so because the proof for $F_n(\omega) = \{x_n\}$ (singleton) is extended in a straightforward manner to the case of arbitrary closed sets $F_n(\omega)$. Specifically, if a sequence of closed sets F_n of $\mathbb{R}^d$ is *locally finite*, i.e., for any $K \in \mathcal{K}(\mathbb{R}^d)$, $\{F_n : F_n \cap K \neq \emptyset\}$ is finite, then $F = \bigcup_{n\geq 1} F_n$ is closed. Indeed, let $x \in \mathbb{R}^d$ and $x \notin F$. Consider the compact $K = \{y \in \mathbb{R}^d : ||x - y|| \leq \varepsilon\}$. Let $F_1, F_2, \ldots, F_m$, say, be the only F_n's that intersect K. Then

$$K^\circ \cap \left(\bigcup_{i=1}^{m} F_i\right)^c$$

is an open neighborhood of x containing no element of F.

Finally, the fact that each realization of a Poisson process on $\mathcal{F}\setminus\{\emptyset\}$ is locally finite follows from the structure of $\mathbb{R}^d$ and the property of the σ-finite measure τ. See [73].

2.8   Exercises

2.1 Let U be a finite set, and $\pi : U \to [0, 1]$.

(i) Let $f : 2^U \to \mathbb{R}$ be defined as

$$f(A) = \left(\prod_{i\in A} \pi(i)\right)\left(\prod_{i\in A^c}(1 - \pi(i))\right).$$

Use a probabilistic argument to show that $\sum_{A\subseteq U} f(A) = 1$.

(ii) For each $A \subseteq U$, let

$$g(A) = \sum_{B\subseteq A}(-1)^{\#(A\setminus B)}\left[1 - \max_{i\in B^c}\pi(i)\right],$$

where $A \setminus B = A \cap B^c$. Show that $\sum_{A\subseteq U} g(A) = 1$.

2.2   The results in Exercise 2.1 can be obtained in a more general setting, without using probabilistic arguments. Referring to Exercise 2.1, the finite set $V = \{\pi(i) : i \in U\}$ is a subset on the real line $\mathbb{R}$.

Let V be a *finite* subset of $\mathbb{R}$ (or, more generally, of a commutative ring with unit 1). For $A \subseteq V$, let

$$f(A) = \left(\prod_{u \in A} u \right) \left(\prod_{u \in A^c} (1 - u) \right),$$

where $A^c = V \setminus A$.

(i) Show that $\forall v \in V$, $\sum_{v \in A \subseteq V} f(A) = v$.

(ii) Since the empty product is 1, prove, by induction on $\#(V)$, that $\sum_{A \subseteq V} f(A) = 1$.

2.3 Let $f : \mathbb{R}^d \to \mathbb{R}^+ = [0, +\infty)$. For $\alpha > 0$, let $A_\alpha = \{x \in \mathbb{R}^d : f(x) \geq \alpha\}$.

(i) Writing $A_\alpha(x)$ for the indicator function $I_{A_\alpha}(x)$, verify that

$$\forall x \in \mathbb{R}^d, \ f(x) = \int_0^{+\infty} A_\alpha(x) d\alpha.$$

(ii) Let f be a multivariate probability function on $\mathbb{R}^d$, with dF as its associated probability measure on $\mathcal{B}(\mathbb{R}^d)$. Let μ denote the Lebesgue measure on $\mathcal{B}(\mathbb{R}^d)$. For $\alpha > 0$, consider the signed-measure $\varepsilon_\alpha = dF - \alpha\mu$. Show that $\forall A \subseteq \mathcal{B}(\mathbb{R}^d)$, $\varepsilon_\alpha(A) \leq \varepsilon_\alpha(A_\alpha)$.

2.4 Let $(\Omega, \mathcal{A}, P)$ be a probability space, and U be a finite set. Let $X : \Omega \to U$. Verify that the condition

$$\{\omega \in \Omega : X(\omega) = A\} \in \Omega, \ \ \forall A \subseteq U,$$

is equivalent to the measurability of X with respect to $\mathcal{A}$ and $\mathcal{B}$, where $\mathcal{B}$ is the power set of 2^U.

2.5 Let $\mathbb{P}$ denote the set of all probability measures on a measurable space $(\Omega, \mathcal{A})$. Let $P \in \mathbb{P}$, and $F, G : \mathcal{A} \to [0, 1]$ with $F(A) = \inf\{P(A) : P \in \mathcal{P}\}$, $G(A) = \sup\{P(A) : P \in \mathcal{P}\}$. Verify that $\forall A \in \mathcal{A}$, $F(A) + G(A^c) = 1$.

2.6 Consider the following concrete probability space $(\Omega, \mathcal{A}, P)$:

- $\Omega = 2^U$, where U is a finite set.

- $\mathcal{A}$ is the σ-field on Ω, which is the power set of 2^U.

- $P = P_f$, where $f : 2^U \to [0, 1]$ is a probability density (mass) function on 2^U (i.e., $f \geq 0$ and $\sum_{A \subseteq U} f(A) = 1$).

Let $\mathbb{R}^U$ the set of all functions from U to $\mathbb{R}$. Let

$$\mathcal{X} = \bigcup_{A \subseteq U} \mathbb{R}^A.$$

For a fixed $t \in \mathbb{R}^U$, define $f_t : \mathcal{X} \to \mathbb{R}$ as follows. For $x \in \mathcal{X}$, $x : A \to \mathbb{R}$ (for some $A \subseteq U$),

$$f_t(x) = \begin{cases} f(A) & \text{if } x = t|_A \\ 0 & \text{otherwise} \end{cases}$$

(where $t|_A$ is the restriction of $t : U \to \mathbb{R}$ to $A \subseteq U$). Verify that

(i) The support of f_t is finite.

(ii) For a σ-field $\sigma(\mathcal{X})$ containing singletons of $\mathcal{X}$, $(\mathcal{X}, \sigma(\mathcal{X}), P_{f_t})$ is a probability space, where P_{f_t} is the probability measure associated with f_t.

2.7 Let $(\Omega, \mathcal{A}, P)$ be a probability space. Let $S : \Omega \to 2^U$ such that

$$\forall A \subseteq U, \quad \{\omega \in \Omega : S(\omega) = A\} \in \mathcal{A}.$$

(i) Let $X : \Omega \to U$ such that $\forall \omega \in \Omega$, $X(\omega) \in S(\omega)$. Verify that $S(\omega) \neq \emptyset$, $\forall \omega \in \Omega$, i.e., S is a *nonempty* random set.

(ii) Suppose that, $\forall u \in U$, $\{\omega \in \Omega : X(\omega) = u\} \in \mathcal{A}$, and

$$P(\omega : X(\omega) \in S(\omega)) = 1.$$

Verify that $\forall A \subseteq U$, $P(S \subseteq A) \leq P(X \in A)$.

2.8 Let $U = \{1, 2, 3, 4\}$ and let X be a random variable taking values in U with probability mass function $g(1) = 0.425$, $g(2) = 0.225$, $g(3) = 0.225$, $g(4) = 0.125$.

Let $S : \Omega \to 2^U \setminus \{\emptyset\}$ with $f(A) = P(S = A)$ given by $f(\{1\}) = 0.4$, $f(\{2\}) = 0.2$, $f(\{3\}) = 0.2$, $f(\{1\}) = 0.1$, $f(\{1, 2, 3, 4\}) = 0.1$, and $f(A) = 0$ for all other subsets A of U.

(i) Show that $P(S \subseteq A) \leq P(X \in A)$, $\forall A \subseteq U$.

(ii) Is S a coarsening of X?

2.9 Let $(\Omega, \mathcal{A}, P)$ be a probability space, and $X : \Omega \to U$, $S : \Omega \to 2^U$, where U is a finite set. Suppose both X and S are random elements. Show that S is coarsening of X if and only if

$$\sum_{A \subseteq U} \sum_{u \in A} P(X = u, S = A) = 1.$$

2.10 Let $(\Omega, \mathcal{A}, P)$ be a probability space, and $(U, \mathcal{U})$ be a measurable space. Let $S : \Omega \to 2^U$, Suppose

$$\forall A \subseteq U, \quad \{\omega \in \Omega : S(\omega) = A\} \in \mathcal{A}.$$

Define $P_* : \mathcal{U} \to [0, 1]$ by $P_*(A) = P(S \subseteq A)$. Verify the following:

(i) If for all ω, $S(\omega) \neq \emptyset$, then $P_*(\emptyset) = 0$.

(ii) $P_*(U) = 1$.

(iii) P_* is monotone increasing.

(iv) P_* is not necessarily additive.

Chapter 3

Finite Random Sets

As in the study of statistical models, we consider first the counterpart of classical probability models, namely, the case of random sets on finite spaces. This framework is simple and is not encumbered with technical details. Many of the concepts presented here have counterparts in the general theory that is dealt with in Chapter 5. The study of random sets in this finite case illustrates the results we can expect to obtain in more abstract spaces.

3.1 Random Sets and Their Distributions

Throughout this chapter, U denotes a *finite* set. The cardinality of a subset A of U is denoted by $|A|$ or $\#(A)$. $(\Omega, \mathcal{A}, P)$ will be a probability space in the background.

DEFINITION 3.1 *A finite random set* with values in 2^U *is a map* $X : \Omega \to 2^U$ *such that* $X^{-1}(\{A\}) = \{\omega \in \Omega : X(\omega) = A\} \in \mathcal{A}$, *for any* $A \subseteq U$.

Remarks.

(i) Clearly, a random set X is a random element when we refer to the measurable space $(2^U, \mathcal{E})$, where $\mathcal{E}$ is the power set of 2^U, i.e., $X^{-1}(\mathcal{E}) \subseteq \mathcal{A}$. This is so because, $\forall \mathbb{A} \in \mathcal{E}, X^{-1}(\mathbb{A}) = \bigcup_{A \in \mathbb{A}} X^{-1}(\{A\})$.

(ii) As in the case of finite random variables, the measurability condition on X is required for us to define rigorously the probability law P_X of X on $\mathcal{E}$ via probability values on "singletons." Specifically, let $f : 2^U \to [0, 1]$ be $f(A) = P(X = A)$, then f is a bona fide *probability density function* on 2^U, i.e., $f \geq 0$ and $\sum_{A \subseteq U} f(A) = P(X \in 2^U) = P_X(2^U) = 1$.

For $\mathbb{A} \in \mathcal{E}, P_X(\mathbb{A}) = \sum_{A \in \mathbb{A}} f(A)$.

Example 1 (Random level set). Let $\varphi : U \to [0, 1]$. Let $\alpha : \Omega \to [0, 1]$ be a random variable, uniformly distributed. Consider $X : \Omega \to 2^U$ where

$$X(\omega) = \{u \in U : \varphi(u) \geq \alpha(\omega)\}.$$

The measurability condition is easily verified (say, by renaming the elements of U so that $u_i \leq u_j \Rightarrow \varphi(u_i) \leq \varphi(u_j)$). Note that this random set is *nested* in the sense that its range is totally ordered by set inclusion. To specify its probability density f on 2^U, it is convenient to use the concept of *distribution functions*. If we let

$$F : 2^U \to [0, 1]$$

where $F(A) = P(X \subseteq A)$, then

$$F(A) = 1 - P(X \cap A^c \neq \emptyset) = 1 - P\left(\omega : \alpha(\omega) \leq \max_{A^c} \varphi(u)\right) = 1 - \max_{A^c} \varphi(u).$$

Now $F(A) = \sum_{B \subseteq A} f(B)$, and hence we need to express f in terms of F. Fortunately this is a familiar situation in combinatorics! See Chapter 4.

As students can guess, we are going to establish a one-to-one correspondence between f and F so that the concept of distribution functions can be used at a more fundamental level, especially when we deal with more general types of random sets (Chapter 5).

Inspired by the situation with random variables, we are going to define axiomatically the concept of distribution functions (for random sets). Now, observe that the distribution function F of a random set X, i.e., $F(A) = P(X \subseteq A)$ is monotone, i.e., $A \subseteq B \Rightarrow F(A) \leq F(B)$, and moreover, F is *2-monotone*, i.e., for any $A, B \subseteq U$,

$$F(A \cup B) \geq F(A) + F(B) - F(A \cap B).$$

But, an *arbitrary* function $F : 2^U \to [0, 1]$ which is 2-monotone might not be necessarily monotone, since for $A \subseteq B$, we have

$$F(B) = F(A \cup (B \setminus A)) \geq F(A) + F(B \setminus A) - F(\emptyset).$$

Thus, unless $F(\emptyset) = 0$, i.e., $F(\emptyset)$ is the minimum value of F, F might not be monotone (and hence it is not a distribution function of some random set).

Now, if F is the distribution function of a random X with density f, then clearly $F(\emptyset) = f(\emptyset) = P(X = \emptyset)$. For $F(\emptyset) = 0$, the random set X should not be allowed to take the value $\emptyset$ in 2^U. While it is possible for an arbitrary random set X to take the value $\emptyset$, we will restrict ourselves *from now on* to *nonempty random sets*, i.e., to random sets X for which $F(\emptyset) = f(\emptyset) = 0$. This will allow us to have a neat definition of abstract distribution functions.

Note that most random sets in applications are nonempty random sets. For example, if $\mathcal{P}$ is a class of probability measures on $(U, 2^U)$, then $F = \inf \mathcal{P}$ (i.e., $\forall A \in U$, $F(A) = \inf\{P(A) : P \in \mathcal{P}\}$), is such that $F(\emptyset) = 0$. Also, if the random set X is a coarsening of some random variable Y, i.e. $P(Y \in X) = 1$, then necessarily X is a nonempty random set.

THEOREM 3.1 Let X be a (nonempty) random set on U. Let $F : 2^U \to [0,1]$ be $F(A) = P(X \subseteq A)$. Then F satisfies the following properties:

(i) $F(\emptyset) = 0$, $F(U) = 1$,

(ii) For any $k \geq 2$, and $A_1, A_2, \ldots, A_k$ subsets of U,

$$F\left(\bigcup_{j=1}^{k} A_j\right) \geq \sum_{\emptyset \neq I \subseteq \{1,2,\ldots,k\}} (-1)^{|I|+1} F\left(\bigcap_{i \in I} A_i\right). \qquad (3.1)$$

Proof. The property (i) is obvious. For (ii), we proceed as follows. For $B \subseteq U$, let

$$J(B) = \{i = 1, 2, \ldots, k \text{ such that } B \subseteq A_i\},$$

i.e., $B \subseteq \bigcap_{i \in J(B)} A_i$.

Clearly,

$$F\left(\bigcup_{j=1}^{k} A_j\right) = \sum_{B \subseteq \bigcup_{i=1}^{k} A_i} f(B) \geq \sum_{B \subseteq U, J(B) \neq \emptyset} f(B).$$

Now, since for any nonempty finite set A, $\sum_{B \subseteq A} (-1)^{|B|+1} = 1$, we can write (apply to $A = J(B)$):

$$\sum_{B \subseteq U, J(B) \neq \emptyset} f(B) = \sum_{B \subseteq U, J(B) \neq \emptyset} \left[\sum_{\emptyset \neq I \subseteq J(B)} (-1)^{|I|+1} \right] f(B) =$$

$$\sum_{\emptyset \neq I \subseteq U} (-1)^{|I|+1} \sum_{B \subseteq U, J(B) \supseteq I} f(B) = \sum_{\emptyset \neq I \subseteq \{1,2,\ldots,k\}} (-1)^{|I|+1} \sum_{B \subseteq \bigcap_{i \in I} A_i} f(B)$$

(by interchanging the order of the summations). $\square$

It turns out that the properties (i) and (ii) of $F(A) = P(X \subseteq A)$ in the above Theorem 3.1 characterize distribution functions of (nonempty) random sets. More precisely, any set function $F : 2^U \to [0,1]$ satisfying (i), (ii) must be a distribution of a (nonempty) random set with values in 2^U.

DEFINITION 3.2 A function $F : 2^U \to [0,1]$ satisfying the properties (i) and (ii) of Theorem 3.1 is called a distribution function on 2^U.

We are going to show that if F is a distribution function on 2^U, then it is the distribution function of some (nonempty) random set X on U, i.e., there exist a probability space $(\Omega, \mathcal{A}, P)$ and a (nonempty) random set X on U such that $F(A) = P(X \subseteq A)$, $\forall A \subseteq U$. For that, it suffices to show that F is of the form

$$F(A) = \sum_{B \subseteq A} f(B)$$

where $f : 2^U \to [0, 1]$ is a probability density on 2^U.

THEOREM 3.2   If $F : 2^U \to [0, 1]$ is such that

(i) $F(\emptyset) = 0$, $F(U) = 1$,

(ii) For any $k \geq 2$, and $A_1, A_2, \ldots, A_k$ subsets of U,

$$F\left(\bigcup_{j=1}^{k} A_j\right) \geq \sum_{\emptyset \neq I \subseteq \{1,2,\ldots,k\}} (-1)^{|I|+1} F\left(\bigcap_{i \in I} A_i\right).$$

then $\forall A \subseteq U$,

$$F(A) = \sum_{B \subseteq A} f(B)$$

where $f : 2^U \to [0, 1]$ is such that $f(\cdot) \geq 0$ and $\sum_{B \subseteq U} f(B) = 1$.

Proof.
   a) Let $f : 2^U \to [0, 1]$ be defined by

$$f(A) = \sum_{B \subseteq A} (-1)^{|A-B|} F(B),$$

where $A - B = A \cap B^c$ and $|\cdot|$ denotes cardinality. Then f is nonnegative.
   Indeed, by (i), $f(\emptyset) = F(\emptyset) = 0$, and by construction $f(\{u\}) = F(\{u\}) \geq 0$. For $A \subseteq U$ with $|A| \geq 2$, say $A = \{u_1, \ldots, u_n\}$, let $A_i = A \setminus \{u_i\}$, $i = 1, 2, \ldots, n$. Then

$$f(A) = F(A) - \sum_{i=1}^{n} F(A_i) + \sum_{i<j} F(A_i \cap A_j) + \ldots + (-1)^{n-1} \sum_{i=1}^{n} F\left(\bigcap_{j \neq i} A_j\right) \geq 0$$

(noting that $\bigcap_{i=1}^{n} A_i = \emptyset$), by (ii), since $A = \bigcup_{i=1}^{n} A_i$.
   b) Now

$$\sum_{B \subseteq A} f(B) = \sum_{B \subseteq A} \sum_{C \subseteq B} (-1)^{|B-C|} F(C) = \sum_{C \subseteq B \subseteq A} (-1)^{|B-C|} F(C).$$

If $C = A$, the last expression is $F(A)$. If $C \neq A$, then $A - C$ has $2^{|A-C|}$ subsets, so there are an even number of subsets B with $C \subseteq B \subseteq A$, exactly half of which have an even number of elements. The half of the numbers $(-1)^{|B-C|}$ are 1 and half are -1. Thus for each $C \neq A$, $\sum_{C \subseteq B \subseteq A} (-1)^{|B-C|} F(C) = 0$ with the summation taken over B. Therefore, $\sum_{B \subseteq A} f(B) = F(A)$.

In particular, $1 = F(U) = \sum_{A \subseteq U} f(A)$, so that f is a probability density on 2^U. $\square$

Remarks.

a) The above proof of Theorem 3.2 is a direct and elementary one. It was based upon the fact that if F is to be of the form

$$F(A) = \sum_{B \subseteq A} f(B)$$

then the function f has to be defined in terms of F as $f(A) = \sum_{B \subseteq A} (-1)^{|A-B|} F(B)$. This is a fact from *combinatorial theory* that will be elaborated in Chapter 4.

b) The condition (ii) in Theorem 3.2 (or Theorem 3.1) is termed *monotonicity of infinite order* or *infinitely monotone*. If F is a probability measure on U, then F is infinitely monotone in view of Poincaré's equalities. In other words, the monotonicity of infinite order is a generalization of Poincaré's equalities for probability measures. Note that distribution functions, as set functions, are, in general, *not additive*. In fact, $\forall A \subseteq U$,

$$P(X \subseteq A) + P(X \cap A^c \neq \emptyset) = 1.$$

A function F that satisfies (ii) in Theorem 3.2 only for some given $k \geq 2$ is called k-*monotone*. Recall that infinite monotonicity requires (ii) for *any* $k \geq 2$. Relations between different k-monotonicities will be investigated in Chapter 4. A set function F which is *2-monotone*, i.e., for for any $A, B \subseteq U$,

$$F(A) + F(B) \leq F(A \cap B) + F(A \cup B),$$

is also referred to as *convex* by the analogy with convex functions in real analysis (Shapley [115]). In the context of *coalitional games* (where U is a set of players, $F(A)$ represents the worth of the coalition A of players), each F is a game, and 2-monotone F (or supermodular) is called a convex game.

As noticed earlier, 2-monotonicity implies monotonicity only if $F(\emptyset)$ is the minimum value of F on 2^U, here $F(\emptyset) = 0$. Thus, distribution functions are monotone set functions.

c) In the finite case, we can also characterize probability laws of (nonempty) random sets by a *dual concept*, namely, *capacity functionals*. If X is a random set on U, then we define its capacity functional as the set function

$$T : 2^U \to [0,1], \quad T(A) = P(X \cap A \neq \emptyset).$$

Since $T(A) = 1 - F(A^c)$, where F is the distribution function of X, T also characterizes the probability law P_X of X in view of Theorem 3.2. Specifically, the axiomatic definition of capacity functionals of random sets is this.

DEFINITION 3.3   *A set function $T : 2^U \to [0,1]$ is a capacity functional of some random set if it satisfies*

(α) $T(\emptyset) = 0$, $T(U) = 1$,

(β) *For any $k \geq 2$, and $A_1, A_2, \ldots, A_k$ in 2^U,*

$$T\left(\bigcap_{j=1}^{k} A_j\right) \leq \sum_{\emptyset \neq I \subseteq \{1,2,\ldots,k\}} (-1)^{|I|+1} T\left(\bigcup_{i \in I} A_i\right).$$

The property (β), dual of monotonicity of infinite order, is referred to as property of *alternating of infinite order*. For a given k, T is said to be *k-alternating*. Like F, the set function T is monotone.

As expected, the definition of capacity functionals for random sets on *infinite spaces* (like $\mathbb{R}$) is more subtle (Chapter 5).

Example 2. Continue with Example 1 where

$$F(A) = 1 - \max_{u \in A^c} \varphi(u) = 1 - T(A^c),$$

we see that the probability density function f of the random set X is given by

$$f(A) = \sum_{B \subseteq A} (-1)^{|A-B|} \left[1 - \max_{u \in A^c} \varphi(u)\right].$$

Example 3 (continuation of Example 1). Example 1 is typical in survey sampling from a finite population U. Recall that a probability sampling plan is a random set X on U with probability density function f on 2^U. In statistical inference in survey sampling, as exemplified by the unbiased estimation of the population total $T(\theta)$, Chapter 2, the *covering function* π of X, $\pi : U \to [0,1]$, $\pi(u) = P(u \in X)$, is essential. In example 1, the given function $\varphi : U \to [0,1]$ plays the role of a predetermined covering function. The general problem in survey sampling is to find a probability sampling plan f admitting π (here

$\pi = \varphi$) as its covering function, i.e., $\pi(u) = \sum\limits_{u \in A} f(A)$, the summation is taken on all A containing a given $u \in U$.

The probability sampling plan f given in example 2 is clearly one solution! Indeed, $\pi(u) = T(\{u\})$, recalling that

$$T(A) = P(X \cap A \neq \emptyset) = \max_{u \in A} \pi(u),$$

and

$$\pi(u) = P(X \cap \{u\} \neq \emptyset) = P(u \in X),$$

i.e., the covering function of a random set X is the restriction of its capacity functional T to singleton sets of U.

Note that the randomized level set of the covering function π is a very special random set (i.e., a special probability sampling design), in the sense that its probability density function f is completely determined by π. In general, the knowledge of the covering function of a random set is *not* sufficient to determine the probability density function. In that sense, the covering function (or one-point coverage function, or first-order probabilities of inclusion) of X (or of f) as well as the many-point coverage function $\pi(A) = \sum\limits_{A \subseteq B} f(B)$, for each $A \subseteq U$ ($= P(A \subseteq S)$), are similar to moments of random variables; see also [105].

The following analysis provides a general way for obtaining all possible probability sampling plans from the specification of a covering function π, i.e., those admitting π as their (common) covering function.

Without loss of generality, let $U = \{1, 2, \ldots, n\}$, and $\pi : U \rightarrow [0, 1]$. If f is a probability density on 2^U, then we write P_f for its associated probability measure on the power set of 2^U, i.e., for $\mathbb{A} \subseteq 2^U$,

$$P_f(\mathbb{A}) = \sum_{A \in \mathbb{A}} f(A).$$

For each $j \in U$, consider the Bernoulli random variable, defined on the probability space $(2^U, P_f)$:

$$I_j : 2^U \rightarrow [0, 1]$$

$$I_j(A) = \begin{cases} 1 \text{ if } j \in A \\ 0 \text{ if } j \notin A \end{cases}$$

with parameter $P_f(I_j = 1) = \pi(j)$, i.e., f is a probability density on 2^U having precisely π as its covering function.

Such an f is determined completely from the joint distribution of the Bernoulli random vector $(I_1, I_2, \ldots, I_n)$. This can be seen as follows. Exhibiting the following bijection between 2^U and $\{0, 1\}^n$: for $A \subseteq U$, we associate $\varepsilon = (\varepsilon_1, \varepsilon_2, \ldots, \varepsilon_n) \in \{0, 1\}^n$ where $\varepsilon_i = 1$ or 0 according to $i \in A$

or not; and conversely, for $\varepsilon = (\varepsilon_1, \varepsilon_2, \ldots, \varepsilon_n) \in \{0,1\}^n$, we associate the set $A_\varepsilon = \{j : \varepsilon_j = 1\}$. Then clearly,

$$f(A_\varepsilon) = P_f(I_1 = \varepsilon_1, I_2 = \varepsilon_2, \ldots, I_n = \varepsilon_n), \quad \forall \varepsilon \in \{0,1\}^n.$$

Since π is given, the Bernoulli random variables I_j, $j = 1, 2, \ldots, n$, have fixed marginal distributions. As such, their joint distributions are determined by n-copulas according to Sklar's theorem (Chapter 1). Specifically, let F_j be the distribution of I_j, namely,

$$F_j(x) = \begin{cases} 0 & \text{if } x < 0 \\ 1 - \pi(j) & \text{if } 0 \le x < 1 \\ 1 & \text{if } x \ge 1 \end{cases}$$

Then the joint distribution function of $(I_1, I_2, \ldots, I_n)$ is of the form

$$G(x_1, x_2, \ldots, x_n) = C[F_1(x), F_2(x), \ldots, F_n(x)]$$

where C is a n-copula.

For example, when $C : [0,1]^n \rightarrow [0,1]$ is the copula $C(y_1, y_2, \ldots, y_n) = \prod_{j=1}^{n} y_j$, we obtain the well-known *Poisson sampling design*:

$$f(A) = \prod_{j \in A} \pi(j) \prod_{i \notin A} (1 - \pi(i)).$$

3.2   Set-Valued Observations

This section and the next two are illustrations of statistical situations where random sets seem appropriate for modeling uncertainty and making inference with imprecise data.

We consider the standard problem of statistics, namely, the estimation of the unknown probability law π_0 of a random variable X with values in U. But unlike the conventional situation, the data is an i.i.d. sample of random sets $S_1, S_2, \ldots, S_n$, drawn from a random S, rather than an i.i.d. sample from X. The relation between X and S is that $X \in S$ almost surely, i.e., X is an a.s. measurable selection (or *selector*) of S. This clearly extends the structure of standard data analysis in statistics. By the strong law of large numbers, the distribution function F of S is estimated consistently by the empirical distribution function

$$F_n(A) = \frac{1}{n} |\{i : S_i \subseteq A\}|,$$

we are led to describe the statistical model in terms of F. Let $\mathbb{P}$ denote the set of all probability measures on U (finite here). The parameter space is some subset $\mathcal{P} \subseteq \mathbb{P}$ containing π_0, where before having additional information to narrow down $\mathcal{P}$, $\mathcal{P}$ is simply the set of probability laws of all possible selectors of copies of S. We would like to specify further the set $\mathcal{P}$. Now observe that since X is a selector of S (the observed sets $S_1, S_2, \ldots, S_n$ contain unobserved values $X_1, X_2, \ldots, X_n$), we have that $F \leq \pi_0$. Indeed, let $(\Omega, \mathcal{A}, P)$ be a probability space on which are defined both S and X. Let $D \in \mathcal{A}$ such that $P(D^c) = 0$ and $X(\omega) \in S(\omega)$ for all $\omega \in D$. Then,

$$F(A) = P(S \subseteq A) = P((S \subseteq A) \cap D) \leq P(X \in A) = \pi_0(A)$$

Thus $\pi_0 \in \mathcal{C}(F) = \{\pi \in \mathbb{P} : F \leq \pi\}$. Borrowing a name from game theory, $\mathcal{C}(F)$ we call the *core* of F or of S. The structure of the core $\mathcal{C}(F)$ is very nice (!), and the question is whether $\mathcal{P} = \mathcal{C}(F)$? This amounts to check the converse of the above fact, namely, given F on 2^U (or equivalently its associated probability measure dF on the power set of 2^U), and $\pi \in \mathbb{P}$ with $F \leq \pi$, can we find a probability space $(\Omega, \mathcal{A}, P)$ and $S : \Omega \to 2^U$, $X : \Omega \to U$ such that $P(X \in S) = 1$ and $PS^{-1} = dF$, $PX^{-1} = \pi$? i.e., is π *selectionable* with respect to F?

The positive answer can be found in [4, 96] in a very general setting, see details in Chapter 5. Here we will discuss this converse problem from an intuitive viewpoint.

The above converse problem is the problem of existence of a measurable selection with given image measure. Let us say a few words about the selection problem in general.

Let S be a set-valued function, defined on some set Ω, with values as subsets of a set U (arbitrary). A selection of S is a function $X : \Omega \to U$ such that $X(\omega) \in S(\omega), \forall \omega \in \Omega$. The existence of a selection is guaranteed by the axiom of choice. In our case, there is more mathematical structure involved, namely, a probability space $(\Omega, \mathcal{A}, P)$ and U together with some σ-field $\mathcal{B}$ on it. We seek selections that are $\mathcal{A} - \mathcal{B}$-measurable as well as "almost sure selections" in the sense that the selection X of S is measurable and $X \in S$ except on a P-null set of Ω. For existence theorems and further details, we refer the reader to [62, 124].

Now, from the given structure (U, π), $(2^U, F)$ with $F \leq \pi$, we consider the probability space

$$(2^U \times [0, 1], \ dF \otimes dx)$$

and the random set $S : 2^U \times [0, 1] \to 2^U$ defined by $S(A, t) = A$ for all $t \in [0, 1]$. The random set S has F as its distribution function. The solution to the converse problem consists of showing that the condition $F \leq \pi$ is sufficient for the existence of a random variable

$$X : 2^U \times [0, 1] \to U$$

such that $(dF \otimes dx)(X \in S) = 1$ and π is the probability law of X on U, i.e., X is a selector of S.

Without going further into technical details at this stage, let us mention that the result is expected since the framework *has the flavor* of an old selection problem, called the *marriage problem* by Halmos and Vaughan [48]. This old problem started out with Hall in 1935 [46] as a problem of choosing distinct representative elements for a class of subsets of some set. The result was generalized by Everett and Whaples [30] and later by Bollobas and Varapoulos [11]. The marriage problem is this.

Consider two finite sets B (boys) and G (girls), of the same cardinality, say. Each boy $b \in B$ is acquainted with a set of girls $S(b)$, so that $S : B \to 2^G$. Suppose we are interested in the question "under what conditions is it possible for each boy to marry one of his acquaintances?" This is a selection problem of a particular type, namely, an injective (one-to-one) selection. Specifically, we seek conditions under which there exists a function $X : B \to G$ with $X(b)$ being the girl in $S(b)$ who is chosen by b for marriage. Clearly, if $b_1 \neq b_2$ then $X(b_1) \neq X(b_2)$!

Suppose $S(b_1) = \{g_1\}$, $S(b_2) = \{g_2\}$, $S(b_3) = \{g_1, g_2\}$ then it is not possible for these boys $\{b_1, b_2, b_3\}$ to marry their acquaintances. Thus, the *necessary condition* for the marriage problem is that any k boys should know, collectively, at least k girls, i.e.,

$$\forall A \subseteq B, |A| \leq \left| \bigcup_{b \in A} S(b) \right|. \tag{3.2}$$

The remarkable result of Hall is that (3.2) is also sufficient.

The "analogy" of the marriage problem with the selection problem in coarse data analysis can be seen as follows. Let U be a finite set (take $B = G = U$). Let μ be the counting probability measure on U, i.e., $A \subseteq U$, $\mu(A) = \dfrac{|A|}{|U|}$. Let S be a (nonempty) random set U, defined on the probability space $(U, 2^U, \mu)$. Let $X : U \to U$ be the map $X(u) = u$, $\forall u \in U$, so that $P_X = \mu$. Clearly X is an a.s. selector of S if $\forall u \in U$, $u \in S(u)$. But then X is also an a.s. of the random set $S' : U \to 2^U \setminus \{\emptyset\}$ where $S'(u) = \{v \in U : u \in S(v)\}$. This implies that $\forall u, v \in U$, $u \in S(v) \Leftrightarrow v \in S'(u)$. The necessary and sufficient condition of *Hall's theorem* takes the form, $\forall A \subseteq U$,

$$|A| \leq \left| \bigcup_{u \in A} S'(u) \right| = |\{u \in U : S(u) \cap A \neq \emptyset\}|,$$

which is equivalent to $\mu(A) \leq \mu(S \cap A \neq \emptyset)$ i.e., $P_X(A) \leq T_S(A)$.

Now, let us get back to our problem of modeling random experiments with set-valued observations. We are going to take a closer look at the core $\mathcal{C}(F)$.

Since here F is monotone of infinite order, $\mathcal{C}(F) \neq \emptyset$. In fact, this follows from a more general result of Shapley [115] for *convex games* on U.

In our context of random sets, i.e., when F is a distribution function (and hence monotone of infinite order), we can show that $\mathcal{C}(F) \neq \emptyset$ by simply constructing a $\pi \in \mathbb{P}$ such that $\pi \geq F$, as follows.

Let f be the density function of F, i.e.,

$$f(A) = \sum_{B \subseteq A} (-1)^{|A \setminus B|} F(B).$$

For a fixed $\emptyset \neq A \subseteq U$, define

$$g_A : U \to \mathbb{R}$$

by

$$g_A(u) = \begin{cases} \displaystyle\sum_{u \in B \subseteq A} \frac{f(B)}{|B|} & \text{for } u \in A \\ \displaystyle\sum_{u \in B} \frac{f(B)}{|B \setminus (A \cap B)|} & \text{for } u \notin A \end{cases}$$

Then $g_A(\cdot) \geq 0$ and

$$\sum_{u \in U} g_A(u) = \sum_{u \in A} g_A(u) + \sum_{u \notin A} g_A(u).$$

But

$$\sum_{u \in A} g_A(u) = \sum_{u \in A} \sum_{u \in B \subseteq A} \frac{f(B)}{|B|} = \sum_{B \subseteq A} \frac{f(B)}{|B|} \sum_{u \in B} 1 = \sum_{B \subseteq A} f(B)$$

and

$$\sum_{u \notin A} g_A(u) = \sum_{B \not\subseteq A} \frac{f(B)}{|B \setminus A \cap B|} \sum_{u \in B \setminus A \cap B} 1 = \sum_{B \not\subseteq A} f(B).$$

Thus,

$$\sum_{u \in U} g_A(u) = \sum_{B \subseteq U} f(B) = 1.$$

Hence $g_A(\cdot)$ is a probability density on U. Let π_A denote the probability measure on U associated with g_A, then for $B \subseteq U$, we have

$$\pi_A(B) = \sum_{u \in B} g_A(u) = \sum_{u \in A \cap B} g_A(u) + \sum_{u \in A' \cap B} g_A(u)$$

$$= \sum_{u \in A \cap B} \sum_{u \in D \subseteq A} \frac{f(D)}{|D|} + \sum_{u \in D \cap A'} \sum_{u \in D \cap A' \neq \emptyset} \frac{f(D)}{|D \cap A'|}$$

$$\geq \sum_{D \subseteq A \cap B} f(D) + \sum_{D \cap A' \neq \emptyset} f(D)$$

Thus, π_A is a probability measure in $\mathbb{P}$ such that $\pi_A \geq F$, i.e., $\pi_A \in \mathcal{C}(F)$. Moreover, the above construction of π_A exhibits the fact that, in particular, $\pi_A(A) = F(A)$, so that for each $A \subseteq U$,

$$F(A) = \inf\{\pi(A) : \pi \in \mathcal{C}(F)\}$$

i.e., F is the lower envelop of its core $\mathcal{C}(F)$.

Remark. If F is the lower envelope of some given set $Q \subseteq \mathbb{P}$, then in general, $Q \subseteq \mathcal{C}(F)$ with possibly strict inclusion.

The above construction leads to a detailed description of $\mathcal{C}(F)$. For $\emptyset \neq A \subseteq U$, and $u \in A$, let the $\alpha(u, A)$ be nonnegative numbers such that $\sum_{u \in A} \alpha(u, A) = f(A)$. For example, take $\alpha(u, A) = \dfrac{f(A)}{|A|}$ but there are many other choices. Such an α is called an *allocation* of F. Then $g_\alpha(u) = \sum_{u \in A} \alpha(u, A)$ is a density. It will be shown, in Chapter 4, that the probability measure $\pi_\alpha(A) = \sum_{u \in A} g_\alpha(u) \geq F(A)$, and if $\pi \geq F$ then $\pi = \pi_\alpha$ for some allocation α, i.e., $\mathcal{C}(F) = \{\pi_\alpha : \alpha \text{ allocations of } F\}$.

Next we mention here the use of random sets as *models* in coarse data analysis.

Let $(\Omega, \mathcal{A}, P)$ be a probability space, and U be a finite set. Let $X : \Omega \to U$, and $S : \Omega \to 2^U$, the power set of U. Since $X(\omega) \in S(\omega)$ implies that $S(\omega) \neq \emptyset$, we have that

$$\{\omega : X(\omega) \in S(\omega)\} \subseteq \Omega \backslash \Omega_o, \quad \text{where} \quad \Omega_o = \{\omega : S(\omega) = \emptyset\}.$$

Thus, $P(X \in S) \leq P(\Omega \backslash \Omega_o)$. Now, $1 = P(\Omega \backslash \Omega_o) + P(\Omega_o)$, which implies that

$$P(X \in S) < 1 \quad \text{if} \quad P(\Omega_o) > 0, \quad (\text{i.e., } P(S = \emptyset) \neq 0).$$

A *coarsening* of X is a *nonempty* random set S on U such that $P(X \in S) = 1$. For a pair (X, S), we observe that

$$\{\omega : X(\omega) \in S(\omega)\} = \bigcup_{A \subseteq U} \bigcup_{x \in A} \{\omega : X(\omega) = x, S(\omega) = A\}.$$

Thus, to check coarsening, we need information about the *joint distribution* of (X, S). Specifically, S is a coarsening of X if and only if

$$\sum_{A \subseteq U} \sum_{x \in A} P(X = x, S = A) = 1.$$

Also, if S is a coarsening of X, then necessarily the probability law P_X has to belong to the *core* $\mathcal{C}(F)$, where F is the distribution function of S.

Clearly, for a given X, there are many possible coarsenings of X. Among coarsenings of X, there are special ones which are "nice" for statistical inference about the unknown distribution P_X of X, when X cannot be directly observed. A coarsening S is said to be a *coarsening at random* (CAR) if

$$P(S = A|X = x) = \text{constant}, \tag{3.3}$$

for any $A \in 2^U \setminus \{\emptyset\}$ and $x \in A$. The condition (3.3) is referred to as the CAR Assumption. The constant is denoted as $\pi(A)$ and is called a *CAR probability* (see Heitjan and Rubin [51]).

Some simple analyses are needed to bring out the statistical meaning of the CAR assumption.

FACT 1   *The condition (3.3) is equivalent to:* for any $A \in 2^U \setminus \{\emptyset\}$, and $x \in A$,

$$P(S = A|X = x) = P(S = A|X \in A). \tag{3.4}$$

Proof. Since S is a coarsening of X, we have $(S = A) \subset (X \in A)$ so that $P(X \in A|S = A) = 1$. But

$$P(X \in A|S = A) = \sum_{x \in A} P(X = x|S = A)$$

$$= \sum_{x \in A} \frac{P(S = A|X = x)P(X = x)}{P(S = A)}.$$

Thus, if (3.3) holds, then

$$1 = \frac{\pi(A)}{P(S = A)} \sum_{x \in A} P(X = x)) = \frac{\pi(A)}{P(S = A)} P(X \in A),$$

i.e.,

$$\pi(A) = \frac{P(S = A)}{P(X \in A)} = P(S = A|X \in A),$$

since, again, $(S = A) \subset (X \in A)$, hence (3.4) holds.

Conversely, assume (3.4). Then clearly (3.3) holds since $P(S = A|X \in A)$ depends only on A. $\square$

FACT 2   *The condition (3.4) is equivalent to:* for any $A \in 2^U \setminus \{\emptyset\}$, and $x \in A$,

$$P(S = A|X = x, X \in A) = P(S = A|X \in A). \tag{3.5}$$

Proof. Since $(X = x) \subset (X \in A)$,

$$P(S = A|X = x, X \in A) = \frac{P(S = A, X = x, X \in A)}{P(X = x, X \in A)}$$

$$= \frac{P(S = A, X = x)}{P(X = x)} = P(S = A|X = x),$$

which is $P(S = A|X \in A)$ by (3.4). □

Remark. The condition (3.5) is a *conditional independence* of the events $(S = A)$ and $(X \in A)$ *given* $(X = x)$. Let $a = (S = A)$, $b = (X \in A)$ and $c = (X = x)$, then a and b are conditionally independent given c if, by definition,

$$P(ab|c) = P(a|c)P(b|c),$$

but that is the same thing as $P(a|bc) = P(a|c)$ or by symmetry, $P(b|ac) = P(b|c)$. □

FACT 3 *The condition (3.3) is equivalent to:*
for any $A \in 2^U \setminus \{\emptyset\}$, and $x \in A$,

$$P(X = x|S = A) = P(X = x|X \in A). \qquad (3.6)$$

Proof. By Fact 1 and Fact 2, we have

$$P(X = x|S = A, X \in A) = P(X = x|X \in A).$$

But $(S = A) \subset (X \in A)$, we obtain (3.6). □

Interpretation of (3.6). The meaning of the CAR assumption is (3.6), which says that as far as the distribution of X is concerned, observing the value A of the random set S is the same as knowing that the unobservable X falls into the set A.

In the above analysis, we see that, when the CAR assumption holds, we have

$$\pi(A) = \frac{P(S = A)}{P(X \in A)} = \frac{f(A)}{P_X(A)}. \qquad (3.7)$$

Let P_X be generated by a density on U, say, g, so that

$$P(X \in A) = \sum_{x \in A} g(x) = P_g(A).$$

Recall also that $\pi(A) = P(S = A|X = x)$, which is zero for $x \notin A$. Consider the conditional density of S given $X = x$, for any $x \in U$. We have

$$\sum_{A \subset U} P(S = A|X = x) = 1 = \sum_{A \ni x} P(S = A|X = x)$$

since, as a coarsening, $P(S = A|X = x) = 0$ for any A that does not contain the given x. When S is a CAR model for X, $P(S = A|X = x) = \pi(A)$, thus, the condition on the families of CAR probabilities is

$$\sum_{A \ni x} \pi(A) = 1 \qquad \text{for any} \quad x \in U. \tag{3.8}$$

FACT 4 *The condition (3.7) is a sufficient condition for the existence of a CAR model for X. Specifically, if a random set S is given by its density $f(A) = P(S = A)$, and a family of CAR probabilities $\pi(A)$, $A \in 2^U \setminus \{\emptyset\}$ satisfying (3.8), such that $Q(A) = f(A)/\pi(A)$ is a probability measure on U, then $Q(\cdot)$ is the probability law of some X such that S is a CAR model for X.*

Proof. Assume (3.7), with $P_X = P_g$. The situation is this. Given two marginal densities f and g, such that (3.7) holds with π satisfying (3.8). We can construct a joint distribution for (S, X) so that

(i) S and X have f and g as their respective marginals,

(ii) S is a coarsening of X, and

(iii) S is a CAR model for X.

The following construction does all that:
  Define $P(X = x|S = A) = g(x)/P_g(A)$ when $x \in A$ and $f(A) > 0$, and zero for $x \notin A$.
  Note that, by (3.7), when $f(A) > 0$, we also have $P_g(A) > 0$ so that the above conditional density of X given $S = A$ is well defined. The joint density of (S, X) is

$$P(S = A, X = x) = P(X = x|S = A)P(S = A) = \frac{g(x)f(A)}{P_g(A)} = \frac{g(x)}{\pi(A)}.$$

Thus, the marginal density of X is

$$\sum_{A \subseteq U} P(S = A, X = x) = \sum_{A \ni x} g(x)\pi(A) = g(x),$$

by (3.8). Next, X is an a.s. selector of S. Indeed, we have

$$\{\omega : X(\omega) \in S(\omega)\} \supseteq \bigcup_{x \in A} \{\omega : X(\omega) = x, S(\omega) = A\}.$$

In view of the construction, $P(X = x, S = A) \neq 0$ only when $x \in A$ and $f(A) > 0$. Now,

$$\sum_{x \in U} \sum_{A \ni x} P(X = x, S = A) = 1 \qquad \text{(joint density)},$$

so that $P(X \in S) = 1$. Finally, since $P(X = x, S = A) = g(x)\pi(A)$ and since $g(.)$ is the marginal density of X, we have

$$P(S = A|X = x) = \frac{P(X = x, S = A)}{P(X = x)} = \pi(A),$$

i.e., the CAR assumption holds.   □

Remark. When the statistical model for X is given as the $\mathcal{C}(F)$, then the problem is to look for the most plausible element of the $\mathcal{C}(F)$ to represent the observations of X; that can be done, e.g., by using the *maximum entropy principle*. In the actual problem of coarse data where X is unobservable, the problem is to see if there is an element of the $\mathcal{C}(F)$ that makes S a CAR model. In any case, an *explicit description* of the $\mathcal{C}(F)$ seems useful.   □

Let $Q \in \mathcal{C}(F)$ then $F \leq Q$ so that $\pi(A) = f(A)/Q(A) \in [0, 1]$. The existence of a CAR model is that of a $Q \in \mathcal{C}(F)$ such that

$$\sum_{A \ni x} \frac{f(A)}{Q(A)} = 1 \qquad \text{for any} \quad x \in U, \tag{3.9}$$

Now

$$Q \in \mathcal{C}(F) \qquad \text{is equivalent to} \qquad Q = Q_\alpha,$$

where $\alpha : U \times 2^U \setminus \{\emptyset\} \to \mathbb{R}^+$ such that for any $A \subseteq U$,

$$\sum_{x \in A} \alpha(x, A) = f(A), \qquad g_\alpha(x) = \sum_{A \ni x} \alpha(x, A),$$

$$\text{and} \qquad Q_\alpha(A) = \sum_{x \in A} g_\alpha(x).$$

Thus, the problem is to find α (if any) such that

$$\sum_{A \ni x} \frac{f(A)}{Q_\alpha(A)} = 1 \qquad \text{for any} \quad x \in U.$$

Gill, Van der Laan, and Robins [36] showed the existence of Q_α (not constructively) by maximizing $E(\log f(S))$. In fact they showed that the maximum of $\sum_A f(A) \log Q(A)$ over $\mathcal{C}(F)$ is attained at some $Q_\circ$ that does satisfy (3.9).

Example. Let $(\Omega, \mathcal{A}, P)$ be a probability space and $U = \{x_1, x_2, x_3, x_4\}$. Let $X : \Omega \to U$ be a random variable with probability measure P_X and $S : \Omega \to 2^U \setminus \{\emptyset\}$ be a random set with density $f(A) = P(S = A)$ given by

$$f(\{x_1\}) = f(\{x_2\}) = \frac{1}{6}, \quad f(\{x_3\}) = \frac{1}{4}, \quad f(\{x_4\}) = \frac{1}{12},$$

$$f(\{x_1, x_2, x_3\}) = \frac{1}{6}, \quad f(U) = \frac{1}{6}, \quad f(A) = 0,$$

where A is any other subset of U. Suppose that S is a CAR model for X. Then
$$P_X(x_1) = P_X(x_2) = \frac{9}{35}, \quad P_X(x_3) = \frac{27}{70} \quad P_X(x_4) = \frac{1}{10}.$$

Thus, by (3.7), the CAR probabilities are given as
$$\pi(\{x_1\}) = \pi(\{x_2\}) = \pi(\{x_3\}) = \frac{35}{54}, \quad \pi(\{x_4\}) = \frac{5}{6},$$
$$\pi(\{x_1, x_2, x_3\}) = \frac{10}{54}, \quad \pi(U) = \frac{1}{6}, \quad \pi(A) = 0,$$

where A is any other subset of U. It can be checked that the families of CAR probabilities satisfy the condition (3.8).

3.3 Imprecise Probabilities

In the context of statistics, the complete knowledge about a random quantity of interest is its probability law. Thus, by partial or imprecise knowledge in a probabilistic sense, we mean the partial specification of a probability law. Typically, the probability law of interest is known only to lie in some known class of probability measures.

For example, suppose we have a box containing 30 red balls and 60 other balls, some of which are white and the rest are black. We are going to draw a ball from this box. The payoffs for getting a red, black, and white ball are $10, $20, and $30, respectively. What is the expected payoff? Of course, there is not enough information to answer this question in the classical way since we do not know the probability distribution of the red, white, and black balls. We do however have a set of probability densities to which the true density must belong. It is given by the following table:

x	red	black	white
f_k	$\dfrac{30}{90}$	$\dfrac{k}{90}$	$\dfrac{60 - k}{90}$

The true density function is one among these 61 densities.

The general situation of imprecise probabilities is this. Let $\mathbb{P}$ denote the class of all probability measures on U. The true probability measure P_0 is only known to belong to a given subclass $\mathcal{P} \subseteq \mathbb{P}$. From the knowledge of $\mathcal{P}$, we can obtain bounds on P_0, namely, $F \leq P_0 \leq T$, where $F = \inf \mathcal{P}$ and $T = \sup \mathcal{P}$. Since $T(A) = 1 - F(A^c)$, it suffices to consider one of the bounds, say, F.

In this section, we consider attractive situations where F is the distribution function of some random set on U, and that in the context of decision-making.

The problem of choosing a "best" course of action in the context of numerical utilities requires a counterpart of expected values of random variables. Since the information about a probability measure on the set of states of nature is incomplete, one is forced to deal with multivalued random elements, that is, random sets, rather than with random elements, so that it is not clear what the counterpart of expectation of a random variable is. Of course, from a mathematical viewpoint, one can consider expected values of random sets in the context of theory of integration of random elements with values in separable Banach spaces. However, from the decision analysis standpoint, where it is necessary to compare expected utilities, such an approach is not feasible due to the lack of a total ordering on the set of values of random sets. Thus, all current approaches to this problem have been directed toward ways of extracting a single expected value from among a set of possible expected values, and providing a rationale for the one extracted. As we will see, since the problem is much less structured than a statistical decision problem, some form of additional, though subjective, information will be used in extracting process.

In its simplest form, a decision problem consists of choosing an action among a collection of relevant actions Γ in such a way that utility is maximized. Specifically, if Θ denotes the collection of possible "states of nature," the true value θ_0 being unknown, then a utility function

$$u : \Gamma \times \Theta \to \mathbb{R}$$

is specified, where $u(a, \theta)$ is a "payoff" when action a is chosen and nature presents θ. In the Bayesian framework, the knowledge about Θ is described by a probability P on Θ. Then the expected value $E_P u(a, \cdot)$ of the utility function $u(a, \theta)$ is used to make a choice as to which action a to take. When P is specified, the "optimal" action is the one that maximizes $E_P u(a, \cdot)$ over $a \in \Gamma$. In practical problems in statistics, one can even gather more information about θ_0 through experimentation. In evidential knowledge problems, this luxury is not available.

The following examples illustrate the case when P is not known completely. To focus on the modeling of incomplete information, we leave aside the space of actions Γ, but we will return to it in Section 3.4.

Example 1. Let $\Theta = \{\theta_1, \theta_2, \theta_3\}$. The "true" probability measure P_0 is known only up to the extent that $P_0\{\theta_1\} = 1/3$ and then of course $P_0\{\theta_2, \theta_3\} = 2/3$. Let $\mathcal{P}$ denote the class of probability measures P having this property. Then $\mathcal{P} = \{P : F \leq P\}$, where F is defined on Θ by $F(A) = \inf\{P(A) : P \in \mathcal{P}\}$. Moreover, it easily checked that $F(A) = \sum_{B \subseteq A} m(B)$, where $m\{\theta_1\} = 1/3$, $m\{\theta_2, \theta_3\} = 2/3$, and $m(A) = 0$ for all other subsets of Θ. Thus, F is a distribution function of some random set on Θ. Note that the mass function m is a density on 2^Θ. For a density f on Θ, P_f is its associated probability. That is, $P_f(A) = \sum_{\theta \in A} f(\theta)$. The class $\mathcal{P}$ consists of those probability measures associated with the densities $\mathcal{F}_m = \{f : F \leq P_f\}$.

In Pfanzagl [99], the following situation is considered. A box contains 30 red balls and 60 black and yellow balls in unknown proportions. A ball is going to be drawn from the box. Suppose the payoff of getting a red, black, and yellow ball are $5, $10, and $20, respectively. What is the expected payoff?

Let $\Theta = \{\theta_1, \theta_2, \theta_3\}$, with θ_1 specifying red, θ_2 black, and θ_3 yellow. The density on Θ is not known, but is one of

Θ	θ_1	θ_2	θ_3
f	$1/3$	$k/90$	$(60-k)/90$

where $k \in \{0, 1, \ldots, 60\}$. This model is strictly smaller than $\mathcal{F}_m$ above since the mass $m\{\theta_2, \theta_3\}$ can only be distributed to θ_2 and θ_3 in proportions of the form $k/90$ and $(60-k)/90$, rather than in any proportions.

Example 2. Let $\Theta = \{1, 5, 10, 20\}$. Suppose that the information about the true probability is that $P\{1\} \geq .4$, $P\{5\} \geq .2$, $P\{10\} \geq .2$, and $P\{20\} \geq .1$. Let $\mathcal{P}$ denote the collection of all P satisfying these conditions. Then we can describe the "model" by $\mathcal{P} = \{P : F \leq P\}$, where for $A \in 2^\Theta$, $F(A) = \inf\{P(A) : P \in \mathcal{P}\}$. If we let $m\{1\} = .4$, $m\{5\} = .2$, $m\{10\} = .2$, $m\{20\} = .1$, and $m(\{1, 5, 10, 20\}) = .1$, then it can be checked that $F(A) = \sum_{B \subseteq A} m(B)$, so that F is a distribution function of some random set on Θ.

Example 3. Let $\Theta = \{1, 5, 10, 20\}$ and $\mathcal{F} = \{f_1, f_2, f_3, f_4\}$:

Θ	1	5	10	20
f_1	.5	.2	.2	.1
f_2	.4	.3	.2	.1
f_3	.4	.2	.3	.1
f_4	.4	.2	.2	.2

If $F(A) = \min_i P_{f_i}(A)$, then it is easy to check that $P(A) = \sum_{B \subseteq A} m(B)$ where m is given as in Example 2. However, $\mathcal{F}$ is strictly contained in $\mathcal{F}_m$. Indeed, the density

Θ	1	5	10	20
g	.425	.225	.225	.125

is clearly in $\mathcal{F}_m$ but not in $\mathcal{F}$. Also, for $\rho \in (0, 1)$, a density f_ρ of the form

Θ	1	5	10	20
f_ρ	$.4 + .1(1 - \rho)$	$.2$	$.2$	$.1 + .1\rho$

is in $\mathcal{F}_m$ but not in $\mathcal{F}$. If we seek a density to "approximate" the unknown true one, then it should be in the model $\mathcal{F}$. The model $\mathcal{F}_m$ should not be substituted for the model $\mathcal{F}$ unless they are the same.

Example 4. This example generalized Example 1. Let $\{\Theta_1, \Theta_2, \ldots, \Theta_k\}$ be a partition of the finite set Θ. Suppose that the information about the true probability measure is that $P_0(\Theta_i) = \alpha_i$. Then this model

$$\mathcal{P} = \{P : P(\Theta_i) = \alpha_i\}$$

can be described as a distribution function of some random set on Θ. That is,

(i) $F(A) = \inf\{P(A) : P \in \mathcal{P}\}$ for $A \in 2^\Theta$ is a distribution function, and

(ii) $\mathcal{P} = \{P : F \le P\}$.

Indeed, let $A \in 2^\Theta$. The "best" lower approximation of A by the Θ_i is

$$\underline{A} = \bigcup_{\Theta_i \subseteq A} \Theta_i.$$

Now it is clear that all the $P \in \mathcal{P}$ agree on $\underline{A}$ and so we may define $G : 2^\Theta \to [0,1]$ by $G(A) = P(\underline{A})$ for any $P \in \mathcal{P}$. We are going to show that

(i) G is infinitely monotone

(ii) $\mathcal{P} = \{P : G \le P\}$.

(iii) $G = F$.

For (i), use the obvious facts that

$$\underline{\bigcup A_i} \subseteq \bigcup \underline{A_i}$$

and

$$\underline{\bigcap A_i} = \bigcap \underline{A_i}.$$

Now, $F(\emptyset) = P(\underline{\emptyset}) = P(\emptyset) = 0$, and $F(\Theta) = P(\underline{\Theta}) = P(\Theta) = 1$. Now it follows that

$$G\left(\bigcup_{i=1}^n \underline{A_i}\right) \ge \sum_{I \subseteq \{1,2,\ldots,n\}} (-1)^{|I|+1} G\left(\bigcap_{i \in I} \underline{A_i}\right).$$

Re (ii), since $\underline{A} \subseteq A$, $G \leq P$ for all $P \in \mathcal{P}$. Conversely, if Q is a probability measure on Θ such that $G \leq Q$, then in particular, $G(\Theta_i) = P(\underline{\Theta_i}) = P(\Theta_i) = \alpha_i \leq Q(\Theta_i)$. Since P and Q are probabilities and $\{\Theta_1, \Theta_2, \ldots, \Theta_k\}$ is a partition, it follows that $Q(\Theta_i) = \alpha_i$ for all i.

For (iii), it suffices to show that for $A \in 2^\Theta$ there is a $P_A \in \mathcal{P}$ such that $P_A(A) = P_A(\underline{A})$. But this is obvious.

Example 5. In expert systems, a typical situation is this. The "evidence" is represented mathematically as a density m on 2^Θ, or equivalently, since Θ is finite, as a distribution function F. The class $\mathcal{F}_m$ is not empty, but it might be immaterial. Besides the situation in this example where the model is given directly by a density function m on 2^Θ, the random set modeling of a probabilistic model such as in Examples 1, 2, and 3 has some advantages. First, when we know only that the true probability P_0 is in $\mathcal{P}$, a lower bound for $P_0(A)$ is $F(A) = \inf\{P(A) : P \in \mathcal{P}\}$. Also, if $u : \Theta \to \mathbb{R}$ is a utility function, a lower bound for the expected utility $E_{P_0}(u)$ is $\inf\{E_P(u) : P \in \mathcal{P}\}$. In general, these quantities are hard to compute. As we will see in Section 3.4, when $\mathcal{P}$ can be modeled by m, $\inf\{E_P(u) : P \in \mathcal{P}\}$ turns out to be fairly easy to compute.

Second, when the maximum entropy principle is used as a way of obtaining a canonical; distribution on Θ, the constraint $\mathcal{F}_m$ turns out to be easier to handle than the original model $\mathcal{P}$. This is due essentially to the fact that densities in $\mathcal{F}_m$ can be related to m in a more "visual" way.

Finally, for decision analysis, where the concept of expected utilities is essential, a model of the form (Θ, m, u) is convenient. Indeed, as shown in Section 3.4, the most general way of defining an expected utility is to select a set function $\varphi : 2^\Theta \to \mathbb{R}$, depending on u, and take

$$E_m(\varphi) = \sum_{A \subseteq \Theta} \varphi(A) m(A),$$

which is an ordinary expectation of a function of a random set.

3.4   Decision Making with Random Sets

We present in this section three approaches to the problem of decision making based on distribution functions of random sets.

Expectation with Respect to a Distribution Function

If u is an integrable random variable defined on a probability space $(U, \mathcal{A}, P)$, then its expected value can be written as

$$E_P(u) = \int_0^\infty P(u^+ > t)dt - \int_0^\infty P(u^- > t)dt =$$

$$\int_0^\infty P(u > t)dt + \int_{-\infty}^0 (P(u > t) - 1)dt.$$

Here, u^+ and u^- are the positive and negative parts of u, respectively. If P is replaced by a capacity F, then E_P is generalized as a Choquet integral (more details in Chapter 6) by

$$E_F(u) = \int_0^\infty F(u > t)dt + \int_{-\infty}^0 (F(u > t) - 1)dt.$$

If $u \geq 0$, then the second term in this definition vanishes. It should be noted that that this definition is obtained just by replacing P by F in

$$\int_0^\infty P(u > t)dt + \int_{-\infty}^0 (P(u > t) - 1)dt$$

rather than by defining it for $u \geq 0$ and then using $u = u^+ - u^-$. This is necessary because E_F is not an additive operator.

When U is finite, the computation of $E_F(u)$ is simple. Suppose that $U = \{\theta_1, \theta_2, \ldots, \theta_n\}$. We may as well take

$$u(\theta_1) \leq u(\theta_2) \leq \ldots \leq u(\theta_n).$$

It is easy to check that

$$E_F(u) = \sum_{i=1}^n u(\theta_i)[F(\{\theta_i, \theta_{i+1}, \ldots, \theta_n\}) - F(\{\theta_{i+1}, \theta_{i+2}, \ldots, \theta_n\})].$$

Now

$$g(\theta_i) = F(\{\theta_i, \theta_{i+1}, \ldots, \theta_n\}) - F(\{\theta_{i+1}, \theta_{i+2}, \ldots, \theta_n\})$$

is a probability density on U, so that the Choquet integral $E_F(u)$ is an ordinary probabilistic expectation, but the density used for this ordinary expectation depends not only on F but on the ordering given U via u. When our probabilistic knowledge about U is modeled by a belief function F, then $E_F(u)$ can be used as a generalized expected value of the utility function u. This seems attractive since our criteria for decision should involve u rather than other more subjective factors. The theorem below shows that, using $E_F(u)$ for decision making leads to the pessimistic strategy of minimizing probabilistic expected values.

Recall that if g is a density on U, then P_g denotes its associated probability measure on 2^U. By $\mathcal{F}_F$, we mean the class of densities g on U such that $P_g \geq F$.

THEOREM *Let U be finite, $u : U \to \mathbb{R}$, and F a distribution function on a random set S. Then there exists a density $g \in \mathcal{F}_F$ such that*

$$E_F(u) = E_{P_g}(u) = \inf\{E_{P_f}(u) : f \in \mathcal{F}_F\}.$$

Proof. As above, let

$$g(\theta_i) = F(\{\theta_i, \theta_{i+1}, \dots, \theta_n\}) - F(\{\theta_{i+1}, \theta_{i+2}, \dots, \theta_n\}).$$

For $A_i = \{\theta_i, \theta_{i+1}, \dots, \theta_n\}$,

$$g(\theta_i) = F(A_i) - F(A_i \setminus \{\theta_i\}) = \sum_{B \subseteq A_i} m(B) - \sum_{B \subseteq A_i \setminus \{\theta_i\}} m(B) = \sum_{\theta_i \in B \subseteq A_i} m(B),$$

which shows that g is in $\mathcal{F}_F$.

For the rest, it suffices to show that for any $t \in \mathbb{R}$ and for all $f \in \mathcal{F}_F$, we have

$$E_{P_f}(u) = \int_0^\infty P_f(u > t)dt + \int_{-\infty}^0 (P_f(u > t) - 1)dt \geq E_{P_g}(u).$$

Indeed, let $(u > t) = \{\theta_i, \theta_{i+1}, \dots, \theta_n\}$. Then by construction of g, we have

$$P_g(u > t) = \sum_{k=i}^n g(\theta_k) = \sum m(B),$$

where the summation is over all subsets of $\{\theta_i, \theta_{i+1}, \dots, \theta_n\}$. If $f \in \mathcal{F}_F$, then $P_f(u > t) = \sum_{k=i}^n f(\theta_k)$, where the set of A's for which $m(A)$ can be distributed to the θ_k for $k \geq i$ is at least as large as that of the set of B's above. □

In Example 2, g turns out to be f_1 as seen from

$$g(1) = F(\{1, 2, 3, 4\}) - F(\{2, 3, 4\}) = 1 - .5 = .5,$$
$$g(2) = F(\{2, 3, 4\}) - F(\{3, 4\}) = .5 - .3 = .2,$$
$$g(3) = F(\{3, 4\}) - F(\{4\}) = .3 - .1 = .2,$$
$$g(4) = F(\{4\}) = .1.$$

$E_F(u) = \sum u(i)g(i) = 5.5$. Further, this expected value is the smallest of all those whose densities are compatible with m_F. This is clear since all the mass to be distributed is assigned to the smallest utility. Namely, $m_F(\{1, 2, 3, 4\})\}) = .1$, it is all assigned to 1, and $u(1)$ is smaller than all the other values of u. This holds in general. Namely, $E_F(u) = E_{P_g}(u)$, where g is a density compatible with m_F with the resulting expectation the smallest.

Expectation of a Function of a Random Set

Current approaches to decision making based on numerical utilities are all centered around the idea of extracting a single value from the evidence to form a reasonable concept of expected utility. To do this, they all rely heavily on various additional subjective assumptions. A plausible naive density for inference is

$$g(u) = \sum_{A \ni u} \frac{f(A)}{|A|}.$$

The rationale behind the choice of the naive distribution g above is the so-called *Laplace insufficient reason principle* [57]. We will elaborate further on this principle in the next subsection in connection with the maximum entropy principle.

The approach of the previous subsection, as well as that of the next subsection based on maximum entropy, leads to the selection of a density $f \in \mathcal{F}_m$ in order to come up with an expected utility $E_f(u)$. Now it is clear that, for an $f \in \mathcal{F}_m$, once can find many set functions $\varphi : 2^U \to \mathbb{R}$ such that

$$E_f(A) = E_m(A) = \sum_{A \in 2^U} \varphi(A) m(A).$$

In fact, define φ arbitrarily on every element of 2^U except for some A for which $m(A) \neq 0$, and define

$$\varphi(A) = \frac{E_f(u) - \displaystyle\sum_{B \neq A} \varphi(B) m(B)}{m(A)}.$$

The point is this. Selecting φ and considering $E_m(\varphi) = \sum \varphi(a) m(a)$ as expected utility seems to be a more general procedure.

A set function φ can be selected for decision making as follows. For $\rho \in [0,1]$, define

$$\varphi_\rho(A) = \rho \max\{u(\theta) : \theta \in A\} + (1 - \rho) \min\{u(\theta) : \theta \in A\}.$$

It turns out that $E_m(\varphi_\rho) = E_f(u)$, where the density f is constructed as follows. Suppose that $U = \{\theta_1, \theta_2, \ldots, \theta_n\}$ is ordered so that

$$u(\theta_1) \leq u(\theta_2) \leq \ldots \leq u(\theta_n).$$

Now f is defined by

$$f(\theta_i) = \rho \sum_{A \in \mathcal{A}} m(A) + (1 - \rho) \sum_{A \in \mathcal{B}} m(A),$$

where

$$\mathcal{A} = \{A : \theta_i \in A \subseteq \{\theta_1, \theta_2, \ldots, \theta_i\}\},$$

and

$$\mathcal{B} = \{A : \theta_i \in A \subseteq \{\theta_i, \theta_{i+1}, \ldots, \theta_n\}\}.$$

The interpretation of this density is this. For each Θ_i, subsets A containing θ_i and made up of elements θ_j with $j \le i$ get the proportion ρ of their mass $m(A)$ given to $f(\theta_i)$, and subsets A containing θ_i and made up of elements θ_j with $j \ge i$ get the proportion $1 - \rho$ of their mass $m(A)$ given to $f(\theta_i)$. If $\rho = 0$, then $E_f(u) = \min\{E_h(u) : h \in \mathcal{F}_m\}$ as in the previous subsection. At the set function level, we see that the set of values $\{u(\theta) : \theta \in A\}$ is transformed into the single number

$$\rho \max\{u(\theta) : \theta \in A\} + (1 - \rho) \min\{u(\theta) : \theta \in A\}.$$

Two remarks are in order here:

(i) This number is a function only of max and min of u.

(ii) The transformation depends on the parameter ρ supplied by the decision maker.

The selection of φ can be carried out in a more general fashion. For each A for which $m(A) > 0$, the decision maker decides subjectively on a vector $(w_1(A), w_2(A), \ldots, w_{|A|}(A))$, with $w_i(A) \ge 0$ and $\sum w_i(A) = 1$. Then $\varphi(A)$ is taken to be $\sum u_i w_i(A)$, where the u_i are the values o the $u(\theta)$ for $\theta \in A$ arranged in increasing order of magnitude. The choice of the vector of w's can be done by the decision maker giving a "degree of optimism" α and replying on some principle such as maximum entropy to specify this vector. For example, choose $(w_1(A), w_2(A), \ldots, w_{|A|}(A))$ to maximize $\sum w_i(A) \log(w_i(A))$ subject to $w_i(A) \ge 0$ and $\sum w_i(A) = 1$, and

$$\sum \frac{|A| - i}{|A| - 1} w_i(A) = \alpha.$$

(For $|A| = 1$, there is nothing to do; $\varphi(A) = u(\theta)$, where $A = \{\theta\}$.)

Once φ is chosen, the ordinary expectation $E_m(\varphi)$ is used as an expected utility for decision purposes. In all of the above, the selection of φ depends on an additional parameter. It all boils down to using some additional subjective opinion of the decision maker to transform sets of values of utilities to numbers, that is, to considering functions of random sets. This leads to strategies different than the extreme ones, either pessimistic or optimistic. The justification of the parameters involved in this process remains.

A very general way to define the concept of expected utility in the framework (U, m, u) is to select a set function $\varphi : 2^U \to \mathbb{R}$ and to form $E_m(\varphi)$. But there should be some rationale for the process used to select φ. As far as the latter is concerned, the maximum entropy principle seems to be a reasonable one in some decision making problems. That is the topic of the next subsection.

Maximum Entropy Distributions

We investigate in some detail a commonly used rationale for selecting a canonical probability distribution from among a set of relevant ones. When there is no information about a finite set of states of nature U, it makes sense to put the elements of U on equal footing, that is, to endow U with a uniform distribution. This is Laplace's *insufficient reason principle* [57]. If the occurrence of elements of U are governed by a probability density f on U, then a measure of uncertainty of the phenomenon is the *entropy* of f, that is,

$$H(f) = - \sum_{\theta} f(\theta) \log(f(\theta)).$$

The entropy of f does not involve the utility function. Since the uniform distribution has the highest entropy among all densities on U, the *insufficient reason principle* is equivalent to the *principle of maximum entropy*. Motivated by the success of statistical mechanics, the principle of maximum entropy has been formalized as an inference procedure [56]. The postulated density of Laplace is one that maximizes $H(f)$ over all densities f on U. More generally, if $\mathcal{F}$ is any set of densities on U, determined by constraints or evidence of the possible laws governing U, then we are led to seek an $f \in \mathcal{F}$ with maximum $H(f)$. A familiar situation is when $\mathcal{F}$ is a collection of densities with a given expected utility value, and in that case, such a density can be found (see, for example, [44]). In the decision problem under the evidence, the constraint set is $\mathcal{F}_m$. The case of total ignorance corresponds to the m with $m(U) = 1$, and then $\mathcal{F}_m$ is the set of all densities on U. Laplace's insufficient reason principle is to select the f having the greatest entropy from the set of all densities on U. A generalization of this principle in the context of evidence is to select a density f having the greatest entropy from $\mathcal{F}_m$, where the "evidence" is presented through m. In Example 1,

X	θ_1	θ_2	θ_3
f	$1/3$	$k/90$	$(60-k)/90$

with $k \in \{0, 1, 2, \ldots, 60\}$. The density in $\mathcal{F}_m$ with maximum entropy is the uniform density, that is, with each color occurring with probability $1/3$. That this density is actually in the model is somewhat of an accident. The mass of θ_1 has been fixed at $1/3$, and the mass of the other two elements can be equally distributed between them, making the situation as uncertain as possible.

If m is a density on 2^U, and the focal elements of m are *disjoint*, then the density in $\mathcal{F}_m$ with maximum entropy is $f(\theta) = \sum_{A \ni \theta} \dfrac{m(A)}{|A|}$. That is, the mass of A is distributed equally to all elements of A. In Example 2, where the focal elements are not mutually disjoint, assigning the remaining mass .1 equally to each of 1, 5, 10, and 20 yields an entropy of 1.2948, while the maximum entropy density is given by $f(1) = .4$, $f(2) = .2$, $f(10) = .2$, and $f(20) = .2$. The mass $m(\{1, 2, 3, 4\})$ can be distributed in any proportion, but to make the uncertainty maximum, the whole mass .1 should be assigned to 20 in view of the other masses .4, .2, and .2 of 1, 5, and 10, respectively. The utility function is irrelevant in this process. To achieve maximum entropy, the mass of $\{1, 5, 10, 20\}$ is not distributed equally among its elements. The correct generalization of Laplace's insufficient reason principle is to maximize the entropy of the $f \in \mathcal{F}_m$, not simply to distribute the mass of a subset of U equally among its elements. In the rest of this section, we will concentrate on algorithms for this maximum entropy problem.

The first situation that we will consider is this. Let $U = \{\theta_1, \theta_2, \ldots, \theta_n\}$, and m be a density on 2^U with $m(\{\theta_i\}) = \alpha_i$ and $m(U) = 1 - \sum \alpha_i = \varepsilon$. The problem here is to apportion ε among the α_i so that the resulting density on U has the largest possible entropy. That is, write $\varepsilon = \sum \varepsilon_i$ with $\varepsilon_i \geq 0$ so that the entropy of the density f given by $f(\theta_i) = \alpha_i + \varepsilon_i$ is maximum. Here is a precise statement of the problem.

PROBLEM For $i = 1, 2, \ldots, n$, let $\alpha_i \geq 0$, $\varepsilon_i \geq 0$, and $\sum \alpha_i + \sum \varepsilon_i = 1$. *Determine the ε_i, which maximize*

$$H(\varepsilon_1, \varepsilon_2, \ldots, \varepsilon_n) = -\sum (\alpha_i + \varepsilon_i) \log(\alpha_i + \varepsilon_i).$$

Offhand, this problem looks like a nonlinear programming problem, The constraint on the variables ε_i is linear, but the function H to be maximized is nonlinear. However, non-linear programming techniques are not needed. The essence of the matter lies in the following simple lemma whose repeated use will effect a solution.

LEMMA *Let x and $c - x$ be positive. Then*

$$L(x) = -[(c - x) \log(c - x) + x \log(x)]$$

is increasing in x if $c - x > x$.

Proof. The derivative $L'(x) = 1 + \log(c - x) - 1 - \log(x) = \log(c - x) - \log(x)$ is positive as long as $c - x > x$. □

Now suppose that we have α_i's and ε_i's satisfying the condition of the problem. We are interested in maximizing the quantity

$$H(\varepsilon_1, \varepsilon_2, \ldots, \varepsilon_n) = -\sum (\alpha_i + \varepsilon_i) \log(\alpha_i + \varepsilon_i).$$

Suppose that $\alpha_i + \varepsilon_i < \alpha_j + \varepsilon_j$ with $\varepsilon_j > 0$. (We may as well take $i < j$ here.) Let δ be such that $\delta > 0$, $\varepsilon_j - \delta > 0$, and $\alpha_i + \varepsilon_i + \delta < \alpha_j + \varepsilon_j - \delta$. Now apply the lemma with $c = \alpha_i + \varepsilon_i + \alpha_j + \varepsilon_j$ and $x = \alpha_i + \varepsilon_i$. Then the lemma asserts that

$$-[(\alpha_i + \varepsilon_i)\log(\alpha_i + \varepsilon_i) + (\alpha_j + \varepsilon_j)\log(\alpha_j + \varepsilon_j)] <$$
$$-[(\alpha_i + \varepsilon_i + \delta)\log(\alpha_i + \varepsilon_i + \delta) + (\alpha_j + \varepsilon_j - \delta)\log(\alpha_j + \varepsilon_j - \delta)].$$

Thus,

$$H(\varepsilon_1, \varepsilon_2, \ldots, \varepsilon_i + \delta, \ldots, \varepsilon_j - \delta, \ldots, \varepsilon_n) > H(\varepsilon_1, \varepsilon_2, \ldots, \varepsilon_n).$$

The upshot of this is that if an appointment $\varepsilon_1, \varepsilon_2, \ldots, \varepsilon_n$ of ε to the α_i's maximizes H, then whenever $\alpha_i + \varepsilon_i < \alpha_j + \varepsilon_j$, we must have $\varepsilon_j = 0$. Now let the α_i's be indexed so that $\alpha_1 \leq \alpha_2 \leq \ldots \leq \alpha_n$. Thus to maximize H, we must have

$$\alpha_1 + \varepsilon_1 = \alpha_2 + \varepsilon_2 = \ldots = \alpha_k + \varepsilon_k \leq \alpha_{k+1} \leq \ldots \leq \alpha_n, \qquad (3.10)$$

with $\varepsilon_{k+i} = 0$ for $i > 0$. Of course, k may be n. There is at most one assignment of the ε_i like this. For any other apportionment $\gamma_1, \gamma_2, \ldots, \gamma_n$ of ε, some $\varepsilon_i < \gamma_i$ and some $\varepsilon_j < \gamma_j$, and in that case, $\alpha_j + \gamma_j < \alpha_i + \gamma_i$ with $\gamma_i > 0$. But (3.10) does not have this property, so there is at most one apportionment satisfying (3.10).

To get ε_i's satisfying (3.10), simply let $\delta_i = \alpha_k - \alpha_i$,$i = 1, 2, \ldots, k$ with k maximum such that $\sum \delta_i \leq \varepsilon$. Now let $\varepsilon_i = \delta_i + (\varepsilon - \sum \delta_i)/k$, $i = 1, 2, \ldots, k$, and $\varepsilon_i = 0$ for $i > k$. Then,

$$\sum_{i=1}^{n} \varepsilon_i = \sum_{i=1}^{n}(\delta_i + (\varepsilon - \sum \delta_i)/k) = \varepsilon,$$

and for $i < k$,

$$\alpha_i + \varepsilon_i = \alpha_i + \delta_i + (\varepsilon - \sum \delta_i)/k = \alpha_i + \alpha_k - \alpha_i + (\varepsilon - \sum \delta_i)/k =$$

$$\alpha_k + (\varepsilon - \sum \delta_i)/k = \alpha_k + \varepsilon_k.$$

Thus there is a unique set of ε_i's satisfying (3.10) and it is the only set of ε_i's that could maximize H. We call this apportionment the *standard apportionment*. At this point we do not know that this apportionment maximizes H. It maximizes H if any apportionment does. That one exists maximizing H follows from a general theorem. The set of points $\{(\varepsilon_1, \varepsilon_2, \ldots, \varepsilon_n) : \varepsilon_i \geq 0, \sum \varepsilon_i = c\}$ for any constant c is a closed and bounded subset of $\mathbb{R}^n$, and thus its image in $\mathbb{R}$ under a continuous function

$$H(\varepsilon_1, \varepsilon_2, \ldots, \varepsilon_n) = -\sum(\alpha_i + \varepsilon_i)\log(\alpha_i + \varepsilon_i)$$

has a maximum.

Thus the standard apportionment maximizes H, and this is the only apportionment that does so. Further, calculating this apportionment as spelled out is routine, and can be programmed easily. Just put the α_i in increasing order, set $\delta_i = \alpha_k - \alpha_i$, $i = 1, 2, \ldots, k$ with k maximum such that $\sum \delta_i \leq \varepsilon$, let $\varepsilon_i = \delta_i + (\varepsilon - \sum \delta_i)/k$, $i = 1, 2, \ldots, k$, and $\varepsilon_i = 0$ for $i > k$.

There is some merit in providing a constructive proof that there is an apportionment maximizing H. We showed that there is at most one such apportionment, and showed how to construct it. But to show that some apportionment provided a maximum for H, we appealed to a general existence theorem. Here is a constructive proof that the standard configuration

$$\alpha_1 + \varepsilon_1 = \alpha_2 + \varepsilon_2 = \ldots = \alpha_k + \varepsilon_k \leq \alpha_{k+1} \leq \ldots \leq \alpha_n$$

maximizes H. First, we remark that the fact that $\sum \alpha_i + \sum \varepsilon_i = 1$ is not crucial in any of the discussion above, but only that the sum is some positive number.

Suppose that $\gamma_1, \gamma_2, \ldots, \gamma_n$ is an apportionment of ε, always with $\alpha_1 \leq \alpha_2 \leq \ldots \leq \alpha_n$. Let i be the smallest index such that $\varepsilon_i \neq \gamma_i$. If $\varepsilon_i > \gamma_i$ then the indices $i_1, i_2, \ldots, i_m$ such that $\varepsilon_{i+j} > \gamma_{i_j}$ satisfy $\sum_j \gamma_{i_j} - \sum_j \varepsilon_{i_j} \geq \varepsilon_i - \gamma_i$.

(Some ε_{i_j} may be 0.) This is simply because $\sum \gamma_j = \sum \varepsilon_j$. Thus we have

$$\alpha_{i_j} + \gamma_{i_j} > \alpha_{i_j} + \varepsilon_{i_j} \geq \alpha_i + \varepsilon_i > \alpha_i + \gamma_i.$$

Let δ_{i_j} be such that $\gamma_{i_j} - \delta_{i_j} \geq 0$, $\alpha_{i_j} + \gamma_{i+j} - \delta_{i_j} \geq \alpha_i + \gamma_i$, and $\alpha_i + \varepsilon_i = \alpha_i + \gamma_i + \sum_j \delta_{i_j}$. By the lemma above, the apportionment obtained from the γ's by replacing γ_{i_j} by $\gamma_{i_j} - \delta_{i_j}$ and γ_i by $\gamma_i + \sum_j \delta_{i_j}$ has larger entropy. But now this new apportionment has the first i terms $\varepsilon_1, \varepsilon_2, \ldots, \varepsilon_i$.

The argument is entirely similar for $\varepsilon_i < \gamma_i$. Thus we may transform in at most n steps any apportionment into our standard one, each step increasing entropy. Thus we have a constructive proof that our standard apportionment yields maximum entropy. We sum up the discussion above in the following theorem.

THEOREM Let $U = \{\theta_1, \theta_2, \ldots, \theta_n\}$, and m be a density on 2^U with $m(\{\theta_i\}) = \alpha_i$ and $m(U) = 1 - \sum \alpha_i$. Then there is exactly one density f on U, which is compatible with m and which has the largest entropy:

$$H(f) = -\sum_\theta f(\theta) \log(f(\theta)).$$

If $\alpha_1 \leq \alpha_2 \leq \ldots \leq \alpha_n$, then the density is given by $f(\theta_i) = \alpha_i + \varepsilon_i$, where $\varepsilon_i \geq 0$, $\sum_{i=1}^k \varepsilon_i = m(U)$, and

$$\alpha_1 + \varepsilon_1 = \alpha_2 + \varepsilon_2 = \ldots = \alpha_k + \varepsilon_k \leq \alpha_{k+1} \leq \ldots \leq \alpha_n.$$

This density is constructed by putting the α_i in increasing order, setting $\delta_i = \alpha_k - \alpha_i$, $i = 1, 2, \ldots, k$ with k maximum such that $\sum \delta_i \le m(U)$, letting $\varepsilon_i = \delta_i + (m(U) - \sum \delta_i)/k$, $i = 1, 2, \ldots, k$, and letting $\varepsilon_i = 0$ for $i > k$.

The discussion above generalizes to the case when instead of having $m(\{\theta_i\}) = \alpha_i$, we have $m(\Theta_i) = \alpha_o$, where $\Theta_1, \Theta_2, \ldots, \Theta_n$ is a partition of U. First we observe that there is an assignment yielding maximum entropy. The set of possible assignments is a closed and bounded subset of $\mathbb{R}^{|U|}$, and entropy is a continuous function of those assignments into $\mathbb{R}$, and hence achieves a maximum. Maximum entropy is achieved by assigning each θ in Θ_i the mass $m(\Theta_i)/|\Theta_i|$, and proceeding as before. To see this, note first that two points in Θ_i must wind up with same probability. If two points in Θ_i are assigned probabilities $\alpha + \varepsilon$ and $\beta + \gamma$, where α and β come from $m(\Theta_i)$ and ε and γ come from $m(U)$, and $\alpha + \beta < \varepsilon + \gamma$, then one of β and γ is positive, and by the lemma above a part of that one may be shifted to $\alpha + \varepsilon$ so that entropy be increased. So for entropy to be maximized, $\alpha + \beta = \varepsilon + \gamma$, and clearly we may take $\alpha = \beta$ and $\varepsilon = \gamma$. So to maximize entropy, we may as well begin by assigning the mass $m(\Theta_i)$ equally among its elements. This puts us in the situation of the above theorem. So there is an assignment yielding maximum entropy, and we have an algorithm for getting it.

Remarks. The general case is this. Let F be a distribution function on 2^U. Let $\mathcal{F}_F$ be the class of all densities g on U such that $F \le P_g$. Find an algorithm to compute the density in $\mathcal{F}_F$ with maximum entropy. Two such algorithms are given by Meyerowitz et al. [75].

It is interesting to note that the maximum entropy probability measure in the core of F is precisely the unique maximal element with respect to Lorenz partial order. See Jaffray [55] and Dutta and Ray [27].

Concluding remarks. Let $U : \Gamma \times U \to \mathbb{R}$ be a utility function in a decision problem. Optimal actions in terms of expected utility require certain information about U. The concept of expected utility value is well formulated as a mathematical expectation when occurrences of states of nature, the elements of U, are specified by a probability density f. Then the decision problem consists of maximizing $E_f u(a, \theta)$ over a in Γ.

If the probabilistic information about U is less precise, for example, that the density on U is known only to be in some collection $\mathcal{F}$ of densities, then some rationale is needed in order to select some density in this collection and proceeding as in the case the density is known. For example, from a minimax viewpoint, one might take a_0 to be the best course of action if

$$E_f u(a_0, \theta) = \sup_{a \in \Gamma} E_f u(a, \theta).$$

The set function

$$F(A) = \inf_{f \in \mathcal{F}} P_f(A)$$

need not be a distribution function of a random set. That is, it does not need to be of the form F_m for some density m on 2^U. The domain of applicability of distribution functions consists of situations in which knowledge about U is in the form F_m, with m arising from an underlying density on 2^U as in the above example, or from F_m being given directly. In the latter case, first m is formally a probability density density of some random set on 2^U, and second, the "constraint" F_m exists as a mathematical entity.

We summarize the potential recipes as follows.

(i) Since the law governing U is a probability density F of a random subset of U rather than a probability measure on U, one can generalize the concept of integral with respect to an additive measure to the case of Choquet's capacities like F. This is

$$E_F u(a, \cdot) = \int_0^\infty F(\theta : u(a, \theta) > t) dt + \int_{-\infty}^0 [F(\theta : u(a, \theta) > t) - 1] dt.$$

As we have seen, the computation in the finite case is simple, and

$$E_F u(a, \cdot) = \inf_{f \in \mathcal{F}_F} E_f u(a, \cdot).$$

In fact, the inf is attained by g_a in $\mathcal{F}_F$ as we constructed in the previous subsection.

(ii) If additional information is available, or the decision maker is willing to incorporate some subjective view into the process, the most general way to formalize this opinion is in the form of set functions. Specifically, since the mass function m on 2^U is a probability density, we imagine a random set S, defined on some probability space $(\Omega, \mathcal{A}, P)$, with values in 2^U, having m as density. Also, let X be a random variable with values in U with density $f_0 \in \mathcal{F}_m$. Since f_0 is unknown, the expected utility $E_{f_0} u(A, X)$ is replaced by $E_m \varphi_A(S)$, where φ_A is a set function $2^U \to \mathbb{R}$ depending on $u(A, \cdot)$, and possibly on some other parameters. The optimal action is the one that maximizes $E_m \varphi_A(S)$ over $A \in \mathcal{A}$.

(iii) If the choice of a canonical h in $\mathcal{F}_m$ is desired, the maximum entropy principle can be called upon. Maximize $H(f)$ over the f in $\mathcal{F}_m$, yielding h, and then maximize $E_h u(a, \cdot)$ over a in Γ.

A Related Maximum Entropy Problem for Random Sets

In standard problems of maximum entropy in probability and statistics, the constraints are in the form of known moments. In the above maximum entropy problem, the constraint is some set of densities on U. Below is a related maximum entropy problem for random sets in which the constraint is in the form of expectation of the size of the random set (when the random set S is finite, it is the expectation of its cardinality).

If S is a random set on $U = \{u_1, u_2, \ldots, u_n\}$ with known $E(|S|) = \theta$ $(1 < \theta < n)$, then the maximum entropy density of S, i.e., $P_S(A) = P(S = A)$,

$A \subseteq U$, is solution of the following.

$$\text{Maximize } \left\{ - \sum_{A \subseteq U} P_S(A) \log P_S(A) \right\}$$

subject to

(i) $P_S(A) \geq 0$, $\sum_{A \subseteq U} P_S(A) = 1$,

(ii) $E(|S|) = \theta$.

Note that $E(|S|) = \sum_{j=1}^{n} j q_j$ where $q_j = \sum P_S(A)$ and the summation is over all $A \subseteq U$ such that $|A| = j$. To simplify the notation, let p_i, $i = 1, 2, \ldots, 2^n - 1 = m$ be the probabilities $P_S(A)$, $A \subseteq U$ (excluding the empty set) and let $a_i \in \{1, 2, \ldots, n\}$, $i = 1, 2, \ldots, 2^n - 1 = m$ be such that $E(|S|)$ is written as $\sum_{i=1}^{m} a_i p_i$.

The optimization problem becomes

$$\text{Maximize } H(S) = - \sum_{i=1}^{m} p_i \log p_i$$

subject to

(i) $p_i \geq 0$, $\sum_{i=1}^{m} p_i = 1$

(ii) $\sum_{i=1}^{m} a_i p_i = \theta$.

Note that if p_i and p_j are probabilities of two sets A and B such that $|A| = |B|$, then $a_i = a_j = |A|$. The summation $\sum_{i=1}^{n} a_i p_i$ is not the expectation of some random variable X with values a_i, $i = 1, \ldots, m$ and distribution p_i, $i = 1, 2, \ldots, m$, since the a_i's are not distinct; it is $E(|S|)$, where $|S|$ takes values in $\{1, 2, \ldots, n\}$ with distribution q_j, $j = 1, \ldots, n$. Let us proceed formally as in the case of random variables.

Using the Lagrange multiplier technique,

$$L(\alpha, \beta, p) = - \sum_{i=1}^{m} p_i \log p_i + \alpha \left[\sum_{i=1}^{m} p_i - 1 \right] + \beta \left[\sum_{i=1}^{m} a_i p_i - \theta \right]$$

$$= \sum_{i=1}^{m} p_i \log \left[\frac{1}{p_i} e^{-\alpha - \beta a_i} \right] + \alpha \sum_{i=1}^{m} p_i + \beta \sum_{i=1}^{m} a_i p_i$$

$$\leq \sum_{i=1}^{m} e^{-\alpha - \beta a_i} - 1 + \alpha \sum_{i=1}^{m} p_i + \beta \sum_{i=1}^{m} a_i p_i$$

(using the fact that $\log x \leq x - 1$) with equality, if and only if $p_i = e^{-\alpha - \beta a_i}$

$$\frac{\partial L}{\partial \alpha}(\alpha, \beta, p) = 0 \Rightarrow \alpha = \log \phi(\beta), \text{ where } \phi(\beta) = \sum_{i=1}^{m} e^{-\beta a_i}$$

$$\frac{\partial L}{\partial \beta}(\alpha, \beta, p) = 0 \Rightarrow \frac{\phi'(\beta)}{\phi(\beta)} = -\theta$$

$$p_i = \frac{1}{\phi(\beta)} e^{-\beta a_i}$$

where β is the unique solution of the equation $\phi'(\beta) + \theta\phi(\beta) = 0$. To see that this last statement is true, consider a finite measure space $(\Omega, \mathcal{A}, \mu)$ with $\mu(\Omega) = 2^n - 1 = m$ and consider the measurable mapping $Y : (\Omega, \mathcal{A}, \mu) \to \{1, 2, \ldots, n\}$ such that $\mu(\omega : Y(\omega) = j) = \binom{n}{j}$, $j = 1, 2, \ldots, n$. We can write

$$\phi(\beta) = \sum_{i=1}^{m} e^{-\beta a_i} = \sum_{i=1}^{n} \binom{n}{i} e^{-i\beta} = \int_{\Omega} e^{-\beta Y(\omega)} d\mu(\omega).$$

Multiplying the equation $[\phi'(\beta) + \theta\phi(\beta) = 0]$ by e^β we get

$$\int_{\Omega} [\theta - Y(\omega)] e^{(\theta - Y(\omega))\beta} d\mu(\omega) = 0.$$

Note that $\theta \neq EY$ (when we normalize μ to obtain a probability measure). Set $g(\omega) = \theta - Y(\omega)$. It is obvious that the measurable function g is bounded μ – a.e., and since we assume $1 < \theta < n$, we have

$$\mu\{\omega : g(\omega) > 0\} > 0, \quad \mu\{\omega : g(\omega) < 0\} > 0.$$

Thus, the function

$$\beta \in \mathbb{R} \to V(\beta) = \int_{\Omega} g(\omega) e^{\beta g(\omega)} d\mu(\omega)$$

is strictly increasing, with $V(+\infty) = +\infty$, $V(-\infty) = -\infty$, hence, $V(\beta) = 0$ has a unique solution.

Interpretation: Since $p_i = \frac{1}{\phi(\beta)} e^{-\beta a_i}$ we see that if $a_i = a_j = k$, say, then $p_i = p_j$, i.e., the canonical distribution $p = (p_1, p_2, \ldots, p_m)$, $m = 2^n - 1$, of the random set S puts the same mass on subsets of U having the same cardinality.

Remark. The random variable, $|S|$, and the measurable function, Y, have the same set of values, namely, $\{1, 2, \ldots, n\}$. By construction, the support of the measure μ_Y $(= \mu Y^{-1})$ is $\{1, 2, \ldots, n\}$. If d is a number such that $1 < d < n$, we have: $\mu\{\omega : d - Y(\omega) > 0\} > 0$ and $\mu\{\omega : d - Y(\omega) < 0\} > 0$. Now it is clear that $E(|S|) = \theta$ is such that $1 \le \theta \le n$, and $\mu\{w : \theta - Y(w) > 0\} > 0$, and $\mu\{w : \theta - Y(w) < 0\} > 0$ if $1 < \theta < n$. The reader might find this analogous to sample surveys, e.g., Hajek [45]. Specifically, in sampling from a finite population, the population $U = \{u_1, u_2, \ldots, u_n\}$ is finite, and by a sampling design, one means a given probability distribution Q on the collection of all 2^n subsets of U. In other words, Q is the induced probability measure of a random set $S : (\Omega, \mathcal{A}, P) \to 2^U$. The inclusion probabilities (first and second order) are precisely the values of the covering functions, i.e.,

$$\pi(u_i) = \sum_{u_i \in A} Q(A) = P(u_i \in S)$$

$$\pi(u_i, u_j) = \sum_{u_i, u_j \in A} Q(A) = P(\{u_i, u_j\} \subset S).$$

The entropy of the random set S (defining the design) is a measure of spread for sampling probabilities and it is well known that every conditional Poisson sampling design maximizes the entropy in the class of designs having the same values $\pi(u_i)$, $i = 1, 2, \ldots, n$ and the same carrier.

3.5   Exercises

3.1   Let U be a finite set, and F be a distribution function on 2^U. Let $T : 2^U \to [0, 1]$ be defined as $T(A) = 1 - F(A^c)$. Verify that T is monotone, and alternating of infinite order. Is T additive?

3.2   Let $U = \{u_1, u_2, \ldots, u_n\}$, and $\pi : U \to [0, 1]$. Consider the n-copula:

$$C : [0, 1]^n \to [0, 1], \quad C(x_1, x_2, \ldots, x_n) = \min\{x_i : i = 1, 2, \ldots, n\}.$$

Find the probability sampling plan corresponding to C.

3.3   Let $U = \{1, 2, \ldots, n\}$. For each $j \in U$, let $I_j : 2^U \to \{0, 1\}$ where

$$I_j(A) = \begin{cases} 1 \text{ if } j \in A \\ 0 \text{ if } j \notin A. \end{cases}$$

Let X be a random set on U with density f on 2^U. Consider I_j as Bernoulli random variables defined on 2^U with probability measure P_f associated with f. Verify that, $\forall j \in U$,

$$P(j \in X) = P_f(\{A : I_j(A) = 1\}).$$

3.4 Consider the space $2^U \times [0,1]$ with the product σ-field (see Appendix) $\mathcal{E} \otimes \mathcal{B}_1$, where $\mathcal{E}$ is the power set of 2^U, and $\mathcal{B}_1$ is the Borel σ-field of $[0,1]$, and product probability measure $dF \otimes dx$, where dx is the Lebesgue measure on $[0,1]$, and dF is the probability measure associated with a given distribution function F on 2^U, i.e., for $\mathbb{A} \subseteq 2^U$,

$$dF(\mathbb{A}) = \sum_{A \in \mathbb{A}} \sum_{B \subseteq A} (-1)^{|A-B|} F(B).$$

Let $X : 2^U \times [0,1] \to 2^U$: $X(A,t) = A$ for all $t \in [0,1]$. Show that X is a random set (i.e., $\mathcal{E} \otimes \mathcal{B}_1$-$\mathcal{E}$-measurable). What is the probability law of X?

3.5 Let $U = \{a, b, c, d\}$ and $X : \Omega \to U$ be a random variable with density

$$g(a) = 0.425, \quad g(b) = g(c) = 0.225, \quad g(d) = 0.125.$$

Let $S : \Omega \to 2^U$ with density

$$f(\{a\}) = 0.4, \quad f(\{b\}) = f(\{c\}) = 0.2, \quad f(\{d\}) = 0.1,$$

$$f(\{a, b, c, d\}) = 0.1, \text{ and } f(A) = 0 \text{ for other } A \subseteq U.$$

i) Verify that the probability law of X is in the core $\mathcal{C}(F)$, where F is the distribution function $F(A) = \sum_{B \subseteq A} f(A)$.

ii) Show that S is not a CAR model for X.

3.6 Let $\mathbb{P}$ denote the class of all probability measures on a finite set U. Let $0 < \varepsilon < 1$, and P_0 be a fixed probability. Let $\mathcal{P} = \{(1 - \varepsilon)P_0 + \varepsilon P : P \in \mathbb{P}\}$. Let $F = \inf\{P : P \in \mathcal{P}\}$.

i) Show that F is the distribution function of some random set on U.

ii) Show that $\mathcal{P} = \mathcal{C}(F)$ (the core of F).

3.7 Let $U = \{a, b, c, d\}$. Let $\mathcal{P}$ denote the class of all probability measures P on U such that

$$P(a) \geq 0.4, \quad P(b) \geq 0.2, \quad P(c) \geq 0.2, \text{ and } P(d) \geq 0.1.$$

Let $F = \inf \mathcal{P}$.

i) Determine the density f associated with F.

ii) Compute $\varphi(A) = f(A)/|A|$ for each A for which $f(A) \neq 0$.

iii) Compute $g : U \to [0, 1]$ where $g(u) = \sum_{A \ni u} \dfrac{f(A)}{|A|}$.

iv) Verify that $P_g \geq F$.

3.8 Let $U = \{u_1, u_2, \ldots, u_n\}$, and F be a distribution function on 2^U. Let $g : 2^U \to [0, 1]$ be

$$g(u_i) = F(\{u_1, u_2, \ldots, u_i\}) - F(\{u_1, u_2, \ldots, u_{i-1}\})$$

(if $i = 1$, $\{u_1, \ldots, u_{i-1}\} = \emptyset$).

  i) Show that g is a density on U.

  ii) Show that $F \leq P_g$.

3.9 (continuation of 3.8). Let σ be a permutation of $\{1, 2, \ldots, n\}$, and g_σ the density on U constructed as in 3.8 corresponding to $\{u_{\sigma(1)}, u_{\sigma(2)}, \ldots, u_{\sigma(n)}\}$. Show that $F = \inf \mathcal{C}(F)$.

3.10 Let S be a coarsening of X on U. Let Q be the probability law of X such that S is a CAR model for X. Let f be the density of S on 2^U. Let $g : U \to [0, 1]$ be $g(u) = \sum_{u \in A} \alpha(u, A)$, where $\alpha(u, A) = \dfrac{Q(\{u\})}{Q(A)} f(A)$. Verify that g is a density on U and $P_g \geq F$, where F is the distribution function of S.

Chapter 4

Random Sets and Related Uncertainty Measures

This chapter is a continuation of Chapter 3 in which the theme of study is *finite random sets*. Here we will discuss related uncertainty measures such as *belief functions, possibility measures, lower and upper probabilities*. A general framework to study set functions is presented in the context of *incidence algebras*.

4.1 Some Set Functions

We are all familiar with probability measures that are special *set functions*, i.e., maps whose domains are classes of subsets of some set. These are *additive* (σ-additive) set functions defined on σ-fields of subsets of some sets and are used to model laws of random phenomena. Problems in the field of *artificial intelligence* exhibit other types of uncertainty as well, due mainly to the subjectivity, imprecision, and vagueness in the data collected.

For example, expert's opinion can be expressed in terms of subjective beliefs or of subjective evaluations; the intuitive qualitative concept of possibility, once quantified, could be useful in general reasoning under uncertainty; the meaning representation in natural languages reveals the uncertainty expressed as fuzziness. General set functions are proposed to model these types of uncertainty. They are generalizations of probability measures, and in general, *nonadditive*. Nonadditive set functions appear in probability and statistics as lower and upper probabilities. The above uncertainties bear, however, some relationship with random sets, which we spell out in this chapter.

Belief Functions

In Chapter 3 we considered the following coarsening scheme. Let U be a *finite* set. Let $(\Omega, \mathcal{A}, P)$ be a probability space and $X : \Omega \to U$ be a random variable of interest. A coarsening of X is a random set S on U such that $P(X \in S) = 1$, i.e., X is an almost sure selector of S. When performing the random experiment with outcome X (or observing a random phenomenon), the outcome X is not observable, but instead, we observe S containing X with probability one. In such a situation, the occurrence of an event $A \subseteq U$ can only be estimated. From a pessimistic viewpoint, we believe that A occurs with degree $P(S \subseteq A)$. Thus, a qualitative concept of belief is defined as a set function $F : 2^U \rightarrow [0, 1]$ by $F(A) = P(S \subseteq A)$, which is nothing else than the distribution function of the random set S. As such, F satisfies the following properties (see Chapter 3):

(i) $F(\emptyset) = 0$, $F(U) = 1$.

(ii) F is monotone of infinite order, i.e., for $n \geq 2$, and $A_1, A_2, \ldots, A_n$ subsets of U,

$$F\left(\bigcup_{i=1}^{n} A_i\right) \geq \sum_{\emptyset \neq I \subseteq \{1,2,\ldots,n\}} (-1)^{|I|+1} F\left(\bigcap_{i \in I} A_i\right).$$

Following Shafer [114], we take these properties as *axioms* for *belief functions*. Clearly these axioms can be considered on an arbitrary set U (finite or not), leading to the *definition of belief functions* on arbitrary U.

Recall that when U is finite or countably infinite, the domain of probability measures on U can be taken as 2^U. For infinite *uncountable* U, e.g., $U = \mathbb{R}$, the situation is more subtle. While some probability measures still can have domain as 2^U, such as the unit mass at some given point $u_0 \in U$(the Dirac measure δ_{u_0} at u_0), where

$$\delta_{u_0}(A) = \begin{cases} 1 & \text{if } u_0 \in A \\ 0 & \text{if } u_0 \notin A \end{cases},$$

in general, the domain of probability measures on U is some σ-field strictly contained in 2^U. For example, probability measures on $\mathbb{R}$ are defined in the Borel σ-field $\mathcal{B}(\mathbb{R})$. This is due essentially to the existence of nonmeasurable sets of $\mathbb{R}$ (see Halmos [47]). In other words, 2^U is too big to accommodate for the axiom of σ-*additivity* of probability measures. Without this constraint of σ-additivity, set functions can be constructed on 2^U. For example, *outer measures* are set functions μ defined on 2^U (with values in $[0, +\infty]$) such that

a) $\mu(0) = 0$

b) μ is monotone: $A \subseteq B \Rightarrow \mu(A) \leq \mu(B)$,

c) μ is countably (or σ-) subadditive: for any sequence $(A_n, n \geq 1)$, subsets of U,

$$\mu\left(\bigcup_{n \geq 1} A_n\right) \leq \sum_{n \geq 1} \mu(A_n).$$

For example, let $U = \mathbb{R}$, and $f : \mathbb{R} \to [0, \infty)$ such that $\int_{\mathbb{R}} f(x)\,dx = 1$. Let $P : \mathcal{B}(\mathbb{R}) \to [0, 1]$ be

$$P(A) = \int_A f(x)\,dx.$$

Then $\mu : 2^{\mathbb{R}} \to [0, +\infty]$ is an outer measure, where $\mu(B) = \inf \sum_{n \geq 1} P(A_n)$, and the infimum is taken over all finite or infinite sequences $A_n \in \mathcal{B}(\mathbb{R})$ such that $B \subseteq \sum_{n \geq} A_n$.

Also, *capacities* are set functions defined on 2^U (see e.g., Meyer [74]).

Now, belief functions need *not* be σ-additive, and therefore could be constructed also on 2^U for *arbitrary* U. For example, let U be an arbitrary set and $f : U \to [0, 1]$ such that $\sup\{f(x) : x \in U\} = 1$. Let $T : 2^U \to [0, 1]$ be

$$T(A) = \sup\{f(x) : x \in A\}.$$

Then T is *maxitive*, i.e., $T(A \cup B) = \max(T(A), T(B))$. As such, T is alternating of infinite order. (See subsection on upper and lower probabilities in this section 4.1.) Thus, its dual $F(A) = 1 - T(A^c)$ is monotone of infinite order, i.e., $F : 2^U \to [0, 1]$ is a belief function.

On the practical side, subjective assignments (say, by "experts") of degrees of belief to "events," measurable or not, are possible from our intuitive viewpoint. As such, belief functions have domain 2^U for arbitrary set U.

Here is a case where belief functions are induced by probability measures on arbitrary measurable space $(U, \mathcal{U})$. Let μ be a probability measure on $\mathcal{U}$. Let $F : 2^U \to [0, 1]$ be the *inner measure* of μ, i.e., for $B \subseteq U$,

$$F(B) = \sup\{\mu(A) : A \in \mathcal{U}, A \subseteq B\}.$$

Then F is a belief function on U. Clearly, $F(\emptyset) = 0$ and $F(U) = 1$. To prove the property of monotonicity of infinite order, we proceed as follows. (For additional details on the following proofs, see Halmos [47], Kampé de Fériet [58], Fagin and Halpern [31], and Halpern [49].)

LEMMA 4.1   For each $B \subseteq U$, $\sup\{\mu(A) : A \in \mathcal{U}, A \subseteq B\}$ is attained.

Proof. For each $n \geq 1$, let $A_n \in \mathcal{U}$, $A_n \subseteq B$ such that

$$\mu(A_n) \geq F(B) - 1/n,$$

where $F(B) = \sup\{\mu(A) : A \in \mathcal{U}, A \subseteq B\}$. Let $A = \bigcup_{n \geq 1} A_n$. Then $A \in \mathcal{U}$ and $A \subseteq B$. Thus,

$$F(B) \geq \mu(A) \geq \mu(A_n) \geq F(B) - 1/n.$$

Since n is arbitrary, it follows that $F(B) = \mu(A)$.   □

LEMMA 4.2 For any $n \geq 1$, let $B_1, B_2, \ldots, B_n$ in 2^U and $A_1, A_2, \ldots, A_n$ in $\mathcal{U}$ with $A_i \subseteq B_i$, $F(B_i) = \mu(A_i)$, $i = 1, 2, \ldots, n$. Then $F\left(\bigcap_{i=1}^{n} B_i\right) = \mu\left(\bigcap_{i=1}^{n} A_i\right)$.

Proof. By Lemma 4.1, there is a $C \in \mathcal{U}$, $C \subseteq \bigcap_{i=1}^{n} B_i$ such that $F\left(\bigcap_{i=1}^{n} B_i\right) = \mu(C)$. Thus, it suffices to show that $\mu(C) = \mu\left(\bigcap_{i=1}^{n} A_i\right)$. Without loss of generality, we can assume that $\bigcap_{i=1}^{n} A_i \subseteq C$ (otherwise, replace C by $C \cup \left(\bigcap_{i=1}^{n} A_i\right)$ by noting that $C \cup \left(\bigcap_{i=1}^{n} A_i\right) \subseteq \bigcap_{i=1}^{n} B_i$ and

$$F\left(\bigcap_{i=1}^{n} B_i\right) = \mu(C) \leq \mu\left(C \cup \left(\bigcap_{i=1}^{n} A_i\right)\right) \leq F\left(\bigcap_{i=1}^{n} B_i\right)$$

so that $\mu\left(C \cup \left(\bigcap_{i=1}^{n} A_i\right)\right) = F\left(\bigcap_{i=1}^{n} B_i\right)$).

Let $D = C \setminus \bigcap_{i=1}^{n} A_i = \bigcup_{i=1}^{n} (C \cap A_i^c)$. Now, for each $i \in \{1, 2, \ldots, n\}$, $D \subseteq C \subseteq B_i$ with $D \cup A_i \subseteq B_i$ implying that $\mu(D \cup A_i) = F(B_i) = \mu(A_i)$, so that $\mu(D \cap A_i^c) = 0$.

Since $D \subseteq \bigcup_{i=1}^{n} A_i^c$, we have

$$\mu(D) = \mu\left(D \cap \left(\bigcup_{i=1}^{n} A_i^c\right)\right) = \mu\left(\bigcup_{i=1}^{n}(D \cap A_i^c)\right) \leq \sum_{i=1}^{n} \mu(D \cap A_i^c) = 0.$$

Thus, $\mu\left(\bigcap_{i=1}^{n} A_i\right) = \mu(C)$. $\square$

COROLLARY F is infinitely monotone.

Proof. For $n \geq 2$, let $B_1, B_2, \ldots, B_n$ in 2^U. By Lemma 4.1, let $A_1, A_2, \ldots, A_n$ in $\mathcal{U}$ such that $A_i \subseteq B_i$ and $\mu(A_i) = F(B_i)$, $i = 1, 2, \ldots, n$. We have, since $\bigcup_{i=1}^{n} A_i \in \mathcal{U}$ and $\bigcup_{i=1}^{n} A_i \subseteq \bigcup_{i=1}^{n} B_i$,

$$F\left(\bigcup_{i=1}^{n} B_i\right) \geq \mu\left(\bigcup_{i=1}^{n} A_i\right) = \sum_{\emptyset \neq I \subseteq \{1,2,\ldots,n\}} (-1)^{|I|+1} \mu\left(\bigcap_{i \in I} A_i\right) =$$

$$\sum_{\emptyset \neq I \subseteq \{1,2,\dots,n\}} (-1)^{|I|+1} F\left(\bigcap_{i \in I} B_i\right),$$

by Lemma 4.2. □

In this chapter, we focus on the case of a *finite* set U.

Example. Consider the measurable space $(U, 2^U)$, with U finite. Let Q be a probability measure on it. Then, for any integer $k \geq 1$, the set function $Q^k : 2^U \to [0,1]$, defined as $Q^k(A) = [Q(A)]^k$, is a belief function. Indeed, it suffices to observe that Q^k is of the form

$$Q^k(A) = \sum_{B \subseteq A} f(B)$$

where $f : 2^U \to [0,1]$ is a probability density (on 2^U). Using the multinomial theorem

$$(x_1 + x_2 + \dots + x_n)^k = \sum \binom{k}{k_1 k_2 \dots k_n} x_1^{k_1} x_2^{k_2} \dots x_n^{k_n}$$

where the summation extends over all nonnegative integral solutions $k_1, k_2, \dots, k_n$ of $k_1 + k_2 + \dots + k_n = k$, and

$$\binom{k}{k_1 k_2 \dots k_n} = \frac{k!}{k_1! k_2! \dots k_n!},$$

the student is asked to verify that the following f will do:

$$f(A) = \begin{cases} 0 \text{ if } A = \emptyset \text{ or } |A| > k \\ \displaystyle\sum_{\substack{n_a \neq 0 \\ \sum n_a = k}} \prod_{a \in A} \frac{k!}{\prod n_a!} [P(a)]^{n_a} \end{cases}$$

by noting that $Q^k(A) = \left[\displaystyle\sum_{a \in A} Q(a)\right]^k$.

Possibility Measures

As mentioned in the previous chapters, the context of coarse data analysis reveals not only that degrees of belief (in the occurrences of events) are taken from a pessimistic point of view, but also that the qualitative concept of possibility can be quantified from an optimistic point of view. Specifically, if S is a coarsening of X, then the *degree of possibility* for the occurrence of an event A is $\Pi(A) = P(S \cap A \neq \emptyset)$. If we start assigning degrees of possibility to points of U, i.e., considering $\pi(\cdot) : U \to [0,1]$ called a *possibility distribution* (Zadeh [128]), then a *possibility measure* associated with π, denoted as $\Pi(\cdot)$, which is a set function $\Pi(\cdot) : 2^U \rightarrow [0,1]$, where $\pi(u) = \Pi(\{u\})$, is defined as: $\Pi(A) = \sup_{u \in A} \pi(u)$. There exists a random S on U such that $\Pi(A) = P(S \cap A \neq \emptyset)$. Indeed, let $\alpha : (\Omega, \mathcal{A}, P) \rightarrow [0,1]$ be a random variable, uniformly distributed. Consider the randomized level-set map:

$$S : \Omega \rightarrow 2^U,$$

$$S(\omega) = \{u \in U : \pi(u) \geq \alpha(\omega)\}.$$

Then, S is a finite random set on U with

$$P(S \cap A \neq \emptyset) = P\left(\omega : \alpha(\omega) \leq \max_{u \in A} \pi(u)\right) = \max_{u \in U} \pi(u).$$

It is interesting to note that the axioms for possibility measures proposed by Zadeh are dual to a special case of *generalized information measures* of Kampé de Fériet (see e.g., Nguyen [82]). Let $(\Omega, \mathcal{A}, P)$ be a probability space. The Wiener-Shannon information measure $J(A) = -c \log P(A)$, where c is a positive constant (e.g., $c = 1/\log(2)$), led to the concept of information provided by a (measurable) partition $A_1, \ldots, A_n$ of U, namely, the *entropy* $-c \sum_{i=1}^{n} P(A_i) \log P(A_i)$. This information measure J, which is a function of a probability measure P, is *composable* in the sense that, if $A \cap B = \emptyset$, then

$$J(A \cup B) = -c \log \left[e^{-J(A)/c} + e^{-J(B)/c}\right].$$

The meaning of information in the context of localization is this. In localizing the unknown quantity, such as the number π, the interval $(3.1, 3.2)$ should provide less information than $(3.14, 3.15)$. Thus, a *general* information J on Ω should be a set function such that $J(\emptyset) = +\infty$, $J(\Omega) = 0$, and $A \subseteq B \Rightarrow J(A) \geq J(B)$. Since $J(A \cup B) \leq \min\{J(A), J(B)\}$, in general, there exist information measures J, which are not functions of probability measures, and possess a special property, namely, $J\left(\bigcup_{i \in I} A_i\right) = \inf_{i \in I} J(A_i)$ for any index set I. For example, $J(A) = \inf_{w \in A} \varphi(w)$, where $\varphi : \Omega \rightarrow \overline{\mathbb{R}^+} = [0, +\infty]$ such that $\inf_{\Omega} \varphi = 0$. These are called information measures of type Inf. Note that Zadeh's *possibility measures* are set functions $\Pi : 2^U \rightarrow [0,1]$ such that:

(i) $\Pi(\emptyset) = 0$, $\Pi(U) = 1$.

(ii) For any family $\{A_i, i \in I\}$ of subsets of U,

$$\Pi\left(\bigcup_{i \in I} A_i\right) = \sup_{i \in I} \Pi(A_i).$$

Possibility measures can be interpreted as limits of probability measures in the *large deviation convergence* sense.

The purpose of the study of large deviations in probability theory (see e.g., Akian [3], Dembo and Zeitouni [20] and the Appendix) is to find the asymptotic rate of convergence of a sequence of probability measures P_n, $n \geq 1$ (or more generally, of a family P_ε, $\varepsilon > 0$) of interest, i.e., $\lim_{n \to \infty} [P_n(A)]^{1/n}$ for some "rare" event A, which is the same as looking at $\lim_{n \to \infty} \frac{1}{n} \log P_n(A)$.

In the simplest setting of classical statistics, let $\{X_n, n \geq 1\}$ be a sequence of independent and identically distributed (i.i.d.) random variables, distributed as X with $E(X) = \mu$, $V(X) = \sigma^2 < \infty$, and let the sample mean be $\bar{X}_n = (X_1 + X_2 + \cdots + X_n)/n$. By the law of large numbers, the distribution of $\bar{X}_n - \mu$ converges to the distribution of the random variable degenerate at 0, whereas an appropriate scaling $\sqrt{n}(\bar{X}_n - \mu)$ leads to a nontrivial limiting distribution, namely, a normal distribution, by the Central Limit Theorem.

While the law of large numbers asserts that the sequence of sample means $\bar{X}_n$ converges to the common mean $EX_1 = \mu$, it is also of interest to find the *rate* at which $P(|\bar{X}_n - \mu| > a)$ goes to zero, as $n \to \infty$, for given a. This type of problems is referred to as *large deviation problems* (e.g., estimation of probability of *large deviations from the mean*). Now, the central limit theorem asserts that, for n sufficiently large,

$$P(Z_n > x) \approx \int_x^\infty \frac{1}{\sqrt{2\pi}} e^{-t^2/2} dt,$$

where $Z_n = \sqrt{n}(\bar{X}_n - \mu)/\sigma$. Let us find the asymptotics of this integral for $x \to \infty$.

Let $x > 0$. By integration by parts of the function $(1/y^2)e^{-y^2/2}$, we have

$$0 \leq \int_x^\infty \frac{1}{y^2} e^{-y^2/2} dy = \frac{1}{x} e^{-x^2/2} - \int_x^\infty e^{-y^2/2} dy$$

so that

$$\int_x^\infty e^{-y^2/2} dy \leq \frac{1}{x} e^{-x^2/2}.$$

Also, by integration by parts of the function $(y^2 - 1)/(1 + y^2)^2$, we have

$$\int_x^\infty \frac{y^2 - 1}{(1 + y^2)^2} e^{-y^2/2} dy = \frac{x}{1 + x^2} e^{-x^2/2} - \int_x^\infty \frac{y^2}{1 + y^2} e^{-y^2/2} dy$$

so that

$$\frac{x}{1 + x^2} e^{-x^2/2} = \int_x^\infty \frac{(y^2 + 1)^2 - 2}{(1 + y^2)^2} e^{-y^2/2} dy$$

$$\leq \int_x^\infty e^{-y^2/2} dy$$

since $((y^2 + 1)^2 - 2)/(1 + y^2)^2) \leq 1$. Thus

$$\frac{1}{\sqrt{2\pi}} \left(\frac{x}{1+x^2} e^{-x^2/2} \right) \leq \frac{1}{\sqrt{2\pi}} \int_x^\infty e^{-y^2/2} dy$$

$$\leq \frac{1}{\sqrt{2\pi}} \frac{e^{-x^2/2}}{x}$$

for any $x > 0$ and hence

$$\frac{1}{\sqrt{2\pi}} \int_x^\infty e^{-y^2/2} dy \approx \frac{1}{\sqrt{2\pi}} \frac{e^{-x^2/2}}{x},$$

as $x \to \infty$. It follows that, for n sufficiently large,

$$P(Z_n \geq a_n) \approx \frac{1}{\sqrt{2\pi}} \frac{e^{-a_n^2/2}}{a_n} = e^{-a_n^2(1+\varsigma_n)/2}$$

as long as $a_n \to \infty$ and for a sequence $\varsigma_n \to 0$ as $n \to \infty$.

A classical result (Cramer) is this. The sequence of sample means $\bar{X}_n$ satisfies the *large deviation principle* in the sense that, for $\varepsilon > 0$, we have

$$P\left(|\bar{X}_n - \mu| > a\right) \leq e^{-nh(a)+\varepsilon}$$

for n sufficiently large, the exponential rate of convergence being $e^{-nh(a)}$.

The general setup is this. Let Ω be a complete, separable metric space, and $\mathcal{A}$ its Borel σ-field. A sequence $(P_n, n \geq 1)$ of probability measures on $(\Omega, \mathcal{A})$ is said to obey the *large deviation principle* (LDP) if there exists a lower semicontinuous function $I : \Omega \to [0, +\infty]$, called the rate function (in fact, $\forall \alpha > 0$, $\{\omega \in \Omega : I(\omega) \leq \alpha\}$ is compact) such that

(i) for each closed set F of Ω,

$$\limsup_{n\to\infty} \frac{1}{n} \log P_n(F) \leq -\inf_{\omega \in F} I(\omega),$$

(ii) for each open set G of Ω,

$$\limsup_{n\to\infty} \frac{1}{n} \log P_n(G) \geq -\inf_{\omega \in G} I(\omega).$$

As Puhalskii [102] pointed out, if we let $\Pi(A) = \sup_{\omega \in A} e^{-I(\omega)}$, then the LDP is formulated as

(i') $\limsup_{n\to\infty} [P_n(F)]^{1/n} \leq \Pi(F)$

(ii') $\limsup_{n\to\infty} [P_n(G)]^{1/n} \leq \Pi(G)$

and, referring to the Portmanteau theorem for weak convergence of probability measures on metric spaces (see Appendix), we can view the set function $\Pi(\cdot)$ as a *limit* of P_n *in the sense* of LDP. Clearly, the set function $\Pi(\cdot)$ is formally a possibility measure. Moreover, its possibility distribution $\Pi(\{\omega\}) = e^{-I(\omega)}$ is upper semicontinuous (with values in $[0, 1]$). As such $\Pi(\cdot)$ characterizes the distribution of a *random closed set* on Ω (see Chapter 5).

Upper and Lower Probabilities

Upper and lower probabilities are set functions that arise mainly in *robust statistics*, see e.g., Huber and Strassen [54]. We have seen several examples in Chapter 3 where incomplete probabilistic information is of the form of classes of probability measures. The term *imprecise probabilities* is also used to describe such situations.

Let $(U, \mathcal{U})$ be a measurable space. Let $\mathbb{P}$ denote the class of all probability measures on $\mathcal{U}$. The law of a random element with values in U is some probability measure $P_o \in \mathbb{P}$. In practice, P_0 might be known to lie in some subclass $\mathcal{P} \subseteq \mathbb{P}$. In the context of hypothesis testing, a class $\mathcal{P} \subseteq \mathbb{P}$ could denote a hypothesis concerning P_0. In any case, let $\mathcal{P} \subseteq \mathbb{P}$. Then the *lower probability* f and the *upper probability* T are defined as

$$F, T : \mathcal{U} \to [0, 1]$$

$$F(A) = \inf\{P(A) : P \in \mathcal{P}\}$$

$$T(A) = \sup\{P(A) : P \in \mathcal{P}\}.$$

Note that $T(A) = 1 - F(A^c)$.

Clearly these set functions are monotone, with $F(\emptyset) = T(\emptyset) = 0$, $F(U) = T(U) = 1$. Depending upon the structure of $\mathcal{P}$, they could satisfy some stronger conditions, such as *monotonicity of order 2*, i.e.,

$$F(A \cup B) \geq F(A) + F(B) - F(A \cap B),$$

or dually, T is *alternating of order 2*, i.e.,

$$T(A \cap B) \leq T(A) + T(B) - T(A \cup B).$$

In particular, F could be *monotone of infinite order* on $\mathcal{U}$, or, dually, T could be *alternating of infinite order* on $\mathcal{U}$, i.e., for any $n \geq 2$, and $A_1, A_2, \ldots, A_n$ in $\mathcal{U}$,

$$T\left(\bigcap_{i=1}^{n} A_i\right) \leq \sum_{\emptyset \neq I \subseteq \{1,2,\ldots,n\}} (-1)^{|I|+1} T\left(\bigcup_{i \in I} A_i\right).$$

We have seen in Chapter 3 that random set models are appropriate when the lower probability associated with $\mathcal{P}$ is monotone of infinite order (or dually, the upper probability associated with $\mathcal{P}$ is alternating of infinite order). More details on this will be given in Section 4.3. Below is a useful result to check the property of alternating of infinite order of set functions.

A general class of set functions appearing e.g., in the theory of extreme stochastic processes (Norberg [93, 95] and Molchanov [77]) consists of set functions T on $(U,\mathcal{U})$ with $T(\emptyset) = 0$, and *maxitive*, i.e.,

$$T(A \cup B) = \max(T(A), T(B)), \quad A, b \in \mathcal{U}.$$

Note that maxitive set functions are monotone and for any $A_1, A_2, \ldots, A_n$,

$$T\left(\bigcup_{i=1}^{n} A_i\right) = \max(T(A_1), T(A_2), \ldots, T(A_n)).$$

It turns out that maxitive set functions are necessarily alternating of infinite order. This result is interesting in its own right.

THEOREM   Let C be a class of subsets of some set Θ, containing $\emptyset$ and stable under finite intersections and unions. Let $T : C \to [0, +\infty)$ be maxitive, i.e.,

$$\forall A, B \in C, \ T(A \cup B) = \max\{T(A), T(B)\}.$$

Then T is alternating of infinite order.

Proof. Clearly T is monotone increasing on C. Note that for any $A_1, \ldots, A_n$ in C, $n \geq 2$,

$$T\left(\bigcup_{i=1}^{n} A_i\right) = \max\{T(A_i), \ i = 1, 2, \ldots, n\}.$$

We need to show that

$$T\left(\bigcap_{i=1}^{n} A_i\right) \leq \sum_{\emptyset \neq I \subseteq \{1,2,\ldots,n\}} (-1)^{|I|+1} T\left(\bigcup_{i \in I} A_i\right). \tag{4.1}$$

Without loss of generalities, we may assume that

$$0 \leq \alpha_n = T(A_n) \leq \alpha_{n-1} = T(A_{n-1}) \leq \ldots \leq \alpha_1 = T(A_1).$$

For $k \in \{1, 2, \ldots, n\}$, let $J(k) = \{I \subseteq \{1, 2, \ldots, n\} : |I| = k\}$. For $I \in J(k)$, let $m(I) = \min\{i : i \in I\}$, and for $i = 1, 2, \ldots, n - k + 1$, let

$$J_i(k) = \{I \in J(k) : m(I) = i\}.$$

Then we have

$$T\left(\bigcup_{j \in I} A_j\right) = \alpha_i$$

for every $I \in J_i(k)$, $i = 1, 2, \ldots, n - k + 1$. Also,

$$|J_i(k)| = \binom{n-i}{k-1} = \frac{(n-i)!}{(k-1)!(n-i-k+1)!}$$

for every $i = 1, 2, \ldots, n - k + 1$. Thus

$$\sum_{I \in J(k)} T\left(\bigcup_{i \in I} A_i\right) = \sum_{i=1}^{n-k+1} \sum_{I \in J_i(k)} T\left(\bigcup_{j \in I} A_j\right)$$

$$= \sum_{i=1}^{n-k+1} \sum_{I \in J_i(k)} \alpha_i = \sum_{i=1}^{n-k+1} \binom{n-i}{k-1} \alpha_i$$

and therefore

$$\sum_{\emptyset \neq I \subseteq \{1,2,\ldots,n\}} (-1)^{|I|+1} T\left(\bigcup_{i \in I} A_i\right) = \sum_{k=1}^{n} (-1)^{k+1} \sum_{I \in J(k)} T\left(\bigcup_{i \in I} A_i\right)$$

$$= \sum_{k=1}^{n} (-1)^{k+1} \sum_{i=1}^{n-k+1} \binom{n-i}{k-1} \alpha_i$$

$$= \sum_{i=1}^{n} \left(\sum_{k=0}^{n-i} \binom{n-i}{k} (-1)^k\right) \alpha_i$$

$$= \alpha_n = T(A_n) \geq T\left(\bigcap_{i=1}^{n} A_i\right)$$

by observing that for any $i = 1, 2, \ldots, n - 1$,

$$\sum_{k=0}^{n-i} \binom{n-i}{k} (-1)^k = (1-1)^{n-i} = 0.$$

□

Here is a couple of examples of maxitive set functions.

Measures of noncompactness. Kuratowski introduced a measure of non-compactness in topology as follows (see e.g., [2, 7]).

Let U be a metric space with metric δ. Recall that the diameter of a subset A of U is

$$\delta(A) = \sup\{\delta(x, y) : x, y \in A\}$$

$(\delta(\emptyset) = 0$, $\delta(A) = \infty$ for A unbounded). Let $\alpha : 2^U \rightarrow \mathbb{R}^+$ be defined as: $\alpha(A) = \inf\{\varepsilon > 0 : A$ can be covered by a finite number of sets of diameter smaller than $\varepsilon\}$.

It is well known that this measure of noncompactness is maxitive (called *semiadditivity* property in topology). Note that for $A \subseteq \mathbb{R}^d$, $\alpha(A) = 0$ or ∞ according to A is bounded or not. But for an infinite dimensional Banach space, $\alpha(\cdot)$ can take any value between 0 and ∞.

Here is an example that will be used to investigate the validity of the Choquet theorem on *non-locally-compact Polish spaces* in Chapter 5.

Let U be the closed unit ball of the Hilbert space ℓ_2, where

$$\ell_2 = \left\{ x = (x_n, \ n \geq 1) : \|x\|^2 = \sum_{n=1}^{\infty} x_n^2 < +\infty \right\}$$

$$U = \{ x \in \ell_2 : \|x\| \leq 1 \}.$$

Let $\mathcal{G}$ denote the class of open sets of U, and

$$B(x, r) = \{ y \in U : \|x - y\| < r \}.$$

For $A \in \mathcal{G}$, let

$$\alpha_n(A) = \inf \left\{ r > 0 : A \subseteq \bigcup_{i=1}^{n} B(x_i, r) \right\}.$$

Since $(\alpha_n(A), \ n \geq 1)$ is a decreasing sequence, we note

$$\alpha(A) = \lim_{n \to \infty} \alpha_n(A), \quad \forall A \in \mathcal{G}.$$

Note that each $\alpha_n(\cdot)$ is not maxitive. Indeed, let

$$\{ B(x_i, r) : r > 0, \ i = 1, 2, \ldots, 2n \}$$

be a family of $2n$ disjoint balls in U. Let $A = \bigcup_{i=1}^{n} B(x_i, r)$, and $B = \bigcup_{i=n+1}^{2n} B(x_i, r)$. Then $\alpha_n(A) = \alpha_n(B) = r$, but $\alpha(A \cup B) \geq 2r$.

However, $\alpha(\cdot)$ is maxitive. Indeed, for $A, B \in \mathcal{G}$, we have

$$\alpha_{2n}(A \cup B) \leq \max\{\alpha_n(A), \alpha_n(B)\}, \quad n \geq 1.$$

Thus,

$$\alpha(A \cup B) \leq \max\{\alpha(A), \alpha(B)\}$$

yielding $\alpha(A \cup B) = \max\{\alpha(A), \alpha(B)\}$.

Fractal dimensions. For concreteness, consider various concepts of dimension in fractal geometry of $\mathbb{R}^d$ (see e.g., [32]).

For $A \neq \emptyset$ and bounded in $\mathbb{R}^d$, let $N_r(A)$ denote the smallest number of sets of diameter r that cover A. Then the upper *box-dimension* of A is defined to be

$$\dim_B(A) = \lim_{r \to 0} \sup \frac{\log N_r(A)}{-\log r}.$$

A more familiar concept of dimension is that of Hausdorff, which is defined in terms of measures. We say that a finite or countable collection of subsets $(U_n, n \geq 1)$ is an r-cover of a set A if it covers A with diameters $\delta(A_n) \leq r$, $\forall n \geq 1$.

For $s \geq 0$ and $r > 0$, let

$$H_r^s(A) = \inf\left\{\sum_{n=1}^{\infty}(\delta(U_n))^s : \{U_n, \ n \geq 1\} \text{ is an } r\text{-cover of } A\right\}.$$

Since $H_r^s(A)$ decreases as r decreases, we let

$$H^s(A) = \lim_{r\to 0} H_r^s(A).$$

The *Hausdorff dimension* of A is defined in terms of these s-Hausdorff measures:

$$\dim_H(A) = \inf\{s : H^s(A) = 0\} = \sup\{s : H^s(A) = \infty\}.$$

Another similar concept of dimension is the *packing dimension*. For $r > 0$, a finite or countable collection of disjoint closed balls B_n of radii at most r with center in A is called a r-packing of A.

For $r > 0$ and $s \geq 0$, let

$$P_r^s(A) = \sup\left\{\sum_{n=1}^{\infty}(\delta(B_n))^s : \{B_n, n \geq 1\} \text{ is an } r\text{-packing of } A\right\}.$$

Again, P_r^s decreases as δ decreases, we let $P_0^s(A) = \lim_{r\to 0} P_r^s(A)$, and the s-packing measure:

$$P^s(A) = \inf\left\{\sum_{n=1}^{\infty} P_0^s(E_n) : A \subseteq \bigcup_{n=1}^{\infty} E_n\right\}.$$

The *packing-dimension* of A is

$$\dim_P(A) = \inf\{s : P^s(A) = 0\} = \sup\{s : P^s(A) = \infty\}.$$

All the three above dimensions, as set functions, are maxitive (called *finitely stable* in the literature of fractal geometry).

Moreover, the Hausdorff and packing dimensions are *countably stable*, i.e.,

$$\dim\left(\bigcup_{n=1}^{\infty} A_n\right) = \sup_{1\leq n<\infty} \dim(A_n)$$

(or *σ-maxitive*), where dim denotes either the Hausdorff or packing dimension. This stronger property implies that countable sets of $\mathbb{R}^d$ have Hausdorff and packing dimensions zero. The proof of the above properties will be given next, as a consequence of a general method for constructing maxitive set functions.

Some limit aspects. We indicate here some limiting procedures leading to the max ($\vee$) operation.

In the study of *generalized information measures* of Kampé de Fériet (see e.g., [82]), the justification of the minimum ($\wedge$) operation as a composition law comes from the study of the convergence of a sequence of *Wiener-Shannon information measures*. Specifically, let P_n be a sequence of probability measures on $(\Omega, \mathcal{A})$, and $\varepsilon_n > 0$ with $\varepsilon_n \to 0$ as $n \to \infty$. We are interested in $\lim_{n\to\infty} (-\varepsilon_n \log P_n(A))$ for $A \in \mathcal{A}$. The reason is this. Suppose this limit exists for each A, then

$$J(A) = \lim_{n\to\infty} (-\varepsilon_n \log P_n(A))$$

can be taken as an information measure. It turns out that $J(\cdot)$ admits the idempotent operation minimum ($\wedge$), i.e., $J(A \cup B) = \min(J(A), J(B))$, if the sequence of *submeasures* $P_n^{\varepsilon_n}(\cdot)$ converges (pointwise) to a (σ)-maxitive set function $I(\cdot)$.

Here is an example where such a situation happens.

Let $(\Omega, \mathcal{A}, P)$ be a probability space.

Let $f : \Omega \to \mathbb{R}^+$, measurable and such that $f \in L^\infty(\Omega, \mathcal{A}, P)$.

Since P is a finite measure, and $f \in L^\infty(\Omega, \mathcal{A}, P)$, it follows that $f \in L(\Omega, \mathcal{A}, P)$ for all $p > 0$. $((\|f\|_p)^p = \int_\Omega |f|^p dP \leq \|f\|_\infty^p \int_\Omega dP = \|f\|_\infty^p$, where $\|f\|_\infty = \inf\{a \geq 0 : P\{\omega : |f(\omega)| > a\} = 0\}$, often written as ess. $\sup_\Omega f$.)

Consider the sequence of probability measures defined by

$$P_n(A) = \frac{\int_A |f|^n dP}{\int_\Omega |f|^n dP}.$$

Then $(P_n(A))^{1/n} = \dfrac{\|f \cdot 1_A\|_n}{\|f\|_n} \to \dfrac{\|f \cdot 1_A\|_\infty}{\|f\|_\infty}$, as $n \to \infty$, (where 1_A is the indicator function of A).

Now, the set function limit

$$\tau(A) = \|f \cdot 1_A\|_\infty$$

is maxitive (in fact σ-maxitive).

This can be seen as follows.

We have

$$\tau(A) = \inf\{t \geq 0 : P(A \cap (f > t)) = 0\}$$
$$= \inf\{t \geq 0 : A \in \mathcal{W}_t\}$$

where

$$\mathcal{W}_t = \{A \in \mathcal{A} : P(A \cap (f > t)) = 0\}$$
$$= \{A \in \mathcal{A} : A \cap (f > t) \in \mathcal{W}\}$$

where $\mathcal{W} = \{A \in \mathcal{A} : P(A) = 0\}$. Note that $\mathcal{W}$ and $\mathcal{W}_t$, $t \geq 0$ are σ-ideals in $\mathcal{A}$, i.e., $\mathcal{W}$ is a nonempty subset of $\mathcal{A}$, stable under countable unions ($A_n \in \mathcal{W} \Rightarrow \bigcup_n A_n \in \mathcal{W}$), $\emptyset \in \mathcal{W}$, and hereditary (if $A \in \mathcal{W}$ and $B \in \mathcal{A}$ with $B \subseteq A$, then $B \in \mathcal{W}$). Moreover, the family of σ-ideals ($\mathcal{W}_t$, $t \geq 0$) is increasing ($s < t \Rightarrow \mathcal{W}_s \subseteq \mathcal{W}_t$). As such, $\tau(A) = \inf\{t \geq 0 : A \in \mathcal{W}_t\}$ is σ-maxitive (inf $\emptyset = \infty$). Indeed, if $A \subseteq B$ then by properties of σ-ideals,

$$\{t : B \in \mathcal{W}_t\} \subseteq \{t : A \in \mathcal{W}_t\}$$

and hence $\tau(\cdot)$ is monotone increasing.

Next, let $A_n \in \mathcal{A}$, then by monotonicity of τ, we have

$$\tau\left(\bigcup_n A_n\right) \geq \sup_n \tau(A_n) \quad (< \infty).$$

Let $s > \sup_n \tau(A_n)$. For any $n \geq 1$, $\tau(A_n) < s \Rightarrow A_n \in \mathcal{W}_s$ since the family ($\mathcal{W}_t$, $t \geq 0$) is increasing, thus, $\bigcup_n A_n \in \mathcal{W}_s$, resulting in $\tau\left(\bigcup_n A_n\right) \leq s$.

Remark. The above proof says that if ($\mathcal{W}_t$, $t \geq 0$) is an increasing family of σ-ideals of $\mathcal{A}$, then the set function on $\mathcal{A}$ defined by $V(A) = \inf\{t \geq 0 : A \in \mathcal{W}_t\}$ is necessarily σ-maxitive. As an application, if we look back at the *Hausdorff dimension*, we see that, with

$$\mathcal{W}_t = \{A : H^t(A) = 0\},$$

the family $\{\mathcal{W}_t, t \geq 0\}$ is an increasing family of σ-ideals in the power set of $\mathbb{R}^d$, and as such, the Hausdorff dimension is σ-maxitive (and hence alternating of infinite order).

Of course, the situation is similar for maxitive set functions. Specifically, if $\tau : \mathcal{A} \to \mathbb{R}^+$, $\tau(\emptyset) = 0$ *and maxitive*, then, for each $t \geq 0$,

$$\mathcal{M}_t = \{A \in \mathcal{A} : \tau(A) \leq t\}$$

is an *ideal* in $\mathcal{A}$ (an ideal is like an σ-ideal except the stability of countable unions is replaced by the stability of finite unions). Moreover, we have $\tau(A) = \inf\{t \geq 0 : A \in \mathcal{M}_t\}$. Conversely, let ($\mathcal{W}_t$, $t \geq 0$) be an increasing family of ideals of $\mathcal{A}$, then the set function on $\mathcal{A}$ defined by

$$\tau(A) = \inf\{t \geq 0 : A \in \mathcal{W}_t\}$$

is maxitive. Note that $(\mathcal{W}_t, \ t \geq 0)$ might be different than $(\mathcal{M}_t, \ t \geq 0)$. They coincide if and only if $(\mathcal{W}_t, \ t \geq 0)$ is "right continuous," i.e., $\mathcal{W}_t = \bigcap\limits_{s \geq t} \mathcal{W}_s$.

Finally, note that if the ideal $\mathcal{W}$ in the construction of

$$\mathcal{W}_t = \{A : A \cap (f > t) \in \mathcal{W}\}$$

is $\{\emptyset\}$, then

$$\tau(A) = \inf\{t \geq 0 : A \cap (f > t) = \emptyset\} = \sup_{x \in A} f(x).$$

Maxitive set functions of the form

$$\tau(A) = \sup_A f(x)$$

where $f : \Omega \to \mathbb{R}^+$ are very special, since they are maxitive in a strong sense, namely, for any index set I, and $(A_i, \ i \in I)$,

$$\tau\left(\bigcup_{i \in I} A_i\right) = \sup_{i \in I} \tau(A_i),$$

when $\tau(\Omega) = 1$, such set functions are called *possibility measures* (see e.g., [23]), or *idempotent probabilities* [102].

4.2 Incidence Algebras

This section provides a fairly general and convenient setting to study set functions. Perhaps the most general type of set functions is the class of real-valued set functions f, defined on some algebra of subsets of some space U, such that $f(\emptyset) = 0$, which are coalitional games in the theory of cooperative games, see e.g., Marinacci [71]. We look at functions defined on 2^U, where U is a *finite* set. Incidence algebras of a locally finite partially ordered set over a commutative ring with identity is a natural setting for the study of combinatorial problems.

DEFINITION 4.1   *Let U be a finite set, and $\mathcal{F} = \{f : 2^U \to \mathbb{R}\}$. For $f, g \in \mathcal{F}$ and $r \in \mathbb{R}$, let*

$$(f + g)(X) = f(X) + g(X)$$
$$(rf)(X) = r(f(X)).$$

PROPOSITION   *$\mathcal{F}$ is a vector space over $\mathbb{R}$.*

The proof is routine and left as an exercise. One basis for $\mathcal{F}$ as a vector space over $\mathbb{R}$ is the set of functions $\{f_Y : Y \subseteq U\}$ defined by $f_Y(Y) = 1$ and $f_Y(X) = 0$ if $X \neq Y$. Thus $\mathcal{F}$ has dimension $2^{|U|}$ over $\mathbb{R}$.

DEFINITION 4.2 *Let $\mathbb{A}$ be the set of functions $\left(2^U\right)^{[2]} \to \mathbb{R}$, where $\left(2^U\right)^{[2]} = \{(X, Y) : X \subseteq Y \subseteq U\}$. On $\mathbb{A}$ define addition pointwise and multiplication by the formula*

$$(\alpha * \beta)(X, Y) = \sum_{X \subseteq Z \subseteq Y} \alpha(X, Z)\beta(Z, Y).$$

$\mathbb{A}$ *with these operations is the* incidence algebra *of 2^U over the field $\mathbb{R}$.*

THEOREM 4.1 $\mathbb{A}$ *is a ring with identity. Its identity is the function given by $\delta(X, X) = 1$ and $\delta(X, Y) = 0$ if $X \neq Y$.*

Proof. Pointwise addition is the operation that we denote by $+$ given by $(\alpha + \beta)(X, Y) = \alpha(X, Y) + \beta(X, Y)$. Let 0 denote the mapping given by $0(X, Y) = 0$ for all $X \subseteq Y$. To show that $\mathbb{A}$ is a ring, we must show the following for all $\alpha, \beta, \gamma \in \mathbb{A}$. Their verifications are left as exercises.

1. $\alpha + \beta = \beta + \alpha$

2. $(\alpha + \beta) + \gamma = \alpha + (\beta + \gamma)$

3. $\alpha + 0 = \alpha$

4. For each α, there exists β such that $\alpha + \beta = 0$

5. $(\alpha * \beta) * \gamma = \alpha * (\beta * \gamma)$

6. $\alpha * (\beta + \gamma) = (\alpha * \beta) + (\alpha * \gamma)$

7. $(\alpha + \beta) * \gamma = (\alpha * \gamma) + (\beta * \gamma)$

8. $\alpha * \delta = \delta * \alpha = \alpha$.

$\square$

Properties 1–4 say that $\mathbb{A}$ with the operation $+$ is an *Abelian group*. The ring $\mathbb{A}$ has an identity δ, but it is not true that every nonzero element α has an inverse. That is, there does not necessarily exist for α an element β such that $\alpha * \beta = \delta = \beta * \alpha$. The following theorem characterizes those elements that have inverses.

THEOREM 4.2   In the ring $\mathbb{A}$, an element α has an inverse if and only if for all X, $\alpha(X, X) \neq 0$. Its inverse is given inductively by

$$\alpha^{-1}(X, X) = \frac{1}{\alpha(X, X)}$$

$$\alpha^{-1}(X, Y) = \frac{-1}{\alpha(X, X)} \sum_{X \subset Z \subseteq Y} \alpha(X, Z)\alpha^{-1}(Z, Y) \text{ if } X \subset Y.$$

Proof. If α has an inverse β, then $(\alpha * \beta)(X, X) = \alpha(X, X)\beta(X, X) = \delta(X, X) = 1$, so that $\alpha(X, X) \neq 0$. Now suppose that for all X, $\alpha(X, X) \neq 0$. We need an element β such that $\beta * \alpha = \alpha * \beta = \delta$. In particular, we need $(\alpha * \beta)(X, Y) = 0$ for $X \subset Y$ and $(\alpha * \beta)(Y, Y) = 1$. We define $\beta(X, Y)$ inductively on the number of elements between X and Y. If That number is 1, that is, if $X = Y$, let $\beta(X, X) = 1/\alpha(X, X)$, which is possible since $\alpha(X, X) \neq 0$. Assume that $\beta(X, Z)$ has been defined for elements X and Z such that the number of elements between the two is $< n$, and suppose that the number of elements between X and Y is $n > 1$. We want

$$0 = (\alpha * \beta)(X, Y)$$
$$= \sum_{X \subseteq Z \subseteq Y} \alpha(X, Z)\beta(Z, Y)$$
$$= \alpha(X, X)\beta(X, Y) + \sum_{X \subset Z \subseteq Y} \alpha(X, Z)\beta(Z, Y).$$

This equation can be solved for $\beta(X, Y)$ since $\alpha(X, X) \neq 0$, yielding

$$\beta(X, Y) = \frac{-1}{\alpha(Y, Y)} \sum_{X \subseteq Z \subset Y} \beta(X, Z)\alpha(Z, Y).$$

Thus $\alpha * \beta = \delta$. Similarly, there is an element γ such that $\gamma * \alpha = \delta$. Then

$$(\gamma * \alpha) * \beta = \delta * \beta = \beta$$
$$= \gamma * (\alpha * \beta) = \gamma * \delta = \gamma.$$

The theorem follows.   □

Elements in a ring that have an inverse are called *units*. There are two very special units in $\mathbb{A}$.

- $\mu(X, Y) = (-1)^{|Y - X|}$ is the *Möbius function*.

- $\xi(X, Y) = 1$ is the *Zeta function*.

The element ξ is easy to define: it is simply 1 everywhere. The element μ is its inverse. These functions are of particular importance.

PROPOSITION   In the ring $\mathbb{A}$, $\mu * \xi = \xi * \mu = \delta$. That is, they are inverses of each other.

Proof.

$$(\mu * \xi)\,(X,Y) = \sum_{X \subseteq Z \subseteq Y} (-1)^{|Z-X|}\xi(Z,Y)$$

$$= \sum_{X \subseteq Z \subseteq Y} (-1)^{|Z-X|}.$$

Now notice that $\sum_{X \subseteq Z \subseteq Y}(-1)^{|Z-X|} = \delta(X,Y)$. Similarly $\xi * \mu = \delta$. $\square$

There is a natural operation on the elements of the vector space $\mathcal{F}$ by the elements of the incidence algebra $\mathbb{A}$. This operation is a common one in combinatorics, and will simplify some of the computations we must make later with belief functions.

DEFINITION 4.3   For $\alpha \in \mathbb{A}$, $f \in \mathcal{F}$, and $X \in 2^U$, let

$$(f * \alpha)\,(X) = \sum_{Z \subseteq X} f(Z)\alpha(Z, X).$$

PROPOSITION $\mathcal{F}$ is a (right) module over the ring $\mathbb{A}$. That is,

1. $f * \delta = f$

2. $(f * \alpha) * \beta = f * (\alpha * \beta)$

3. $(f + g) * \alpha = f * \alpha + g * \alpha$

4. $f * (\alpha + \beta) = f * \alpha + f * \beta$.

The proof of this proposition is left as an exercise. It is a straightforward calculation. Notice that for $f \in \mathcal{F}$, $f * \xi * \mu = f * \mu * \xi = f$.

With the operation $f * \alpha$, elements of $\mathbb{A}$ are linear transformations on the real vector space $\mathcal{F}$. So $\mathbb{A}$ is a ring of linear transformations on $\mathcal{F}$. Since U is finite, $\mathcal{F}$ is finite dimensional and of dimension $|2^U|$, so $\mathbb{A}$ is isomorphic to a subring of the ring of $|2^U| \times |2^U|$ real matrices. With a basis ordered properly, these matrices are upper triangular. Such a matrix has an inverse if and only if its diagonal entries are nonzero. This corresponds to an element $\alpha \in \mathbb{A}$ having an inverse if and only if $\alpha(X, X) \neq 0$. Following are some observations, elementary but significant.

- For each $r \in \mathbb{R}$, we identify r with the constant map $r \in \mathcal{F}$ defined by $r(X) = r$ for all $X \in 2^U$.

- $\mathbb{A}$ is an *algebra* over $\mathbb{R}$ via the embedding $\mathbb{R} \to \mathbb{A} : r \mapsto r\delta$, where $r\delta\,(X, Y) = r\,(\delta\,(X, Y))$. That is, $\mathbb{A}$ is a vector space over $\mathbb{R}$ and $r\delta * \alpha = \alpha * r\delta$. Note that $(r\delta * \alpha)\,(X, Y) = r\,(\alpha\,(X, Y))$.

- For $r \in \mathbb{R}$, $f \in \mathcal{F}$ and $\alpha \in \mathbb{A}$, $r(f * \alpha) = (rf) * \alpha = f * (r\alpha)$.

- If α is a unit in $\mathbb{A}$, then $\mathcal{F} \to \mathcal{F} : f \to f * \alpha$ and $\mathcal{F} \to \mathcal{F} : f \to f * \alpha^{-1}$ are one-to-one maps of $\mathcal{F}$ onto $\mathcal{F}$, and are inverses of one another.

- $\mathcal{F} \to \mathcal{F} : f \to f * \mu$ and $\mathcal{F} \to \mathcal{F} : f \to f * \xi$ are one-to-one maps of $\mathcal{F}$ onto $\mathcal{F}$, and are inverses of one another. This case is of particular interest.

- $f * \mu$ is called the *Möbius inverse* of f, or the *Möbius inversion* of f.

We begin now with some facts that will be of particular interest in the study of belief functions and other uncertainty measures.

DEFINITION 4.4   Let k be ≥ 2. An element $f \in \mathcal{F}$ is monotone of order k if for every nonempty subset $\mathcal{S}$ of 2^U with $|\mathcal{S}| \leq k$,

$$f\left(\bigcup_{X \in \mathcal{S}} X\right) \geq \sum_{\varnothing \neq \mathcal{T} \subseteq \mathcal{S}} (-1)^{|\mathcal{T}|+1} f\left(\bigcap_{X \in \mathcal{T}} X\right)$$

f *is* monotone of infinite order *if it is monotone of order k for all k.*

Of course, monotone of order k implies monotone of smaller order. Our first goal is to identify those f that are Möbius inversions of maps that are monotone of order k. This is the same as identifying those f such that $f * \xi$ is monotone of order k. There is an alternate form for the right-hand side of the inequality above, which is convenient to have.

LEMMA   Let $f : 2^U \to \mathbb{R}$. Let $\mathcal{S}$ be a subset of 2^U. Let $\Gamma = \Gamma\,(\mathcal{S})$ be the set of subsets that are contained in at least one X in $\mathcal{S}$. Then

$$\sum_{\varnothing \neq \mathcal{T} \subseteq \mathcal{S}} (-1)^{|\mathcal{T}|+1} (f * \xi) \left(\bigcap_{X \in \mathcal{T}} X\right) = \sum_{X \in \Gamma} f(X).$$

Proof.

$$\sum_{\varnothing \neq \mathcal{T} \subseteq \mathcal{S}} (-1)^{|\mathcal{T}|+1} (f * \xi) \left(\bigcap_{X \in \mathcal{T}} X\right) = \sum_{\varnothing \neq \mathcal{T} \subseteq \mathcal{S}} (-1)^{|\mathcal{T}|+1} \sum_{Y \subseteq \cap_{X \in \mathcal{T}} X} f(Y)$$

This last expression is a linear combination of $f(Y)'s$, for Y a subset of some elements of $\mathcal{S}$. Fix Y. We will find the coefficient of $f(Y)$. Let $\mathcal{T}_Y$ be the subset of $\mathcal{S}$ each of whose elements contains Y. Then for Y,

$$\sum_{\substack{\varnothing \neq T \subseteq \mathcal{S} \\ Y \subseteq \cap_{X \in T} X}} (-1)^{|T|+1} f(Y) = \sum_{\varnothing \neq T \subseteq \mathcal{T}_y} (-1)^{|\mathcal{T}_Y|+1} f(Y) = f(Y).$$

The result follows.   □

Of course, the result could have been stated as

$$\sum_{\varnothing \neq T \subseteq \mathcal{S}} (-1)^{|T|+1} f\left(\bigcap_{X \in T} X\right) = \sum_{X \in \Gamma} (f * \mu)(X).$$

The set Γ plays an important role in what follows. Let $X \subseteq U$ with $|X| \geq 2$. Let $\mathcal{S} = \{X - \{x\} : x \in X\}$. Then

- $\bigcup_{Y \in \mathcal{S}} Y = X$

- Every subset Y not X itself is uniquely the intersection of the sets in a subset of $\mathcal{S}$. In fact
$$Y = \bigcap_{x \notin Y} (X - \{x\}).$$

- The set Γ for this $\mathcal{S}$ is precisely the subsets Y of X not X itself.

We will use these facts below.

THEOREM 4.3 $f * \xi$ is monotone of order k if and only if for all A, C with $2 \leq |C| \leq k$, $\sum_{C \subseteq X \subseteq A} f(X) \geq 0$.

Proof. Suppose $f * \xi$ is monotone of order k, and $2 \leq |C| \leq k$. For $C \subseteq A$, let $\mathcal{S} = \{A - \{u\} : u \in C\}$. Then $A = \bigcup_{V \in \mathcal{S}} V$, $|\mathcal{S}| = |C|$, and

$$(f * \xi)(A) = (f * \xi)\left(\bigcup_{V \in \mathcal{S}} V\right) \geq \sum_{Y \in \Gamma} f(Y)$$

where Γ is as in the lemma. The elements of Γ are those subsets of A that do not contain some element of C, and thus are precisely those that do not contain C. Thus

$$(f * \xi)(A) = \sum_{Y \subseteq A} f(Y) = \sum_{C \subseteq X \subseteq A} f(X) + \sum_{Y \in \Gamma} f(Y)$$

Thus $\sum_{C \subseteq X \subseteq A} f(X) \geq 0$.

Now suppose that $\displaystyle\sum_{C\subseteq X\subseteq A} f(X) \geq 0$ for all A, C with $2 \leq |C| \leq k$. Let

$\varnothing \neq S = \{A_1, A_2, \ldots, A_k\}$. Then letting $A = \displaystyle\bigcup_{i=1}^{k} A_i$,

$$(f * \xi)(A) = \sum_{Y\subseteq A} f(Y) = \sum_{X\subseteq A, X\notin\Gamma} f(X) + \sum_{Y\in\Gamma} f(Y).$$

We need to show that $\displaystyle\sum_{X\subseteq A, X\notin\Gamma} f(X) \geq 0$. To do this, we will write it as

disjoint sums of the form $\displaystyle\sum_{C\subseteq X\subseteq A} f(X)$ with $2 \leq |C| \leq k$.

For $i = 1, 2, \ldots, k$, let

$$E_i = A - A_i = \{x_{i1}, \ldots, x_{in_i}\}$$
$$A_0 = A - \cup E_i$$
$$E_{ij} = \{x_{ij}, \ldots, x_{in_i}\}.$$

For each $B \notin \Gamma$, let m_i be the smallest integer such that $x_{im_i} \in B$, $i = 1, 2, \ldots, k$. Let

$$C = \{x_{1m_1}, x_{2m_2}, \ldots, x_{km_k}\}$$
$$A_C = E_{1m_1} \cup \ldots \cup E_{km_k} \cup A_0.$$

Then the intervals $[C, A_C]$ consist of elements of Γ and each $B \in \Gamma$ is in exactly one of these intervals. The theorem follows.   □

Again, the result could have been stated as f is monotone of order k if and only if $\displaystyle\sum_{C\subseteq X\subseteq A} (f * \mu)(X) \geq 0$ for all A, C with $2 \leq |C| \leq k$. Taking $A = C$, we get

COROLLARY   If $f * \xi$ is monotone of order k, then $f(X) \geq 0$ for $2 \leq |X| \leq k$.

The following corollary is of special note.

COROLLARY $f * \xi$ is monotone of infinite order if and only if $f(X) \geq 0$ for $2 \leq |X|$.

Proof. Suppose that $f(X) \geq 0$ for all X such that $2 \leq |X|$. Let S be a nonempty set of subsets of U. We need

$$(f * \xi)\left(\bigcup_{X\in S} X\right) \geq \sum_{\varnothing\neq T\subseteq S} (-1)^{|T|+1} (f * \xi)\left(\bigcap_{X\in T} X\right).$$

Let Γ be as in the lemma. Using the fact that Γ contains all the subsets Y such that $f(Y) < 0$, we have

$$(f * \xi)\left(\bigcup_{X \in \mathcal{S}} X\right) = \sum_{Y \subseteq \bigcup_{X \in \mathcal{S}} X} f(Y) \geq \sum_{Y \in \Gamma} f(Y)$$

$$= \sum_{\varnothing \neq \mathcal{T} \subseteq \mathcal{S}} (-1)^{|\mathcal{T}|+1} (f * \xi)\left(\bigcap_{X \in \mathcal{T}} X\right)$$

So $f * \xi$ is monotone of infinite order.

Now suppose that $f * \xi$ is monotone of infinite order. Let $|X| \geq 2$. We need $f(X) \geq 0$. Let $\mathcal{S} = \{X - \{x\} : x \in X\}$. Then using the fact that Γ is the set of all subsets of X except X itself, we have

$$(f * \xi)\left(\cup_{Y \in \mathcal{S}} Y\right) = (f * \xi)(X)$$
$$= \sum_{Z \subseteq X} f(Z)$$
$$\geq \sum_{Z \in \Gamma} f(Z)$$
$$= \sum_{Z \subseteq X} f(Z) - f(X).$$

Therefore, $0 \geq -f(X)$, or $f(X) \geq 0$. $\quad\square$

Some additional easy consequences of the theorem are these.

COROLLARY *The following hold.*

1. *Constants are monotone of infinite order. In fact, $(r * \mu)(X) = 0$ if $X \neq \varnothing$, and $(r * \mu)(\varnothing) = r$.*

2. *If f and g are monotone of order k, then so is $f + g$.*

3. *If f is monotone of order k and $r \geq 0$, then rf is monotone of order k.*

4. *A function f is monotone of order k if and only if for $r \in \mathbb{R}$, $f + r$ is monotone of order k.*

A connection with ordinary monotonicity is the following.

THEOREM 4.4 *If f is monotone of order 2, then f is monotone if and only if $f(\varnothing)$ is the minimum value of f.*

Proof. If f is monotone, then clearly $f(\varnothing)$ is its minimum value. Suppose that $f(\varnothing)$ is minimum. Let $Y \subseteq X$. Then $X = Y \cup Z$ with $Y \cap Z = \varnothing$. By 2-monotonicity, $f(X) = f(Y \cup Z) \geq f(Y) + f(Z) - f(\varnothing)$. Since $f(Z) \geq f(\varnothing)$, $f(X) \geq f(Y)$. $\quad\square$

COROLLARY If f is monotone of order 2 and not monotone, then $f(\{x\}) < f(\varnothing)$ for some $x \in U$.

Proof. Suppose that f is monotone of order 2 and not monotone. By the theorem, there is an $X \in 2^U$ with $f(X) < f(\varnothing)$. Let X be such an element with $|X|$ minimum. If $|X| = 1$ we are done. Otherwise, $X = Y \cup \{x\}$ with $f(Y) \geq 0$. By 2-monotonicity, $f(X) \geq f(Y) + f(\{x\}) - f(\varnothing)$. Thus $f(X) - f(Y) \geq f(\{x\}) - f(\varnothing)$, and the left side is negative. Hence $f(\{x\}) < f(\varnothing)$. $\quad\square$

By choosing f appropriately, it is easy to get $f * \xi$ that are monotone of infinite order and not monotone, for example, so that $(f * \xi)(U)$ is not the biggest value of $f * \xi$. Just make $f(x)$ very large negatively, and $f(X)$ of the other subsets positive. Then $(f * \xi)(U) < (f * \xi)(U - \{x\})$, so that $f * \xi$ is not monotone.

One cannot state the definition of ordinary monotonicity in the form of Definition 4.4. Having the $|S| = 1$ imposes no condition at all. However,

THEOREM 4.5 $f * \xi$ is monotone if and only if for all A, C with $1 = |C|$,
$$\sum_{C \subseteq X \subseteq A} f(X) \geq 0.$$

Proof. It is clear that $f * \xi$ is monotone if and only if for $A = B \cup \{a\}$, $(f * \xi)(B) \leq (f * \xi)(A)$. This latter holds if and only if

$$\sum_{X \subseteq B} f(X) \leq \sum_{X \subseteq A} f(X) = \sum_{X \subseteq B} f(X) + \sum_{\{b\} \subseteq X \subseteq A} f(X)$$

if and only if $0 \leq \displaystyle\sum_{\{b\} \subseteq X \subseteq A} f(X)$. $\quad\square$

We turn now to the study of functions that are of particular interest to us, namely, belief functions. Recall the following.

DEFINITION 4.5 *A function* $F : 2^U \to [0, 1]$ *is a belief function if*

1. $F(\varnothing) = 0$

2. $F(U) = 1$

3. F is monotone of infinite order.

Note the following.

- A belief function is monotone of order k for any k.

- A belief function is monotone by Theorem 4.4.

There is an intimate connection between belief functions on 2^U and densities on 2^U. We establish that connection now.

DEFINITION 4.6 A density *on* 2^U *is a function* $f : 2^U \rightarrow [0,1]$ *such that* $\sum\limits_{X \subseteq U} f(X) = 1$.

THEOREM 4.6 Let f be a density on 2^U with $f(\varnothing) = 0$. Then

1. $(f * \xi)(\varnothing) = 0$

2. $(f * \xi)(U) = 1$

3. $f * \xi$ is monotone of infinite order.

Proof. The first two items are immediate. Since densities are nonnegative, by corollary on the criterion for monotone of infinite order, $f * \xi$ is monotone of infinite order. □

Now we get the precise correspondence between belief functions and densities.

THEOREM 4.7 F is a belief function if and only if $F * \mu$ is a density with value 0 at $\varnothing$.

Proof. If $F * \mu$ is a density with value 0 at $\varnothing$, then the previous theorem gets $(F * \mu) * \xi = F$ to be a belief function. Assume that F is a belief function. Then

$$\sum_{X \subseteq U} (F * \mu)(X) = ((F * \mu) * \xi)(U) = F(U) = 1.$$

We need $g * \mu \geq 0$.

$$(F * \mu)(\varnothing) = F(\varnothing)\mu(\varnothing, \varnothing) = F(\varnothing) = 0.$$

For $\{x\}$,

$$(F * \mu)(\{x\}) = F(\varnothing)\mu(\varnothing, \{x\}) + F(\{x\})\mu(\{x\}, \{x\}) = F(\{x\}) \geq 0.$$

Since g is monotone of infinite order, $(F * \mu)(X) \geq 0$ for $|X| \geq 2$. □

COROLLARY Let $\mathcal{D}$ be the set of densities on 2^U with value 0 at $\varnothing$, and $\mathcal{B}$ let be the set of belief functions on 2^U. Then $\mathcal{D} \xrightarrow{\xi} \mathcal{B}$ is a one-to-one correspondence with inverse μ.

COROLLARY A belief function F is a probability measure if and only if $F * \mu$ is a density on U, that is, if and only if $(F * \mu)(X) = 0$ for all $|X| \geq 2$.

Proof. Densities f on U give measures F on 2^U via $F(X) = \sum_{x \in X} f(x)$. Any measure on 2^U comes about this way. $\square$

Note on Möbius Transforms of Set Functions

An interesting and natural question, not only for possible applications, but also as a mathematical problem of its independent interest is whether there exists Möbius transforms in the non-locally-finite case. In [71], a formulation of such a counterpart is presented. The question is this. Suppose U is infinite, and $\mathcal{U}$ is an infinite algebra of subsets of U. Let $g : \mathcal{U} \to \mathbb{R}$, and denote the set of all such functions by V. With pointwise addition and multiplication by scalars, V is a vector space over $\mathbb{R}$.

First, consider the case where U is finite. Then the Möbius transform of $g \in V$ plays the role of coefficients in the development of g with respect to some linear basis of V. Specifically,

$$g(A) = \sum_{\varnothing \neq B \in \mathcal{U}} \alpha_g(B) u_B(A) \tag{4.2}$$

where $u_B \in V$ for any $B \in \mathcal{U}$,

$$u_B(A) = \begin{cases} 1 \text{ if } B \subseteq A \\ 0 \text{ otherwise} \end{cases}$$

and

$$\alpha_g(B) = \sum_{D \subseteq B} (-1)^{|B-D|} g(D).$$

Now, if we view the Möbius transform α_g of g as a signed measure on the finite set $\mathcal{U}$, for $\mathcal{A} = \{A_1, \ldots, A_n\} \subseteq \mathcal{U}$, and define $\alpha_g(\mathcal{A}) = \sum_{i=1}^n \alpha_g(A_i)$, then the Möbius transform of a set function is a signed measure on $\mathcal{U}$ satisfying (4.2).

We need another identification. For each $A \in \mathcal{U}$, we identify A with the principal filter generated by A, namely, $p(A) = \{B \in \mathcal{U} : A \subseteq B\}$. But, each principal filter p generated by A corresponds uniquely to an element u_p where $u_p(B)$ is equal to 1 if $B \in p$ and zero otherwise. Thus in the case of a finite algebra $\mathcal{U}$, the Möbius transform α_g of a set function $g : \mathcal{U} \to \mathbb{R}$ is a signed measure on the space

$$\mathbb{F} = \{u_p : p \text{ a principal filter of } \mathcal{U}\}$$

satisfying (4.2). Its existence is well known from combinatorial theory. As we will see, its existence is due to the fact that the "composition norm" of g, namely $\|g\| = \sum_{A \in \mathcal{U}} |\alpha_g(A)|$, is finite.

Now consider the case where the algebra $\mathcal{U}$ of subsets of U is arbitrary. From the identifications above, it seems natural to define the Möbius transform $g : \mathcal{U} \to \mathbb{R}$ by generalizing the space $\mathbb{F}$ and the relation (4.2) to an integral representation, as well as specifying sufficient conditions for their existence. In other works, the Möbius transform of g, when it exists, will be a signed measure living on some appropriate space.

The space $\mathbb{F}$ is generalized as follows. Observe that an element u_p in $\mathbb{R}$ is a special case of a set-function v_p on $\mathcal{U}$ where p is a proper filter of $\mathcal{U}$, that is $p \neq \mathcal{U}$, $v_p = 1$ if $A \in p$ and 0 otherwise. Thus $\mathbb{F}$ is generalized to

$$\mathbb{G} = \{v_p : p \text{ a proper filter of } \mathcal{U}\}.$$

Note that the space $\mathbb{G}$ can be topologized appropriately. By a measure on $\mathbb{G}$, we mean a Borel measure on $\mathbb{G}$. For a set-function $g" : \mathcal{U} \to \mathbb{R}$, we define its composition norm by

$$\|g\| = \sup\{\|g|_{\mathcal{F}}\| : \mathcal{F} \text{ a finite subalgebra of } \mathcal{U}\},$$

where $g|_{\mathcal{F}}$ denotes the restriction of g to $\mathcal{F}$.

The basic representation theorem of Marinacci [71] is this. If $\|g\| < \infty$, then there is a unique regular and bounded signed measure α_g on $\mathbb{G}$ such that for $A \in \mathcal{U}$,

$$g(A) = \int_{\mathbb{G}} v_p(A) d\alpha_g(v_p).$$

It is clear that this is an extension of (4.2). Thus, via identifications, the signed measure α_g on $\mathbb{G}$ can be viewed in the infinite case as the Möbius transform of the set-function g with bounded norm.

4.3   Cores of Capacity Functionals

Let S be a nonempty random set on a finite set U. The probability law of S is determined either by the distribution function $F(A) = P(S \subseteq A)$ or by its *capacity functional* $T(A) = P(S \cap A \neq \emptyset) = 1 - F(A^c)$. Recall that, in the finite case, T satisfies

(i) $T(\emptyset) = 0$, $T(U) = 1$

(ii) T is alternating of infinite order, i.e., $\forall n \geq 2$, $\forall A_1, A_2, \ldots, A_n$ in 2^U,

$$T\left(\bigcap_{i=1}^{n} A_i\right) \leq \sum_{\emptyset \neq I \subseteq \{1,2,\ldots,n\}} (-1)^{|I|+1} T\left(\bigcup_{i \in I} A_i\right).$$

Let $\mathbb{P}$ denote the class of all probability measures on U. The core of T, denoted by $\mathcal{C}(T)$, is $\mathcal{C}(T) = \{P \in \mathbb{P} : P \leq T\}$, where $P \leq T$ means $\forall A \subseteq U$, $P(A) \leq T(A)$. The motivation to look at $\mathcal{C}(T)$ in statistical inference with set-valued observations is this. Consider the situation in *coarse data analysis*. The random variable X with values in U is unobservable. Instead, a coarsening S of X is observable. X is an almost sure selector of S which is a (non-empty) random set on U with unknown capacity functional T. The true (unknown) probability law of X is an element of $\mathcal{C}(T)$, which plays the role of a "parameter space." However $\mathcal{C}(T)$ is unknown. It can be consistently estimated from a random sample $S_1, S_2, \ldots, S_n$ drawn from S. Indeed, let T_n denote the empirical capacity functional based on $S_1, S_2, \ldots, S_n$, i.e.,

$$T_n(A) = \frac{1}{n} \#\{1 \leq j \leq n : S_j \cap A \neq \emptyset\}.$$

Then by the strong law of large numbers, $T_n(A) \to T(A)$, with probability 1, as $n \to +\infty$. Moreover, the empirical probability dF_n based on the unobservable $X_1, X_2, \ldots, X_n$ from X belongs to $\mathcal{C}(T_n)$ almost surely. As such, inference about the true probability law of X can be based upon the approximation of $\mathcal{C}(T)$ by $\mathcal{C}(T_n)$, for n sufficiently large.

We proceed to investigate the structure of the core of a capacity functional T, denoted by $\mathcal{C}(T)$, or equivalently, the core of its associated distribution function F (of a nonempty random set on the finite set U), denoted by $\mathcal{C}(F)$, where $\mathcal{C}(F) = \{P \in \mathbb{P} : F \leq P\}$. In the following f always denoted the Möbius transform of F, which is a probability density on 2^U with $f(\emptyset) = 0$.

Given f, there is a natural way to construct densities on U. These are *allocations*.

DEFINITION 4.7   *Let f be a density on 2^U with $f(\emptyset) = 0$. An* allocation *of f is a function $\alpha : U \times (2^U \setminus \{\emptyset\}) \to [0,1]$ such that $\forall A \subseteq U$, $\sum_{u \in A} \alpha(u, A) = f(A)$.*

Each allocation α gives rise to a density g_α on U, namely, $g_\alpha(u) = \sum_{u \in A} \alpha(u, A)$ where the sum is over all sets A containing u. Indeed, $g_\alpha(\cdot) \geq 0$, and

$$\sum_{u \in U} g_\alpha(u) = \sum_{u \in U} \sum_{u \in A} \alpha(u, A) = \sum_{A \subseteq U} \sum_{u \in A} \alpha(u, A) = \sum_{A \subseteq U} f(A) = 1.$$

Example 1. Let $p(\cdot) : U \to [0,1]$ be a density such that $p(u) \neq 0$, $\forall u \in U$. We write $P(A) = \sum\limits_{u \in A} p(u)$ for $A \subseteq U$. Then clearly

$$\alpha(u, A) = \frac{f(A)}{P(A)} p(u)$$

is an allocation. In particular, if $p(\cdot)$ is the uniform probability density on U, i.e., $p(u) = \frac{1}{|U|}$, $\forall u \in U$, then

$$\alpha(u, A) = \frac{f(A)}{|A|}.$$

Example 2. Let $\{U_1, U_2, \ldots, U_n\}$ be a partition of U with $U_i \neq \emptyset$, $i = 1, 2, \ldots, k$. Let P be a probability measure on U and $p(\cdot)$ its associated density on U, i.e., $p(\cdot) : U \to [0,1]$, $p(u) = P(\{u\})$.
   Let $f : 2^U \to [0,1]$ be

$$f(A) = \begin{cases} P(U_i) \text{ if } A = U_i \\ 0 \quad \text{ if } A \text{ is not one of the } U_i\text{'s} \end{cases}$$

Then f is a density on 2^U with $f(\emptyset) = 0$. Let

$$\alpha(u, A) = \begin{cases} p(u) \text{ if } u \in A = U_i \\ 0 \quad \text{ otherwise.} \end{cases}$$

Then clearly α is an allocation of f.

Example 3. Let U with $|U| = n$ and f be a density on 2^U with $f(\emptyset) = 0$. Let $U = \{u_1, u_2, \ldots, u_n\}$ be an *ordering* of the set U. Let $\alpha(u, A) = f(A)$ if $u = u_j \in A$ with $j = \max\{i : u_i \in A\}$, and zero otherwise. The associated density g_α is

$$g_\alpha(u_i) = \sum_{u_i \in A} \alpha(u_i, A) = \sum_{u_i \in A \subseteq \{u_1, u_2, \ldots, u_i\}} f(A) =$$

$$F(\{u_1, u_2, \ldots, u_i\}) - F(\{u_1, u_2, \ldots, u_{i-1}\})$$

for $i = 1, 2, \ldots, n$ (if $i = 1$, then $F(\{u_1, \ldots, u_{i-1}\}) = F(\emptyset) = 0$) where F is the corresponding distribution function with density f, i.e., $F(A) = \sum\limits_{B \subseteq A} f(B)$.

   Let U with $|U| = n$, and $\mathbb{P}$ be the set of all probability measures on U. Since each $P \in \mathbb{P}$ is uniquely determined by its density p, we identify P with a vector $p = (p_1, p_2, l \ldots, p_n)$, where $p_i = P(\{u_i\})$. Thus, we identify $\mathbb{P}$ with the unit simplex $\mathbb{S}_n$ of $\mathbb{R}^n$, where

$$\mathbb{S}_n = \{p = (p_1, p_2, \ldots, p_n) \in [0,1]^n, \sum_{i=1}^{n} p_i = 1\}.$$

We recall some facts from *convex analysis* here (for more background, see a text like [52]).

A subset A of $\mathbb{R}^n$ is *convex* if $ax + (1-a)y \in A$ whenever $x, y \in A$ and $a \in [0,1]$. The simplex $\mathbb{S}_n$ is a convex subset of $\mathbb{R}^n$. A *convex combination* of the elements $x_1, \ldots, x_k$ in $\mathbb{R}^n$ is an element of $\mathbb{R}^n$ of the form $\sum_{i=1}^{k} a_i x_i$ with $a_i \geq 0$ and $\sum_{i=1}^{k} a_i = 1$. A subset A of $\mathbb{R}^n$ if convex if and only if A contains every convex combination of its elements. Let A be a (nonempty, closed) convex subset of $\mathbb{R}^n$. A point $x \in A$ is called an *extreme point* of A is $x = ay + (1-a)z$ with $y, z \in A$ and $0 < a < 1$ imply $x = y = z$. If A is a compact convex subset of $\mathbb{R}^n$, then each element of A is a finite convex combination of extreme points of A, noting that compact convex subsets have extreme points.

In the following, F denotes a distribution function (of a nonempty random set) on 2^U with density f on 2^U. The *capacity functional* T is dual to F as $T(A) = 1 - F(A^c)$. The cores of F and of T are the same:

$$\{P \in \mathbb{P} : P \leq T\} = \text{core of } T = \{P \in \mathbb{P} : F \leq P\} = \text{core of } F,$$

denoted as $\mathcal{C}(F)$.

It can be checked that $\mathcal{C}(F)$, by identification, is a compact convex subset of the simplex $\mathbb{S}_n$ ($|U| = n$).

There are $n!$ different orderings of the elements of a set U with $|U| = n$. Specifically, let Σ denote the set of all permutations of $\{1, 2, \ldots, n\}$. For $\sigma \in \Sigma$, the elements of U are indexed as $\{u_{\sigma(1)}, u_{\sigma(2)}, \ldots, u_{\sigma(n)}\}$. For each σ, we associate a density g_σ on U as follows:

$$g_\sigma(u_{\sigma(1)}) = F(\{u_{\sigma(1)}\}), \text{ and for } i \geq 2,$$

$$g_\sigma(u_{\sigma(i)}) = F(\{u_{\sigma(1)}, u_{\sigma(2)}, \ldots, u_{\sigma(i)}\}) - F(\{u_{\sigma(1)}, \ldots, u_{\sigma(i-1)}\}).$$

For example, to simplify notations, take $\sigma(i) = i$, $\forall i = 1, \ldots, n$, we write $U = \{u_1, u_2, \ldots, u_n\}$, and dropping σ from our writing (but not from our mind!). The associated probability measure P is defined as $P(A) = \sum_{u \in A} g(u)$, $A \subseteq U$, then $F(A) \leq P(A)$ when $A = U$. Suppose $A \neq U$. Since $U \setminus A = A^c \neq \emptyset$, let $j = \min\{i : u_i \in A^c\}$. For $B = \{u_1, u_2, \ldots, u_j\}$, we have $A \cap B = \{u_1, u_2, \ldots, u_{j-1}\}$, $A \cup B = A \cup \{u_j\}$. Since F is monotone of order 2, we have $F(A \cup B) \geq F(A) + F(B) - F(A \cap B)$, i.e.,

$$F(A \cup \{u_j\}) \geq F(A) + F(\{u_1, u_2, \ldots, u_j\}) - F(\{u_1, u_2, \ldots, u_{j-1}\}),$$

or $g(u_j) \leq F(A \cup \{u_j\}) - F(A)$. But $g(u_j) = P(\{u_j\}) = P(A \cup \{u_j\}) - P(A)$, so that

$$F(A) - P(A) \leq F(A \cup \{u_j\}) - P(A \cup \{u_j\}).$$

On the right-hand side of the above inequality, viewing $A \cup \{u_j\}$ as another set A', and using the same argument, we have

$$F(A \cup \{u_j\}) - P(A \cup \{u_j\}) = F(A') - P(A') \leq F(A \cup \{u_j, u_k\}) - P(A \cup \{u_j, u_k\}),$$

where $u_k \neq u_j$. Continuing this process, we arrive at

$$F(A) - P(A) \leq F(A \cup \{u_j\}) - P(A \cup \{u_j\}) \leq$$

$$F(A \cup \{u_j, u_k\}) - P(A \cup \{u_j, u_k\}) \leq \ldots \leq F(U) - P(U) = 1 - 1 = 0,$$

i.e., for $\forall A \subseteq U$, $F(A) \leq P(A)$, so that $P \in \mathcal{C}(F)$.

For each permutation σ of $\{1, 2, \ldots, n\}$, we obtain an element of $\mathcal{C}(F)$ as above, denoted as P_σ. There are $n!$ (not necessarily distinct) P_σ. These elements of $\mathcal{C}(F)$ are very special. For example, as indicated in Example 3, they all come from allocations, namely, for every σ, allocate $f(A)$ to the element of A with highest rank. Secondly, we have, $\forall A \subseteq U$,

$$F(A) = \inf\{P(A) : P \in \mathcal{C}(F)\},$$

and the infimum is attained for some $P_\sigma(A)$. Indeed, by definition of $\mathcal{C}(F)$,

$$F(A) \leq \inf\{P(A) : P \in \mathcal{C}(F)\},$$

Now for given A, choose σ so that $A = \{u_1, u_2, \ldots, u_k\}$, say. Then

$$P_\sigma(A) = \sum_{i=1}^{k} [F(\{u_1, \ldots, u_i\}) - F(\{u_1, \ldots, u_{i-1}\})] =$$

$$F(\{u_1, \ldots, u_k\}) = F(A).$$

Thus, $F(A) \geq \inf\{P(A) : P \in \mathcal{C}(F)\}$.

But the most important fact about these P_σ is that they are the only *extreme points* of $\mathcal{C}(F)$ (when $\mathcal{C}(F)$ is identified as a convex subset of the simplex $\mathbb{S}_n$ in $\mathbb{R}^n$), a remarkable result of Shapley [115]. We state Shapley's result in the following special form:

THEOREM (Shapley) *Let F be the distribution function of some non-empty random set U with $|U| = n$. Then $\mathcal{C}(F)$ is a compact convex polyhedron with at most $n!$ extreme points that are precisely the P_σ's.*

Interested students could find the proof of Shapley's theorem in a more general form (for convex games) in his paper in 1971.

With Shapley's theorem, we are now in a position to describe the structure of $\mathcal{C}(F)$.

THEOREM $\mathcal{C}(F)$ *consists of probability measures coming from allocations.*

Proof. Let $\mathbb{A}$ denote the subset of $\mathbb{P}$ consisting of all probability measures P_α on U coming from allocations α (of the Möbius inverse f of F). Let α be an allocation of f. Then, $\forall A \subseteq U$,

$$F(A) = \sum_{B \subseteq A} f(B) = \sum_{B \subseteq A} \sum_{u \in B} \alpha(u, B) \leq \sum_{u \in A} \sum_{u \in B} \alpha(u, B) = P_\alpha(A).$$

Thus, $\mathbb{A} \subseteq \mathcal{C}(F)$.

Conversely, by Shapley's theorem, $\mathcal{C}(F)$ has the P_σ's as extreme points, and thus $\mathcal{C}(F)$ is the set of convex combinations of these extreme points. But the P_σ's are elements of $\mathbb{A}$ so that, since $\mathbb{A}$ is clearly convex, $\mathbb{A}$ contains all convex combinations of its elements, in particular, convex combinations of these P_σ's, which is $\mathcal{C}(F)$.   □

Remarks.

(i) In the context of *coarse data modeling*, a nonempty random set S (observable) or a finite set U is viewed as the coarsening of some random variable X with values (unobservable) in U, i.e., X is an a.s. selector of S. The unknown probability law P_X of X lies precisely in $\mathcal{C}(T)$ where T is the capacity functional of S. In other words, as far as statistical inference about P_X is concerned (from, say, i.i.d. observations $S_1, S_2, \ldots, S_n$ of S), the *parameter space* is $\mathcal{C}(T)$. This requires a proof to the fact that a probability measure μ on U is *sectionable* with respect to T if and only if μ comes from an allocation of T. Recall that (see Chapter 3) μ is said to be sectionable w.r.t. T if there exists a probability space $(\Omega, \mathcal{A}, P)$, a random set S' on U, defined on Ω, a random variable X which is also defined on it the same Ω, such that $P_X = \mu$, $T_S = T_{S'}$ (we simply write T as the capacity functional of S when no confusion is possible, otherwise, T_S is the capacity functional of S), and $P(X \in S') = 1$.

The theorem essentially says this. $\mathcal{C}(T)$ is the set of all possible probability laws of a.s. selectors of S and these probability laws can be described in terms of allocations of T. We will provide, to this effect, a *probabilistic proof* of the above theorem in a more general setting of random sets on topological spaces in Chapter 5. Unlike the above proof of the theorem the probabilistic proof will need neither Shapley's theorem nor some convex analysis!

(ii) We can show directly that $F(A) = \inf\{P(A) : P \in \mathbb{A}\}$ as follows. For $A \subseteq U$, let α_A be the allocation such that $\alpha_A(u, B) = 0$ whenever $u \in A$ and

B not contained in A. Let P denote the probability measure associated with α_A, then

$$P(A) = \sum_{u \in A} \sum_{u \in B} \alpha_A(u, B) = \sum_{u \in A} \sum_{B \subseteq A, u \in B} \alpha_A(u, B) =$$

$$\sum_{B \subseteq A} \sum_{u \in B} \alpha_A(u, B) = \sum_{B \subseteq A} f(B) = F(A).$$

(iii) In general, $\mathcal{C}(F)$ is infinite, unless F is a probability measure, in which case, F is the sole element of its core. At the other extreme, let F be the "vacuous prior" on U, i.e.,

$$F(A) = \begin{cases} 1 \text{ if } A = U \\ 0 \text{ if } A \neq U, \end{cases}$$

then $\mathcal{C}(F) = \mathbb{P}$.

(iv) F is always the lower envelope of its core, i.e.,

$$F(A) = \inf\{P(A) : P \in \mathcal{C}(F)\}, \quad \forall A \subseteq U.$$

It is now true that the lower envelope of any subset $\mathcal{P} \subseteq \mathbb{P}$ is a distribution function (belief function). It seems to be a difficult problem to give a reasonable classification of sets of probability measures on $(U, 2^U)$ whose lower envelope is a distribution function of a random set on U.

(iv) Let $\mathcal{P} \subseteq \mathbb{P}$ and $F = \inf \mathcal{P}$. If the *lower probability* F is a distribution function of some random set, then $\mathcal{F} \subseteq \mathcal{C}(F)$ and $F = \inf \mathcal{C}(F)$. However, it might happen that either $\mathcal{P} = \mathcal{C}(F)$ or $\mathcal{P}$ is strictly contained in $\mathcal{C}(F)$.

Example 4. (Shapley value). As a set function, a distribution function F can be viewed as a *coalitional game* (e.g., Owen [97]). The *Shapley value* of the game F is the center of gravity of the extreme points P_σ of $\mathcal{C}(F)$, i.e., it is a probability measure μ on U ($|U| = n$) whose density on U is given by

$$h(u) = \frac{1}{n!} \sum_{\sigma \in \Sigma} g_\sigma(u),$$

where g_σ are densities defined, in terms of F¡ as

$$g_\sigma(u_{\sigma(i)}) = F(\{u_{\sigma(1)}, u_{\sigma(2)}, \ldots, u_{\sigma(i)}\}) - F(\{u_{\sigma(1)}, \ldots, u_{\sigma(i-1)}\}),$$

$u \in U$, $i = 1, 2, \ldots, n$ (for $\sigma(1)$, $F(\emptyset) = 0$, so that $g_\sigma(u_{\sigma(1)}) = F(\{u_{\sigma(1)}\})$).
As an element of $\mathcal{C}(F)$, μ comes from some allocation α. To see this, we need to write $h(\cdot)$ in terms of the Möbius transform f of F.

Using the relation $F(A) = \sum\limits_{B \subseteq A} f(B)$, a direct computation leads to (the details of the manipulations are left as an exercise for students): $\forall u \in U$, $h(u) = \sum\limits_{u \in A} \dfrac{f(A)}{|A|}$ where, as understood, the sum is over A. Thus, the Shapley value μ comes from the allocation

$$\alpha(u, A) = \frac{f(A)}{|A|}.$$

It is interesting to observe that the Shapley value μ, viewing as a point $\mu = (\mu_1, \mu_2, \ldots, \mu_n)$ in the simplex $\mathbb{S}_n$, maximizes the function $H : \mathbb{S}_n \to \mathbb{R}$,

$$H(p) = \sum\limits_{\emptyset \neq A \subseteq U} f(A) \log \left[\prod\limits_{u \in A} p(u) \right]^{1/|A|}$$

where $p = (p_1, p_2, \ldots, p_n) \in \mathbb{S}_n$ and $p_i = p(u_i)$, $i = 1, 2, \ldots, n$.

Example 5. (Allocations for the CAR model). Recall the CAR model in coarse data analysis. Let X be a random variable, defined on a probability space $(\Omega, \mathcal{A}, Pr)$ with values in the finite set U with $|U| = n$. A coarsening of X is a nonempty random set S with values in $2^U \setminus \{\emptyset\}$ such that X is an almost sure selector of S, i.e., $Pr(X \in S) = 1$. Let f be the density on 2^U of S. The random set S is called a coarsening at random (CAR) of X if its distribution is related to the probability law P of X in some special way. The existence of CAR is proved in Gill et al. [36]. Specifically, there exist CAR probabilities $\pi(A)$, $A \subseteq U$, and a density $g(\cdot)$ on U such that

$$\sum\limits_{u \in A} \pi(A) = 1, \quad \forall u \in U \tag{4.3}$$

where the sum is over A, and

$$f(A) = P_g(A)\pi(A), \quad \forall u \in U, \tag{4.4}$$

where, as usual, $P_g(A) = \sum\limits_{u \in A} g(u)$. Define

$$\alpha(u, A) = \begin{cases} g(u) \dfrac{f(A)}{P_g(A)} & \text{if } P_g(A) \neq 0 \\ 0 & \text{if } P_g(A) = 0 \end{cases}$$

In view of (4.4), we see that $\alpha(\cdot, \cdot)$ is an allocation. Now, if $g(u) = 0$, then $\sum\limits_{i \in A} \alpha(u, A) = 0$, and if $g(u) \neq 0$, then $P_g(A) \neq 0$ for any $A \ni u$, thus,

$$\sum\limits_{u \in A} \alpha(u, a) = \sum\limits_{u \in A} g(u) \frac{f(A)}{P_g(A)} = g(u) \sum\limits_{u \in A} \pi(A) = g(u),$$

in view of (4.3). Thus the CAR model of X, i.e., P_g, comes from the above $\alpha(\cdot, \cdot)$ allocation (and hence belongs to $\mathcal{C}(F)$! where F is the distribution function with density f on 2^U).

Note that, like the Shapley value, the CAR solution P_g, viewed as a point in the simplex $\mathbb{S}_n$, maximizes the following entropy function:

$$G : \mathbb{S}_n \to \mathbb{R},$$

$$G(p) = \sum_{\emptyset \neq A \subseteq U} f(A) \log \left[\frac{P(A)}{|A|} \right],$$

where $P(A) = \sum_{u \in A} P(u)$, $p = (p_1, p_2, \ldots, p_n) \in \mathbb{S}_n$, $p_i = P(u_i)$, $i = 1, 2, \ldots, n$.

(See Exercises.)

4.4   Exercises

4.1 Let P and Q be two probability measures on the finite set U.

i) Show that P^3 is a distribution function of some random set on U.

ii) Is $F(A) = P(A)Q(A)$, $A \subseteq U$, a distribution function of some random set on U?

4.2 Let $(\Omega, \mathcal{A}, P)$ be a probability space. A *measurable kernel* of a set $B \subseteq \Omega$, is a measurable set $A \in \mathcal{A}$ such that $A \subseteq B$ and whenever $D \in \mathcal{A}$ and $D \subseteq B \setminus A$, we have $P(D) = 0$.

i) Let A be a measurable kernel, where $B \in 2^\Omega$. Verify that if $G \in \mathcal{A}$ and $G \subseteq B$, then $P(G) \leq P(A)$.

ii) For $B \in 2^\Omega$, define $P_* : 2^\Omega \to [0, 1]$ by

$$P_*(B) = \sup\{P(A) : A \in \mathcal{A}, A \subseteq B\}.$$

Show that if $A \in \mathcal{A}$ and $A \subseteq B$ with $P(A) = P_*(B)$, then A is a measurable kernel of B.

iii) Show that every $B \in 2^\Omega$ has a measurable kernel.

4.3 Let $F, T : 2^U \to [0, 1]$, such that $F(A) = 1 - T(A^c)$. Suppose that F is *monotone of order n*, i.e., for some $n \geq 2$, and any $A_1, \ldots, A_n$,

$$F\left(\bigcup_{i=1}^n A_i \right) \leq \sum_{\emptyset \neq I \subseteq \{1,2,\ldots,n\}} (-1)^{|I|+1} F\left(\bigcap_{i \in I} A_i \right).$$

Show that T is then *alternating of order n*, i.e.,

$$T\left(\bigcap_{i=1}^{n} A_i\right) \leq \sum_{\emptyset \neq I \subseteq \{1,2,\ldots,n\}} (-1)^{|I|+1} T\left(\bigcup_{i \in I} A_i\right).$$

4.4 Let $(U, \mathcal{U})$ be a measurable space. An *ideal J* in $\mathcal{U}$ is a nonempty subset of $\mathcal{U}$, which is stable under finite unions and hereditary, i.e., if $A \in J$ and $B \in \mathcal{U}$ with $B \subset A$, then $B \in J$. A family of ideals $\{J_t, t \geq 0\}$ is said to be *increasing* if $s < t \Rightarrow J_s \subset J_t$.

i) Show that any maxitive set function T on $\mathcal{U}$ is of the form

$$T(A) = \inf\{t \geq 0 : A \in J_t\}$$

where $\{J_t, t \geq 0\}$ is an increasing family of ideals in U.

ii) Let $\{N_t, t \geq 0\}$ be an increasing family of ideals in U which is "right continuous" in the sense that

$$N_t = \bigcap_{s > t} N_s.$$

Let $T(A) = \inf\{t \geq 0 : A \in N_t\}$. Let $\{J_t, t \geq 0\}$ be an increasing family of ideals associated with the maxitive set function T (from i)). Show that $J_t = N_t$, $\forall t \geq 0$.

4.5 Let F be a distribution function on U with $|U| = n$. Let Σ denote the set of all permutations of $\{1, 2, \ldots, n\}$. Consider the densities g_σ on U, $\sigma \in \Sigma$,

$$g_\sigma(u_{\sigma(i)}) = F(\{u_{\sigma(1)}, u_{\sigma(2)}, \ldots, u_{\sigma(i)}\}) - F(\{u_{\sigma(1)}, \ldots, u_{\sigma(i-1)}\})$$

for $i = 1, 2, \ldots, n$, $U = \{u_{\sigma(1)}, u_{\sigma(2)}, \ldots, u_{\sigma(n)}\}$. Show that, $\forall u \in U$,

$$\frac{1}{n!} \sum_{\sigma \in \Sigma} g_\sigma(u) = \sum_{u \in A} \frac{f(A)}{|A|},$$

where the sum is over A containing u, and f is the Möbius inverse of F.

4.6 Let $U = \{u_1, u_2, \ldots, u_n\}$. Let

$$\mathbb{P} = \left\{p = (p_1, p_2, \ldots, p_n) : p_i > 0, \sum_{i=1}^{n} p_i = 1\right\}.$$

For $A \subseteq U$, $P(A) = \sum_{u_i \in A} p_i$, where $p_i = P(\{u_i\})$. Let $\varphi(p) = \sum_{\emptyset \neq A \subseteq U} f(A) \log\left(\frac{P(A)}{|A|}\right)$ where $f : 2^U \setminus \{\emptyset\} \to [0, 1]$ is a probability density.

Use the Lagrange multiplier technique to show that the CAR model is a point in $\mathbb{P}$ maximizing $\varphi(\cdot)$.

4.7 Continuing with the setting of Exercise 4.6, show that the Shapley value of a point in $\mathbb{P}$ is a point in $\mathbb{P}$ maximizing the function Ψ defined on $\mathbb{P}$ as

$$\Psi(p) = \sum_{\emptyset \neq A \subseteq U} f(A) \log \left[\prod_{u_i \in A} p_i \right]^{1/|A|}.$$

4.8 Let $U = \{1, 2, 3\}$ and $f : 2^U \setminus \{\emptyset\} \to [0, 1]$ be

$$f(\{1\}) = f(\{2\}) = f(\{1, 2\}) = 1/9, \quad f(\{3\}) = 0,$$

$$f(\{1, 3\}) = f(\{2, 3\}) = 1/6, \text{ and } f(\{1, 2, 3\}) = 1/3.$$

Find the CAR model and the Shapley value.

4.9 Let $U = \{u_1, u_2, \ldots, u_n\}$ and let S be a (nonempty) random set on U with density f. Show that if the CAR model is equal to the Shapley value p, then $\forall A$ with $f(A) > 0$ and $u \in A$, $P(A) = p(u)|A|$.

4.10 Let $U = \{1, 2, \ldots, n\}$ be a set of "players." A *cooperative game* is a set function $T : 2^U \to \mathbb{R}$ such that $T(\emptyset) = 0$, where for $A \subseteq U$, $T(A)$ is the amount of utility that the members of the coalition A can obtain from the game. For $\emptyset \neq A \subseteq U$, let T_A denote the game

$$T_A(B) = \begin{cases} 1 \text{ if } A \subseteq B \\ 0 \text{ if not} \end{cases}$$

Show that there exist $2^n - 1$ real numbers α_A, $A \subseteq U$, such that

$$T(\cdot) = \sum_{A \subseteq U} \alpha_A T_A(\cdot)$$

4.11 Let $\mathbb{Z}$ denote the set of integers. The partial order on $\mathbb{Z} \times \mathbb{Z}$ is defined as $(x, y) \leq (x', y')$ iff $x \leq x'$ and $y \leq y'$.
Determine the Möbius function of $\mathbb{Z} \times \mathbb{Z}$.

4.12 Let $U = \{u_1, u_2, \ldots, u_n\}$, and $T : 2^U \to [0, 1]$, $T(\emptyset) = 0$, $T(U) = 1$, alternating of order 2. Let $\mathbb{P}$ denote the set of all probability measures on U, and $\mathcal{C}(T) = \{P \in \mathbb{P} : P \leq T\}$. Define $g : U \to [0, 1]$ by $g(u_1) = T(\{u_1\})$, and for $i \geq 2$,

$$g(u_i) = T(\{u_1, u_2, \ldots, u_i\}) - T(\{u_1, \ldots, u_{i-1}\}).$$

Let $P_g : 2^U \to [0, 1]$ be $P_g(A) = \sum_{u \in A} g(u)$.

  i) Show that $P_g \in \mathcal{C}(T)$.

  ii) Show that $\forall A \subseteq U$, $T(A) = \sup\{P(A) : P \in \mathcal{C}(T)\}$.

4.13 Let f be a density on 2^U with $f(\emptyset) = 0$. Let $F, T : 2^U \to [0, 1]$ where $F(A) = \sum_{B \subseteq A} f(B)$ and $T(A) = 1 - F(A^c)$. Show that T is maxitive if and only if $\{A \subseteq U : f(A) > 0\}$ is totally ordered.

Chapter 5

Random Closed Sets

This chapter is concerned with the foundations of random sets taking values as closed subsets of $\mathbb{R}^d$ (or more generally of a Hausdorff, locally compact and second countable topological space).

5.1 Introduction

In Chapters 3 and 4 we considered random sets on a *finite* set U, i.e., random elements taking subsets of U as values. These correspond to games of chance in classical probability theory. While the theory of discrete random variables (or vectors), i.e., random elements with values in an at most countable subset D of $\mathbb{R}^d$, follow in a straightforward manner from finite random variables, the situation is different for random sets. This is so because if U is discrete (infinitely countable, e.g., the set of integers $\mathbb{Z}$), its power set 2^U is infinitely uncountable, and its power set $(2^U, \subseteq)$ is not locally finite. Note that we are talking about *random sets on a discrete space U*, i.e., random elements with values in 2^U. If we consider *discrete random sets*, i.e., random sets taking values in a discrete subset $\mathbb{D}$ of 2^U, then of course, their probability laws are determined by their densities on $\mathbb{D}$. Specifically, let X be a discrete random set on U, with density f on $\mathbb{D}$ (i.e., $f : \mathbb{D} \to [0,1]$, $\sum_{A \in \mathbb{D}} f(A) = 1$), then $\forall \mathbb{A} \subseteq 2^U$,

$$P(X \in \mathbb{A}) = \sum_{A \in \mathbb{D} \cap \mathbb{A}} f(A).$$

Thus, for example, a random set X on $\mathbb{Z}$ is a discrete random set if X only takes *finite* subsets of $\mathbb{Z}$ as values, otherwise it is not.

Random sets on discrete spaces appear typically in image analysis. For example, black and white images in two-dimensional space $\mathbb{R}^2$ are modeled as random sets taking values in the power set of $\mathbb{Z} \times \mathbb{Z}$. As usual, in cases such as these, in order to specify probability laws, we need to consider probability measures on some appropriate σ-field on the power set of $\mathbb{Z} \times \mathbb{Z}$. For applications, it is necessary to be able to specify probability measures in some simpler way, similar to the ways to specify probability measures on $\mathcal{B}(\mathbb{R}^d)$ by distribution functions via the Lebesgue-Stieltjes theorem (see Appendix).

We will address this problem in a specific context, namely for random sets on topological spaces.

Now a random vector $X : \Omega \to \mathbb{R}^d$ can be viewed as random set taking values as singletons $\{X\}$ in the power set of $\mathbb{R}^d$. Singletons are closed sets in $\mathbb{R}^d$. Thus, a random vector is a random set taking closed sets as values. Similarly, point processes are such random sets (see chapter 2). If $X : \Omega \to \mathbb{R}^+$ is a non-negative random variable, then $S(\omega) = [0, X(\omega)]$ is a random set on $\mathbb{R}^+$, taking values in the class of closed sets of $\mathbb{R}^+$. The expected value of X is

$$EX = \int_0^{+\infty} P(X \geq t)\, dt = \int_0^{+\infty} \pi(t)\, dt$$

where $\pi(t)$ is $P(t \in S)$. As pointed out in Chapter 2, confidence regions are *random closed sets* (i.e., random sets taking closed subsets of $\mathbb{R}^d$ as values), and the above observation led to the general Robbins' formula for the computation of expected volume $\Lambda(S)$ of S (where Λ denotes the Lebesgue measure on $\mathcal{B}(\mathbb{R}^d)$) via π, without using the distribution of S. However, in general, when using random sets to model patterns with emphasis on shape, it is necessary to define random sets rigorously, i.e., as bona fide random elements.

Let U be a set and $\mathbb{B} \subseteq 2^U$ (and $(\Omega, \mathcal{A}, P)$ be always a probability space in the background!). To define random elements, defined on Ω, we need to specify a σ-field $\sigma(\mathbb{B})$ of subsets of $\mathbb{B}$. We will proceed to do so in a fairly general setting, namely when U is a Hausdorff, locally compact and second countable space (see Appendix). In fact, for concreteness, we will take $U = \mathbb{R}^d$. As spelled out above, random closed sets are general enough for applications. As such $\mathbb{B}$ will be taken as $\mathcal{F}(\mathbb{R}^d)$, or simply $\mathcal{F}$, the class of all closed subsets of $\mathbb{R}^d$. The two remaining questions are:

- What is the σ-field $\sigma(\mathcal{F})$?

- How to specify probability measures on $\sigma(\mathcal{F})$?

Fortunately, the answers to these basic questions are given in Mathéron [73]. This chapter is devoted to the foundations of random closed sets as laid down by Mathéron. At the end of the chapter, for interested students, we discuss *Choquet's theorem* on *non-locally compact* Polish spaces.

5.2 The Hit-or-Miss Topology

Let $\mathcal{F}$ denote the class of all closed subsets of $\mathbb{R}^d$. A natural way to specify a σ-field $\sigma(\mathcal{F})$ on $\mathcal{F}$ is to equip $\mathcal{F}$ with an appropriate topology and to take $\sigma(\mathcal{F})$ as its Borel σ-field. Of course one can topologize $\mathcal{F}$ in many different ways. We look for a topology $\mathcal{T}$ on $\mathcal{F}$ which is appropriate in the sense that probability laws on random closed sets can be characterized by simpler objects just like the case of random vectors (via Lebesgue-Stieltjes theorem), since after all, random closed sets are generalizations of random vectors.

If we view a *finite* set U as a topological space with its discrete topology (i.e., all subsets of U are open sets), then a finite random set X is a random closed set, and its probability law as a probability measure on the σ-filed of 2^U, namely the power set of 2^U, is uniquely determined via its capacity functional:

$$T : 2^U \to [0,1], \quad T(A) = P(X \cap A \neq \emptyset).$$

Now we observe that $\{\omega : X(\omega) \cap A \neq \emptyset\} = X^{-1}(\mathcal{F}_A)$, where

$$\mathcal{F}_A = \{B \subseteq U : B \cap A \neq \emptyset\}.$$

Thus, when $U = \mathbb{R}^d$, and $X : \Omega \to \mathcal{F}$, we would like to have subsets of $\mathcal{F}$ of the form $\mathcal{F}_A = \{F \in \mathcal{F} : F \cap A \neq \emptyset\}$, for $A \subseteq \mathbb{R}^d$, to belong to the Borel σ-filed $\sigma(\mathcal{F})$ generated by the topology $\mathcal{T}$. But then the set complement of $\mathcal{F}_A$, namely $\mathcal{F}^A = \{F \in \mathcal{F} : F \cap A = \emptyset\}$, should be in $\sigma(\mathcal{F})$ too.

For given $A \subseteq \mathbb{R}^d$, $\mathcal{F}_A$ is the class of closed subsets which *hit* A, while $\mathcal{F}^A$ is the class of closed subsets which *miss* A. These considerations lead to the following way to define an appropriate topology on $\mathcal{F}$.

Just like open intervals in $\mathbb{R}$ form a base for its topology (i.e., open sets of $\mathbb{R}$ are taken to be unions of open intervals), the space of closed sets $\mathcal{F}$ of $\mathbb{R}^d$, can be topologized in a similar fashion.

In the following, $\mathcal{F}$, $\mathcal{G}$, $\mathcal{K}$ denote the classes of closed, open, and compact subsets of $\mathbb{R}^d$, respectively.

For $A \subseteq \mathbb{R}^d$, $\mathcal{F}_A = \{F \in \mathcal{F} : F \cap A \neq \emptyset\}$, $\mathcal{F}^A = \{F \in \mathcal{F} : F \cap A = \emptyset\}$. Let

$$\mathbb{B} = \{\mathcal{F}^K_{G_1,G_2,\dots,G_n} : K \in \mathcal{K}, G_i \in \mathcal{G}, n \neq 0\}$$

where

$$\mathcal{F}^K_{G_1,G_2,\dots,G_n} = \mathcal{F}^K \cap \mathcal{F}_{G_1} \cap \dots \cap \mathcal{F}_{G_n}.$$

Note that in $\mathbb{B}$, when $n = 0$, $\mathcal{F}^K_{G_1,G_2,\dots,G_n}$ means $\mathcal{F}^K$. Also, for $G \in \mathcal{G}$, we have

$$\mathcal{F}_G = \mathcal{F}^\emptyset_G, \quad \mathcal{F}_\emptyset = \emptyset, \quad \mathcal{F} = \mathcal{F}^\emptyset.$$

It is left as an exercise for students to check that $\mathbb{B}$ is qualified as a base for a topology $\mathcal{T}$ on $\mathcal{F}$, its generated topology is called the *hit-or-miss topology* of $\mathcal{F}$. The associated Borel σ-field on $\mathcal{F}$ is denoted as $\mathcal{B}(\mathcal{F})$.

DEFINITION 5.1 *Let $(\Omega, \mathcal{A}, P)$ be a probability space. By a random closed set on $\mathbb{R}^d$, we mean a map $X : \Omega \to \mathbb{R}^d$ which is $\mathcal{A}$-$\mathcal{B}(\mathcal{F})$-measurable. The probability law of X is the probability measure P_X on $\mathcal{B}(\mathcal{F})$ where $P_X = PX^{-1}$.*

Before giving examples of random closed sets and their probability laws, let us mention few important facts about the measurable space $(\mathcal{F}, \mathcal{B}(\mathcal{F}))$ which is the range of random closed sets. In general, random elements take values in metric spaces, such as $\mathbb{R}^d$, $C([0,1])$. Weak convergence (i.e., convergence in distribution) of random elements is well-established for metric spaces (see Appendix). For random closed sets, the range $\mathcal{F}$ with the hit-or-miss topology turns out to be a *metric space*. Specifically, the space $\mathcal{F}(\mathbb{R}^d)$ with the hit-or-miss topology is *compact*, Hausdorff, and *second countable* (i.e., the hit-or-miss topology has a countable base); see Mathéron [73]. As such, $\mathcal{F}$ is metrizable. This fact is important for the consideration of weak convergence of random closed sets in the next chapter 6.

Remark. By definition, a random closed set X in $\mathbb{R}^d$ is a map from Ω to $\mathcal{F}(\mathbb{R}^d)$ which is $\mathcal{A}\text{-}\mathcal{B}(\mathcal{F})$-measurable, i.e., $\forall \mathbb{F} \in \mathcal{B}(\mathcal{F})$, $X^{-1}(\mathbb{F}) \in \mathcal{A}$. Similarly to the measurability of random vectors, it can be checked that X is $\mathcal{A}\text{-}\mathcal{B}(\mathcal{F})$-measurable if and only if $X^{-1}(\mathcal{F}_F) \in \mathcal{A}$, $\forall F \in \mathcal{F}(\mathbb{R}^d)$. In fact, the last condition is equivalent to (see an exercise):

(i) $X^{-1}(\mathcal{F}_G) \in \mathcal{A}$, $\forall G \in \mathcal{G}(\mathbb{R}^d)$,

or

(ii) $X^{-1}(\mathcal{F}_K) \in \mathcal{A}$, $\forall K \in \mathcal{K}(\mathbb{R}^d)$.

Example 1 (Randomized level-sets). Let $\varphi : \mathbb{R}^d \to [0,1]$ be upper semi-continuous (u.s.c.), i.e., $\forall a \in \mathbb{R}$, $\{x \in \mathbb{R}^d : \varphi(x) \geq a\}$ is a closed set in $\mathbb{R}^d$.

Let $\alpha : \Omega \to [0,1]$ be a random variable, uniformly distributed. Define $S : \Omega \to \mathcal{F}(\mathbb{R}^d)$ by

$$S(\omega) = \{x \in \mathbb{R}^d : \varphi(x) \geq \alpha(\omega)\}.$$

Then clearly S is a random closed set on $\mathbb{R}^d$

Example 2. Let S be a random closed set on $\mathbb{R}^d$. Let Λ be the Lebesgue measure on $\mathcal{B}(\mathbb{R}^d)$. Then $\Lambda(S)$ is a non-negative random variable (see chapter 2, section 2.2).

5.3   Capacity Functionals

Let $S : (\Omega, \mathcal{A}, P) \to (\mathcal{F}(\mathbb{R}^d), \mathcal{B}(\mathcal{F}))$ be a random closed set on $\mathbb{R}^d$. The probability law on S is the probability P_S on $\mathcal{B}(\mathcal{F})$ where $P_S = PS^{-1}$.

When S is singleton-valued, we can view S as a random vector. Now, as it is well known, probability laws of random vectors are characterized by distribution functions, via the Lebesgue-Stieltjes theorem, which are defined on $\mathbb{R}^d$. We wish to be able to characterize probability laws of random closed sets on $\mathbb{R}^d$, generalized random vectors, in some similar way. Note that for applications, such a possibility is very important for specifying probability measures on $\mathcal{B}(\mathcal{F})$.

We have seen that (Chapter 3) a set function T on 2^U, for U *finite*, such that

(i) $T(\emptyset) = 0$, $0 \leq T \leq 1$,

(ii) T is alternating of infinite order, i.e., $\forall n \geq 2$, $A_1, A_2, \ldots, A_n$ subsets of U,

$$T\left(\bigcap_{i=1}^{n} A_i\right) \leq \sum_{\emptyset \neq I \subseteq \{1,2,\ldots,n\}} (-1)^{|I|+1} T\left(\bigcup_{i \in I} A_i\right)$$

characterizes a probability measure Q on the power set of 2^U, i.e., a probability law of some random set S on U, via $T(A) = Q(\mathcal{F}_A)$, $\forall A \subseteq U$, where $\mathcal{F}_A = \{B \subseteq U : B \cap A \neq \emptyset\}$.

More specifically, there exist a probability space $(\Omega, \mathcal{A}, P)$, a random element $S : \Omega \to 2^U$, such that

$$\forall A \subseteq U, \quad P(S \cap A \neq \emptyset) = P_S(\mathcal{F}_A) = Q(\mathcal{F}_A) = T(A).$$

On the other hand, let $X : (\Omega, \mathcal{A}, P) \to (\mathbb{R}^d, \mathcal{B}(\mathbb{R}^d))$ be a random vector, then its probability law P_X on $\mathcal{B}(\mathbb{R}^d)$ is determined from the values of P_X on the class $\mathcal{K}$ of compact subsets of $\mathbb{R}^d$. For ease of reference, here is the proof of this fact.

Sample spaces like $\mathbb{R}^d$ are *metric spaces*, and as such, probability measures on their Borel σ-fields are somewhat special, in the sense that each P on $\mathcal{B}(\mathbb{R}^d)$ is determined by its values on *closed sets* or *open sets* of $\mathbb{R}^d$, a property which is called *regularity*. Specifically, any probability measure P on $(\mathbb{R}^d, \mathcal{B}(\mathbb{R}^d))$ or, more generally, on any *metric space* $(U, \mathcal{B}(\rho))$, where ρ denotes a metric on U, and $\mathcal{B}(\rho)$ denotes its Borel σ-field, is *regular*, i.e., for any $A \in \mathcal{B}(\rho)$,

$$P(A) = \sup\{P(F) : F \in \mathcal{F}, F \subseteq A\} = \inf\{P(G) : G \in \mathcal{G}, A \subseteq G\},$$

where $\mathcal{F}$ and $\mathcal{G}$ denote the classes of closed and open sets of $\mathbb{R}^d$, respectively.

Observe that, by definition of supremum and infimum, the above regularity condition is equivalent to:

if $A \in \mathcal{B}(\rho)$, and $\varepsilon > 0$, then there exist $F_\varepsilon \in \mathcal{F}$ and $G_\varepsilon \in \mathcal{G}$ such that $F_\varepsilon \subseteq A \subseteq G_\varepsilon$ and $P(G_\varepsilon \setminus F_\varepsilon) < \varepsilon$.

The above assertion is proved as follows.

Let $\mathcal{A}$ denotes the class of $A \in \mathcal{B}(\rho)$ which are regular in the above sense. Since clearly $U \in \mathcal{A}$, $\mathcal{A}$ is not empty! $\mathcal{A}$ is also clearly closed under complementation. Now let $A_n \in \mathcal{A}$. Let F_n, G_n such that $F_n \subseteq A_n \subseteq G_n$ with

$$P(G_n \backslash F_n) < \frac{\varepsilon}{2^{n+1}}.$$

Take $G = \bigcup_{n \geq 1} G_n \in \mathcal{G}$. Next, since

$$\bigcup_{n \leq N} F_n \nearrow \bigcup_{n \geq 1} F_n \quad \text{as} \quad N \to \infty,$$

it follows from the continuity of P that there exists N such that

$$P\left[\left(\bigcup_{n \geq 1} F_n\right) \backslash \left(\bigcup_{n \leq N} F_n\right)\right] < \varepsilon/2.$$

Take $F = \bigcup_{n \leq N} F_n \in \mathcal{F}$, then $F \subseteq \bigcup_{n \geq 1} A_n \subseteq G$, and

$$P\left[\left(\bigcup_{n \geq 1} G_n\right) \backslash \left(\bigcup_{n \leq N} F_n\right)\right]$$

$$\leq \sum_{n \geq 1} P(G_n \backslash F_n) + P\left[\left(\bigcup_{n \geq 1} F_n\right) \backslash \left(\bigcup_{n \leq N} F_n\right)\right] < \varepsilon.$$

Thus $\bigcup_{n \geq 1} A_n \in \mathcal{A}$ and hence $\mathcal{A}$ is a σ-field. On the other hand, $\mathcal{F} \subseteq \mathcal{A}$, since if $A \in \mathcal{F}$ and $\varepsilon > 0$, then take $F = A$, and observe that, for $\rho(x, A) = \inf\{\rho(x, y) : y \in A\}$, the map $x \to \rho(x, A)$ is continuous, so that $G_n = \{y : \rho(x, A) < 1/n\}$ is open and $\searrow A$. Thus, there exists N such that $P(G_N \backslash A) < \varepsilon$, noting that $A \subseteq G_N$. Therefore, $\mathcal{A} = \mathcal{B}(\rho)$.

In fact, for spaces like $\mathbb{R}^d$, probability measures on their Borel σ-fields are even more special, namely that they are determined by their values on *compact sets*. This is a consequence of the following: if P is a probability measure on a complete, separable metric space $(U, \mathcal{B}(\rho))$, then P is *tight*, i.e. for each $\varepsilon > 0$, there exists $K_\varepsilon \in \mathcal{K}$ (class of compact sets of U) such that $P(K_\varepsilon) > 1 - \varepsilon$. Indeed, let $\varepsilon > 0$. For each $n \geq 1$, let $S_n(x) = \{y \in U : \rho(x, y) < 1/n\}$. The family $S_n(x)$, $x \in U$, is a covering of U, and since U is separable, there is a countable sub-collection, say, $\{S_n(x_j), j \geq 1\}$, such that

$$U = \bigcup_{j \geq 1} S_n(x_j) = \bigcup_{j \geq 1} \bar{S}_n(x_j).$$

The sequence

$$\bigcup_{j \leq m} \bar{S}_n(x_j) \in \mathcal{F} \nearrow U \quad \text{as} \quad m \to \infty$$

implies that there is j_0 such that

$$P\left(\bigcup_{j \leq j_0} \bar{S}_n(x_j)\right) > 1 - \frac{\varepsilon}{2^n}.$$

Let $K_\varepsilon = \bigcap_{n \geq 1} B_n$, where $B_n = \bigcup_{j \leq j_0} \bar{S}_n(x_j)$. Then, for each n, $K_\varepsilon \subseteq B_n$, and hence K_ε is compact (a topological fact, see e.g., Billingsley [10] or Parthasarathy [98]), and

$$P(K_\varepsilon^c) \leq \sum_{n \geq 1} P(B_n^c) < \varepsilon.$$

Now for $A \in \mathcal{B}(\rho)$ and $\varepsilon > 0$. Since P is regular, there is an $F_\varepsilon \in \mathcal{F}$, $F_\varepsilon \subseteq A$ such that

$$P(A) - P(F_\varepsilon) = P(A \backslash F_\varepsilon) < \varepsilon/2.$$

Since P is tight, for each n, let $K_n \in \mathcal{K}$ such that $P(K_n^c) \leq 1/n$ so that $P(K_m^c) < \varepsilon/2$ for, say, $m > [2/\varepsilon]$. Take $K_\varepsilon = F_\varepsilon \cap K_m \in \mathcal{K}$. Then $K_\varepsilon \subseteq F_\varepsilon \subseteq A$ and $P(A \backslash K_\varepsilon) < \varepsilon$, i.e.,

$$P(A) = \sup\{P(K) : K \in \mathcal{K}, K \subseteq A\}.$$

From the above fact we see that it suffices to consider P_X on $\mathcal{K}$. Let $T : \mathcal{K} \to [0,1]$,

$$T(K) = P(X \in K) = P_X(K).$$

Then T determines completely the probability law P_X of the random vector X. The set function T satisfies the following properties (exercise):

(i) $T(\emptyset) = 0$, $0 \leq T \leq 1$,

(ii) If $K_n \searrow K$ (i.e., $K_{n+1} \subseteq K_n$, $K = \bigcap_{n \geq 1} K_n$), then $T(K_n) \searrow T(K)$.

(iii) T is monotone increasing ($K_1 \subseteq K_2 \Rightarrow T(K_1) \leq T(K_2)$) and for $n \geq 2$, and $K_1, K_2, \ldots, K_n$ in $\mathcal{K}$,

$$T\left(\bigcap_{i=1}^n K_i\right) \leq \sum_{\emptyset \neq I \subseteq \{1,2,\ldots,n\}} (-1)^{|I|+1} T\left(\bigcup_{i \in I} K_i\right).$$

Now let $S = \{X\}$ be a singleton-valued random closed set on $\mathbb{R}^d$. We have

$$P_X(K) = P(X \in K) = P(\{X\} \cap K \neq \emptyset) = P(S \cap K \neq \emptyset).$$

Thus, for *general* random closed set S on $\mathbb{R}^d$, we could consider the set function T_S, defined on $\mathcal{K}$, by

$$T_S(K) = P(S \cap K \neq \emptyset) = P_S(\mathcal{F}_K),$$

where $\mathcal{F}_K = \{F \in \mathcal{F} : F \cap K \neq \emptyset\}$. When no confusion is possible, we write simply T for T_S.

Then T satisfies the following properties.

(a) $0 \leq T(\cdot) \leq 1$ with $T(\emptyset) = P(\mathcal{F}_\emptyset) = P(\emptyset) = 0$.

(b) Writing in terms of T the probabilities $P\left(\mathcal{F}^K_{K_1,\ldots,K_n}\right)$, for $K, K_1, \ldots, K_n \in \mathcal{K}, n \geq 1$,

$$P\left(\mathcal{F}^K_{K_1,\ldots,K_n}\right) = \Delta_n(K; K_1,\ldots,K_n),$$

where

$$\Delta_1(K; K_1) = T(K \cup K_1) - T(K)$$
$$\Delta_2(K; K_1, K_2) = \Delta_1(K; K_1) - \Delta_1(K \cup K_2; K_1)$$
$$\vdots$$
$$\Delta_n(K; K_1,\ldots,K_n) = \Delta_{n-1}(K; K_1,\ldots,K_{n-1})$$
$$-\Delta_{n-1}(K \cup K_n; K_1,\ldots,K_{n-1}).$$

As such, $\Delta_n \geq 0$ for all $n \geq 1$. Note that we also let $\Delta_0(K) = 1 - T(K)$.

(c) If $K_n \searrow K$ then $T(K_n) \searrow T(K)$.

This can be seen as follows. We have

$$\mathcal{F}_{K_n} \searrow \bigcap \mathcal{F}_{K_n} \supseteq \mathcal{F}_{\cap K_n} = \mathcal{F}_K.$$

But since $K_n \searrow K$, the sequence of compact sets $\{A_n = F \cap K_n, n \geq 1\}$, for any $F \in \bigcap_{n \geq 1} \mathcal{F}_{K_n}$, has the finite intersection property, i.e.,

$$\bigcap_{i=1}^{n} A_n = F \cap \left(\bigcap_{i=1}^{n} K_i\right) = F \cap K_n \neq \emptyset, \quad n \geq 1,$$

and hence

$$\bigcap_{n \geq 1} A_n = F \cap \left(\bigcap_{n \geq 1} K_n\right) \neq \emptyset,$$

i.e., $F \in \mathcal{F}_{\cap K_n} = \mathcal{F}_K$. Thus, by continuity of P, we get the announced property.

Remark. Clearly the setting of (b) is similar to the way we impose conditions of multi-dimensional distribution functions to characterize probability measures on $(\mathbb{R}^d, \mathcal{B}(\mathbb{R}^d))$, namely probabilities of d-dimensional rectangles. The property $\Delta_n \geq 0$ for all $n \geq 0$ in (b) can be equivalently written as follows.

(b*) T is monotone increasing on $\mathcal{K}$, i.e.,

$$K_1 \subseteq K_2 \quad \Longrightarrow \quad T(K_1) \leq T(K_2),$$

and for any $K_1, \ldots, K_n$ in $\mathcal{K}$, $n \geq 2$,

$$T\left(\bigcap_{i=1}^{n} K_i\right) \leq \sum_{\emptyset \neq I \subseteq \{1,2,\ldots,n\}} (-1)^{|I|+1} T\left(\bigcup_{i \in I} K_i\right).$$

This condition (b*) is called the *alternating of infinite order* property of T. Indeed, observe that

$$\Delta_n(K; K_1, K_2, \ldots, K_n)$$

$$= \sum_{\emptyset \neq I \subseteq \{1,2,\ldots,n\}} (-1)^{|I|+1} T\left(K \cup \left(\bigcup_{i \in I} K_i\right)\right) - T(K).$$

Thus for any $n \geq 1$, $\Delta_n \geq 0$ is the same as

(α) For any $K, K_1, \ldots, K_n \in \mathcal{K}$, $n \geq 1$,

$$T(K) \leq \sum_{\emptyset \neq I \subseteq \{1,2,\ldots,n\}} (-1)^{|I|+1} T\left(K \cup \left(\bigcup_{i \in I} K_i\right)\right).$$

But (α) is equivalent to

(β) T is monotone increasing on $\mathcal{K}$ and for any $K_1, \ldots, K_n \in \mathcal{K}$, $n \geq 2$, we have

$$T\left(\bigcap_{i=1}^{n} K_i\right) \leq \sum_{\emptyset \neq I \subseteq \{1,2,\ldots,n\}} (-1)^{|I|+1} T\left(\bigcup_{i \in I} K_i\right). \tag{5.1}$$

Indeed, assume (α), then T is monotone increasing (take $n = 1$ in (α)). For $K_1, \ldots, K_n$ with $n \geq 2$, take $K = \bigcap_{i=1}^{n} K_i$ in (α) yielding (5.1). Conversely, assume (β). Then (α) holds for $n = 1$ by monotonicity of T. For $n \geq 2$, apply (β) to $A_i = K \cup K_i$, $i = 1, 2, \ldots, n$ yielding

$$T\left(\bigcap_{i=1}^{n} A_i\right) \leq \sum_{\emptyset \neq I \subseteq \{1,2,\ldots,n\}} (-1)^{|I|+1} T\left(\bigcup_{i \in I} A_i\right)$$

$$= \sum_{\emptyset \neq I \subseteq \{1,2,\ldots,n\}} (-1)^{|I|+1} T\left(K \cup \left(\bigcup_{i \in I} K_i\right)\right).$$

But $K \subseteq \bigcap_{i=1}^{n} A_i$ and hence

$$T(K) \leq T\left(\bigcap_{i=1}^{n} A_i\right),$$

again, by monotonicity of T.

As in the case of distribution functions, we arrive at

DEFINITION 5.2   *A set-function $T : \mathcal{K} \to \mathbb{R}$ is called a* capacity functional *if it satisfies:*

(i) $0 \leq T \leq 1$, $T(\emptyset) = 0$.

(ii) T *is* alternating of infinite order.

(iii) *If $K_n \searrow K$ then $T(K_n) \searrow T(K)$.*

It turns out that capacity functionals characterize the probability measures on $(\mathcal{F}, \mathcal{B}(\mathcal{F}))$, a remarkable result due to Choquet. This can be viewed as the counter-part of Lebesgue/Stieltjes theorem.

CHOQUET THEOREM   *If $T : \mathcal{K} \to \mathbb{R}$ is a capacity functional, then there exists a unique probability measure P on $\mathcal{B}(\mathcal{F})$ such that*

$$P\left(\mathcal{F}_K\right) = T(K), \qquad \text{for all} \quad K \in \mathcal{K}.$$

Although a complete *probabilistic proof* of Choquet theorem is given in Matheron [73], we sketch below the main ideas for students to gain some insights. The students should recognize the analogy between "intervals" of $\mathbb{R}$ or rectangles of $\mathbb{R}^d$ and various subsets of $\mathcal{F}$ in the below approach.

As in the case of distribution functions on $\mathbb{R}^d$, we seek a *field* $\mathcal{A}$ of subsets of $\mathcal{F}$ such that the σ-field generated by $\mathcal{A}$, $\sigma(\mathcal{A}) = \mathcal{B}(\mathcal{F})$ and on $\mathcal{A}$ we can define unambiguously an σ-additive probability measure in terms of T, and then use Caratheodory theorem to conclude (see Appendix). First note that T defined on $\mathcal{K}$ can be extended to $\mathcal{P}(\mathbb{R}^d)$ as follows.

For $G \in \mathcal{G}$, define

$$T(G) = \sup\{T(K) : K \in \mathcal{K}, K \subseteq G\},$$

and for $A \in \mathcal{P}(\mathbb{R}^d)$, define

$$T(A) = \inf\{T(G) : G \in \mathcal{G} \ A \subseteq G\}.$$

Note that such extended T is left continuous on $\mathcal{P}(\mathbb{R}^d)$, i.e., if $A_n \in \mathcal{P}(\mathbb{R}^d)$ and $A_n \nearrow A$ then $T(A_n) \nearrow T(A)$. Also, the alternating of infinite order is preserved on $\mathcal{P}(\mathbb{R}^d)$, i.e., the above $\Delta_n \geq 0$ for any $n \geq 0$, when the compacts are replaced by $A, A_1, ..., A_n$ in $\mathcal{P}(\mathbb{R}^d)$. To get an idea, consider $n = 2$ and Borel sets of $\mathbb{R}^d$. We have

$$\Delta_2(A; A_1, A_2) = T(A \cup A_1) - T(A) - T(A \cup A_1 \cup A_2) + T(A \cup A_2).$$

Since Borel sets are T-*capacitable* (i.e., approximable by compact sets, a property similar to tightness of probability measures, see e.g., Meyer [74]), if we let $K_n, K_n^{(1)}, K_n^{(2)}$ be compacts subsets of A, A_1, A_2, respectively, and tending upwards to them, then

$$T(K_n \cup K_n^{(1)}) - T(K_n) - T(K_n \cup K_n^{(1)} \cup K_n^{(2)}) + T(K_n \cup K_n^{(2)})$$
$$\nearrow T(A \cup A_1) - T(A) - T(A \cup A_1 \cup A_2) + T(A \cup A_2)$$

so that

$$\Delta_2(A; A_1, A_2) = \lim_{n \to \infty} \Delta_2(K_n; K_n^{(1)}, K_n^{(2)}) \geq 0.$$

Let $\mathcal{C}$ be the class of subsets of $\mathbb{R}^d$ of the form

$$\{V = K \cup G, \ K \in \mathcal{K}, \ G \in \mathcal{G}\},$$

and $\mathcal{A}$ be the class of subsets of $\mathcal{F}$ of the form

$$\{\mathcal{F}_{V_1,\ldots,V_n}^V : V, \ V_i \in \mathcal{C}\}.$$

Then $\mathcal{A}$ is a *semi-field* of subsets of $\mathcal{F}$ (i.e., $\mathcal{A}$ contains $\emptyset$ and $\mathcal{F}$, closed under finite intersection, and if $A \in \mathcal{A}$, then A^c is the union of a finite collection of elements of $\mathcal{A}$), and $\sigma(\mathcal{A}) = \mathcal{B}(\mathcal{F})$. Define $P : \mathcal{A} \to [0,1]$ by

$$P\left(\mathcal{F}_{V_1,\ldots,V_n}^V\right) = \Delta_n(V; V_1, \ldots, V_n),$$

then P is well-defined, additive and $P(\mathcal{F}) = 1$, since $\mathcal{F} = \mathcal{F}^\emptyset$ and

$$P(\mathcal{F}^\emptyset) = \Delta_0(\emptyset) = 1 - T(\emptyset) = 1.$$

As such, it suffices to verify that P can be approximated by some "compact" subclass of $\mathcal{A}$ (see Appendix).

Few technical comments are in order here. The results on fields remain valid for semi-fields. The proof of Choquet theorem relies heavily on topological properties of $\mathbb{R}^d$ as a *locally compact, Hausdorff and second countable space*. In other words, Choquet theorem is valid on such general spaces. However, we will make a remark on infinitely dimensional polish spaces after presenting an example.

Example. Let $\phi : \mathbb{R}^d \to [0,1]$ be upper semi-continuous (u.s.c.), i.e., for any $s \in \mathbb{R}$, the set $\{x \in \mathbb{R}^d : \phi(x) \geq s\}$ is closed. Define

$$T : \mathcal{K} \longrightarrow \mathbb{R}, \qquad T(K) = \sup_{x \in K} \phi(x).$$

Then T is a capacity functional.

A simple *probabilistic proof* will be left as an exercise, but here is a direct proof!

Proof. Since $\phi : \mathbb{R}^d \to [0, 1]$, the property (i) in the definition of a capacity functional is obvious, noting also that the supremum over the empty set is zero. For (iii), let $K_n \searrow K$ in $\mathcal{K}$. We have

$$T(K) \leq \inf_n T(K_n) = s,$$

say. Let $\varepsilon > 0$ and consider

$$A_n = \{x \in \mathbb{R}^d : \phi(x) \geq s - \varepsilon\} \cap K_n.$$

By hypothesis, the set $\{x \in \mathbb{R}^d : \phi(x) \geq s - \varepsilon\}$ is closed, and so is A_n. The A_n's are contained in K_1, and hence $\bigcap_{n \geq 1} A_n \neq \emptyset$ (since the A_n's have the finite intersection property, and noting that $A_n \neq \emptyset$ for each n). We have $A_n \subseteq K_n$ for all n, and hence

$$T(K) \geq \sup_{x \in \cap A_n} \phi(x) \geq s - \varepsilon$$

(noting that T is obviously monotone increasing). Thus (iii) follows.

We reserve to prove the property (ii), namely alternating of infinite order property of T to the end because of an interesting phenomenon. The set-function $T(K) = \sup_{x \in K} \phi(x)$ is very special. It is *maxitive* in the sense that, for any K_1, K_2,

$$T(K_1 \cup K_2) = \max\{T(K_1), T(K_2)\}.$$

Such set-functions are necessarily alternating of infinite order. (See chapter 4).
□

Notes on the Core of Capacity Functionals of Random Closed Sets

In Chapter 4, using Shapley's theorem in game theory, we have shown that probability measures in the core $\mathcal{C}(T)$ of a capacity functional T of a *finite* random set S (i.e., a random element, defined on $(\Omega, \mathcal{A}, P)$, with values in 2^U, with U being a *finite* set) all come from allocations of the probability density f on 2^U (where f is the Möbius inverse of the dual F of T, i.e., $F(A) = 1 - T(A^c)$). It turns out that this result can be obtained without using Shapley's theorem. In other words, the description of $\mathcal{C}(T)$ can be obtained from a *probabilistic proof*. Specifically, as pointed out in chapter 3, if X is an almost sure (a.s.) selection of S, then its probability law P_X belongs to $\mathcal{C}(T)$, i.e., $P_X \leq T$ on 2^U. The converse holds: $\mathcal{C}(T)$ consists precisely of all probability laws of a.s. selections of S. Here is a probabilistic proof of this result in which the *necessity condition* is interesting since it will involve a concept of *ordered coupling* for general random closed sets.

THEOREM *Let T be a capacity functional on a finite set U. Let $f : 2^U \to [0, 1]$ be the Möbius inverse of the dual F of T. Then a probability measure μ on U is in $\mathcal{C}(T)$ is and only if μ comes from an allocation of f.*

Probabilistic proof.

(i) *Sufficiency.* Consider the probability space $(\Omega, \mathcal{A}, P)$ where

$$\Omega = (2^U \setminus \{\emptyset\}) \times U,$$

$\mathcal{A}$ = power set of Ω, and $P((A, u)) = \alpha(u, A)$, where α is an allocation of f. Define $S : \Omega \to 2^U \setminus \{\emptyset\}$ and $X : \Omega \to U$ by

$$S(A, u) = A, \quad X(A, u) = u, \quad \forall (u, A) \in \Omega.$$

Then, clearly,

$$P(\omega : S(\omega) = A) = P(A \times U) = \sum_{u \in U} P((A, u)) = f(A),$$

i.e. S has f (or T) as its distribution. Also, if μ is the probability measure on U defined by $\mu(\{u\}) = \sum_{u \in A} \alpha(u, A)$ (the sum is over all A containing u), then μ is the probability law of X, since

$$P(\omega : X(\omega) = u) = P((2^U \setminus \{\emptyset\}) \times \{u\}) =$$

$$\sum_{\emptyset \neq A \subseteq U} P((A, u)) = \sum_{\emptyset \neq A \subseteq U} \alpha(u, A).$$

Moreover, X is a P-a.s. selection of S, since

$$P(\omega : X(\omega) \in S(\omega)) = P((A, u) : u \in A) =$$

$$\sum_{A \subseteq U} \sum_{u \in A} \alpha(u, A) = \sum_{A \subseteq U} f(A) = 1.$$

Thus, $\mu \leq T$, i.e., $\mu \in \mathcal{C}(T)$.

(ii) *Necessity.* Suppose $\mu \in \mathcal{C}(T)$. Then according to *Norberg's theorem* (stated after this proof), there exists a probability space $(\Omega, \mathcal{A}, P)$, on which are defined a random set $S : U \to 2^U \setminus \{\emptyset\}$ and a random variable $X : \Omega \to U$ such that $P(X \in S) = 1$, and S and X have as distributions T and μ, respectively.

Define $\alpha : U \times (2^U \setminus \{\emptyset\}) \to \mathbb{R}^+$ by:

$$\alpha(u, A) = P(\omega : S(\omega) = A, X(\omega) = u).$$

Clearly α is an allocation of f, i.e., $\alpha(u, A) = 0$ for $u \notin A$, and $\forall A \subseteq U$, $\sum_{u \in A} \alpha(u, A) = f(A)$, and

$$\mu(\{u\}) = P(\omega : X(\omega) = u) = \sum_{u \in A} \alpha(u, A).$$

$\square$

Ordered Coupling of Random Sets (Norberg [96])

Given two capacity functionals on $\mathcal{K}(\mathbb{R}^d)$, ν and ν', we say that there exists an ordered coupling for ν and ν' if there exists a common probability space $(\Omega, \mathcal{A}, P)$ on which are defined two random closed sets S and S' such that $S' \subseteq S$, P-almost surely, where S and S' have ν and ν' as their distributions, respectively. When the random closed set S' takes only singletons as values (identified as a random vector on $\mathbb{R}^d$), we have S' as an P-a.s. selector of S. Note that this setting covers the case of finite sets as a special case when considering the discrete topology (every subset of the *finite* set U is open). A necessary and sufficient condition for the existence of random closed sets on $\mathbb{R}^d$ is given by Norberg [96]. We state below the result in our context of almost sure selectors.

THEOREM (Norberg [96]) *Let μ be a probability measure on $\mathcal{B}(\mathbb{R}^d)$ and T be a capacity functional on $\mathcal{K}(\mathbb{R}^d)$. Then the following are equivalent:*

(a) *$\mu \in \mathcal{C}(T)$, i.e., $\mu \leq T$ on $\mathcal{K}(\mathbb{R}^d)$.*

(b) *There exists a common probability space $(\Omega, \mathcal{A}, P)$ on which are defined a random closed set $S : \Omega \to \mathcal{F}(\mathbb{R}^d)$, and a random vector $X : \Omega \to \mathbb{R}^d$ such that $P(X \in S) = 1$, and S and X have T and μ as their distributions, respectively.*

5.4   Notes on the Choquet Theorem on Polish Spaces (optional)

Since the natural domain of probability theory is polish spaces (i.e., completely metrizable separable spaces), it is necessary to see whether or not Choquet theorem holds on non-locally compact polish spaces.

In this preliminary work towards a definite answer to the question, we provide an example showing that in some *non-locally compact* polish spaces, the version of Choquet theorem on open sets (which is equivalent to that on compact sets, in the case of locally compact spaces) fails if we equip the space of all closed sets with the topology induced by the Hausdorff metric.

In the case of a locally compact separable Hausdorff space, an equivalent version of Choquet theorem is this. Let $S : \mathcal{G} \to [0, 1]$ be a set function. Then there exists a unique probability measure P on $\sigma(\mathcal{F})$ such that

$$P(\mathcal{F}_G) = S(G) \qquad \text{for every} \quad G \in \mathcal{G} \tag{5.2}$$

if and only if

(α) $S(\emptyset) = 0$, $0 \leq S \leq 1$,

(β) S is alternating of infinite order on $\mathcal{G}$, and

(γ) if $G_n \nearrow G$ on $\mathcal{G}$ $\left(\text{i.e., } G_n \subset G_{n+1} \text{ for every } n \in \mathbb{N} \text{ and } G = \bigcup_{n=1}^{\infty} G_n\right)$,
then $T(G_n) \nearrow T(G)$.

As stated earlier, the need to consider non-locally-compact polish spaces (or more generally, metric spaces) is because these spaces form the most general domain of probability theory, e.g., the space $C[0,1]$ of continuous functions on the unit interval $[0,1]$, see e.g., Billingsley [10]. Also, capacities on infinite dimensional spaces are considered for control theory of distributed systems, see, e.g., Li and Yong [67].

To be specific, we take E to be the closed unit ball of the Hilbert space

$$\ell_2 = \left\{ x = (x_n) : \|x\| = \left(\sum_{n=1}^{\infty} x_n^2\right)^{1/2} < \infty \right\},$$

that is,

$$E = \left\{ x = (x_n) \in \ell_2 : \|x\| = \left(\sum_{n=1}^{\infty} x_n^2\right)^{1/2} \leq 1 \right\}.$$

Note that E is a bounded, infinite dimensional Polish space. Moreover, E is *not locally compact at any point* (that is no point of E has a compact neighborhood). As a consequence, if we equip the space $\mathcal{F}$ of closed subsets of E with the miss-or-hit topology, then the space $\mathcal{F}$ is *no longer Hausdorff*. In fact, we have the following stronger result.

LEMMA *If E is not locally compact at any point, then the miss-or-hit topology on $\mathcal{F}$ has the following property:*

Any two nonempty open sets in $\mathcal{F}$ have a nonempty intersection.

Proof. Assume that E is not locally compact at any point. Let $\mathcal{U}$ and $\mathcal{V}$ be two nonempty open sets in $\mathcal{F}$. By the definition of the miss-or-hit topology there exist $\mathcal{F}_{G_1,\ldots,G_n}^K \subset \mathcal{U}$ and $\mathcal{F}_{O_1,\ldots,O_m}^C \subset \mathcal{V}$, where $K, C \in \mathcal{K}$, and $G_i, O_j \in \mathcal{G}$ for $i = 1,\ldots,n, \ j = 1,\ldots,m$. Since $K \cup C$ is compact and E is not locally compact at any point, for each $i = 1,\ldots,n$, and $j = 1,\ldots,m$, there exists $y_i \in G_i\backslash(K \cup C)$ and $z_j \in O_j\backslash(K \cup C)$. It is easy to see that

$$\{y_1, \ldots, y_n, z_1, \ldots, z_m\} \in \mathcal{F}_{G_1,\ldots,G_n}^K \cap \mathcal{F}_{O_1,\ldots,O_m}^C \subset \mathcal{U} \cap \mathcal{V}.$$

Therefore $\mathcal{U} \cap \mathcal{V} \neq \emptyset$. $\square$

Remark. By the above lemma, if E is an open set in an infinite dimensional Banach space, then any two nonempty open sets in $\mathcal{F}$ have nonempty intersection. Observe that an open set in a Banach space is not complete with respect to the norm, but it can be metrizable by a topologically equivalent metric so that it will be complete with respect to the new metric.   □

In considering the Choquet theorem on non-locally-compact spaces, we have a choice between a version on $\mathcal{K}$ or a version on $\mathcal{G}$. As stated, the two versions are equivalent in the locally compact case. However, it is not clear whether they remain equivalent in the non-locally-compact case.

In the following example, we consider the version of the Choquet theorem on open sets. In fact, we will provide an example showing that there is a functional S on $\mathcal{G}$ satisfying the conditions (α), (β) and (γ) such that no probability measure on $(\mathcal{F}, \sigma(\mathcal{F}))$ satisfying (5.2), where $\sigma(\mathcal{F})$ is the Borel σ-field defined below.

Example. Let E denote the closed unit ball of the Hilbert space ℓ_2. As stated earlier, this space E is not locally compact at any point. Therefore by lemma, the space $\mathcal{F}$ of all closed subsets of E equipped with the miss-or-hit topology is no longer Hausdorff. Thus, if we insist on having $\mathcal{F}$ as a Hausdorff topological space, we need to use another topology. An appropriate topology for the space of all closed subsets of a metric space is, perhaps, the topology induced by the Hausdorff metric. Our result says that the Choquet theorem does not hold for this space E if we use the topology induced by the Hausdorff metric instead of the miss-or-hit topology.

Since E is bounded, it is possible to define the Hausdorff metric on $\mathcal{F}$ (including empty set $\emptyset$). Indeed,

$$\rho(A, B) = \begin{cases} \max \left\{ \sup_{x \in A} \|x - B\|, \sup_{x \in B} \|x - A\| \right\} & \text{if } A \neq \emptyset, B \neq \emptyset \\ 0 & \text{if } A = B = \emptyset \\ 2 & \text{otherwise,} \end{cases}$$

where

$$\|x - A\| = \inf\{\|x - y\| : y \in A\}.$$

It is easy to see that d is indeed a metric on $\mathcal{F}$. Let $\sigma(\mathcal{F})$ denote the Borel σ-field of $\mathcal{F}$ (with respect to the Hausdorff metric).

Remark. We observe that the space $\mathcal{F}$ equipped with the Hausdorff metric is not separable. In fact, for each $n \in \mathbb{N}$, let e_n denote the nth standard unit vector of ℓ_2, (i.e., $\mathbf{e}_n$ has 1 at its nth position and zeros elsewhere). Then the sequence $\{\mathbf{e}_n : n \in \mathbb{N}\} \subset E$ satisfies

$$\|\mathbf{e}_n - \mathbf{e}_m\| = 1 \qquad \text{for every } n \neq m.$$

Let $2^{\mathbb{N}}$ denote the family of all nonempty subsets of $\mathbb{N}$. For every $S \in 2^{\mathbb{N}}$, let $A_S = \{x_n : n \in S\}$. It is easy to see that $\rho(A_S, A_T) = 1$ for every $S \neq T$. Since $\{A_S : S \in 2^{\mathbb{N}}\}$ is uncountable, it follows that $\mathcal{F}$ is not separable.

In general, it is easily shown that if E is any set in an infinite dimensional Banach space that contains at least one interior point, then the family of all closed subsets in E equipped with the Hausdorff metric is not separable.

We are going to define a set-function $S : \mathcal{G} \to [0,1]$ satisfying the previous conditions (α), (β), and (γ), equivalent conditions of (5.2). Denote

$$B(x,r) = \{y \in E : ||x - y|| < r\}.$$

For every $G \in \mathcal{G}$, let

$$S_n(G) = \inf\{r > 0 : G \subset B(x_1, r) \cup \ldots \cup B(x_n, r)\}.$$

It is easy to see that $\{S_n(G)\}$ is a decreasing sequence. Define

$$S(G) = \lim_{n \to \infty} S_n(G) \qquad \text{for every } \ G \in \mathcal{G}.$$

Observe that the *Kuratowski measure of noncompactness* is defined on any subset A of a metric space and depends on A metrically, i.e., two topologically equivalent metrics on E may induce two different Kuratowski measures of noncompactness on E. For instance, let $R^n_\infty = R^n \cup \{\infty\}$ denote the one-point compactification of R^n. Let d be any compatible metric on R^n_∞. Then $S_d(A) = 0$ for any $A \subset R^n$. However $S_{||\cdot||}(A) = \infty$ if A is unbounded (with respect to the norm $|| \cdot ||$ in R^n).

S **satisfies** (α). Obviously, $S(\emptyset) = 0$. The fact that $S(E) = 1$ is a special case of the following more general result.

CLAIM 1 $S(G) = 1$ for any open set G of E containing $\{e_n : n \in \mathbb{N}\}$.

Proof. Assume on the contrary that $S(G) < 1$. Take r so that $S(G) < r < 1$. By definition of S there exists $N \in \mathbb{N}$ such that $G \subset \bigcup_{i=1}^{N} B(x_i, r)$. Let $x_i = (x_i(n))$ for $i = 1, \ldots, N$. Observe that

$$||x_i - e_n||^2 = \sum_{k=1}^{n-1} |x_i(k)|^2 + |x_i(n) - 1|^2 + \sum_{k=n+1}^{\infty} |x_i(k)|^2 \geq |x_i(n) - 1|^2,$$

for every $i = 1, \ldots, N$. Since $x_i(n) \to 0$ as $n \to \infty$, it follows that for each $i = 1, \ldots, N$,

$$||x_i - e_n|| > r \qquad \text{for infinitely many} \ \ n.$$

Therefore, for each $i = 1, ..., N$,

$$e_n \notin B(x_i, r) \qquad \text{for infinitely many} \ \ n.$$

Thus, $\bigcup_{i=1}^{N} B(x_i, r)$ contains only finitely many e_n, and cannot cover G. This contradiction indicates that $S(G) = 1$. $\square$

S **satisfies** (β). Observe that S is *maxitive*, i.e.,

$$S(U \cup V) = \max\{S(U), S(V)\} \qquad \text{for every } U, V \in \mathcal{G}.$$

Indeed, for every $U, V \in \mathcal{G}$ we have

$$S_{2n}(U \cup V) \leq \max\{S_n(U), S_n(V)\} \qquad \text{for every } n \in \mathbb{N}.$$

Therefore

$$S(U \cup V) \leq \max\{S(U), S(V)\} \qquad \text{for every } U, V \in \mathcal{G},$$

which yields

$$S(U \cup V) = \max\{S(U), S(V)\} \qquad \text{for every } U, V \in \mathcal{G}.$$

Thus, S is alternating of infinite order.

S **satisfies** (γ). We need to elaborate the condition (γ). If E is a locally compact separable Hausdorff space, then E has a one-point compactification $E_\infty = E \cup \{\infty\}$. Since E is separable and Hausdorff, E_∞ is metrizable. Let d be any compatible metric on E_∞. Then, if $G_n \nearrow G$ in E, we have

$$\sup_{x \in G} \rho(x, G_n) \to 0 \qquad \text{as } n \to \infty, \tag{5.3}$$

where $\rho(x, A) = \inf\{\rho(x, y) : y \in A\}$.

Indeed, assume that (5.3) fails. Then there exist $\delta > 0$, a sequence $\{x_k\} \subset G$, and a subsequence $\{G_{n(k)}\} \subset \{G_n\}$ such that

$$\rho(x_k, G_{n(k)}) \geq \delta \qquad \text{for every } k \in \mathbb{N}.$$

Since E_∞ is compact, by passing to a subsequence if necessary, we may assume that $x_k \to x \in \bar{G}$, where $\bar{G}$ denotes the closure of G in E_∞. It follows that

$$\rho(x, G_{n(k)}) \geq \delta \qquad \text{for every } k \in \mathbb{N}.$$

Since $\{G_n\}$ is increasing,

$$\bigcup_{k=1}^{\infty} G_{n(k)} = \bigcup_{n=1}^{\infty} G_n = G.$$

Therefore we get

$$\rho(x, G_n) \geq \delta \qquad \text{for every } k \in \mathbb{N},$$

which contradicts the fact that $x \in \bar{G} = \overline{\cup_{n=1}^{\infty} G_n}$. Consequently, our claim is justified.

For non-locally-compact spaces, the condition (5.3) does not follow from $G_n \nearrow G$. Thus (5.3) needs to be added. In other words, the condition (γ) is strengthened to be

$$G_n \nearrow G \quad \text{and} \quad \sup_{x \in G} \rho(x, G_n) \to 0 \quad \text{as } n \to \infty. \tag{5.4}$$

We are going to show that (5.4) implies $S(G_n) \nearrow S(G)$. Since S is monotone, $\lim_{n \to \infty} S(G_n) \leq S(G)$. We claim that

$$\lim_{n \to \infty} S(G_n) = S(G).$$

Suppose it is not the case, say,

$$\lim_{n \to \infty} S(G_n) < \alpha < S(G),$$

then we take $\delta > 0$ such that $\alpha + \delta < S(G)$. By (5.4), there exists $N \in \mathbb{N}$ such that

$$G \subset B(G_N, \delta), \quad \text{where} \quad B(G_N, \delta) = \{x \in E : ||x - G_N|| < \delta\}.$$

Since $\{G_n\}$ in increasing,

$$S(G_N) \leq \lim_{n \to \infty} S(G_n) < \alpha.$$

By definition

$$G_N \subset G(x_1, \alpha) \cup \ldots \cup B(x_k, \alpha) \quad \text{for sufficiently large } k \in \mathbb{N},$$

which implies
$$G \subset B(x_1, \alpha + \delta) \cup \ldots \cup B(x_k, \alpha + \delta)$$

for sufficiently large $k \in \mathbb{N}$. It follows that $S(G) \leq \alpha + \delta < S(G)$, a contradiction. Consequently, $S(G_n) \to S(G)$.

Now we are going to show that no probability measure P on $\mathcal{F}$ satisfies the condition (5.2). Of course for (5.2) to make sense, we need to verify first that

CLAIM 2 $\mathcal{F}_G \in \sigma(\mathcal{F})$ *for any open set G of E.*

Proof. Recall that
$$\mathcal{F}_G = \{A \in \mathcal{F} : A \cap G \neq \emptyset\}.$$

We are going to prove that $\mathcal{F}_G$ is open in $\mathcal{F}$ (in the Hausdorff metric) for any open set G of E. Indeed, let $A \in \mathcal{F}_G$. Then $A \cap G \neq \emptyset$. Since G is an open set of E, there exist $r > 0$ and $x \in A$ such that $B(x, r) \subset G$. We claim that $B(A, r) \subset \mathcal{F}_G$, where $B(A, r)$ denotes the ball in $\mathcal{F}$ (in the Hausdorff metric) with center at A and radius r. Observe that if $B \in B(A, r)$, then there exists

$y \in B$ such that $||x - y|| < r$, which implies $y \in B(x, r) \subset G$. Therefore $B \cap G \neq \emptyset$. The claim is proved. $\square$

Let $\{a_n : n \in \mathbb{N}\}$ be a countable dense subset of E. For every $n \in \mathbb{N}$, let

$$B_n = conv\{e_1, \ldots, e_n, a_1, \ldots, a_n\} \quad \text{and} \quad B = \bigcup_{n=1}^{\infty} B_n.$$

Observe that B_n is a compact convex set of E for every $n \in \mathbb{N}$.

CLAIM 3 *For each $n \in \mathbb{N}$, there exists an open set $G_n \supset B_n$ such that*

$$S(G_n) < 2^{-n-1}.$$

Proof. Let $\varepsilon = 2^{-n-1}$. Since B_n is compact, there exist $B(x_i, \varepsilon)$, $i = 1, \ldots, k$, such that $B_n \subset \bigcup_{i=1}^{k} B(x_i, \varepsilon)$. Denote $G_n = \bigcup_{i=1}^{k} B(x_i, \varepsilon)$. Then we have $S_k(G_n) \leq \varepsilon$, which implies $S(G_n) \leq \varepsilon = 2^{-n-1}$. $\square$

Now, assume on the contrary that there is a probability measure P on $\sigma(\mathcal{F})$ such that $P(\mathcal{F}_G) = S(G)$ for every $G \in \mathcal{G}$. By Claim 3, for each $n \in \mathbb{N}$ there exists an open set $G_n \supset B_n$ such that $T(G_n) < 2^{-n-1}$. Denote $G = \bigcup_{n=1}^{\infty} G_n$. Then G is an open set containing B. Therefore, from Claim 3 and Claim 1 we obtain

$$P(F_G) = P\left(F_{\bigcup_{n=1}^{\infty} G_n}\right) = P\left(\bigcup_{n=1}^{\infty} F_{G_n}\right)$$

$$\leq \sum_{n=1}^{\infty} P(F_{G_n}) = \sum_{n=1}^{\infty} S(G_n)$$

$$< \sum_{n=1}^{\infty} 2^{-n-1} = 2^{-1} < 1 = S(G),$$

i.e., (5.2) is violated. Consequently, our result is established. $\square$

Remark. Obviously, we are interested in finding a "positive version" of the Choquet theorem on non-locally-compact spaces, but finally a "negative version" has come to us before we get a "positive one"! The negative result in this note is not quite "a counterexample" to the Choquet theorem on polish spaces. It simply says this: if we try to introduce probability measures for closed sets on non-locally-compact spaces by looking at capacities defined on open sets, then it is not the right way to proceed when the space of closed sets $\mathcal{F}$ is topologized by the Hausdorff metric. As a consequence, we should try other venues, such as using other topologies for $\mathcal{F}$, or considering capacities on compact sets $\mathcal{K}$ rather than on open sets $\mathcal{G}$. As far as we know, the problem remains open.

5.5 Exercises

5.1 Let $\{A_i, i \in I\}$ be a class of subsets of $\mathbb{R}^d$. Show that

(i) $\mathcal{F}_{\bigcup_{i \in I} A_i} = \bigcup_{i \in I} \mathcal{F}_{A_i}$; $\mathcal{F}^{\bigcup_{i \in I} A_i} = \bigcap_{i \in I} \mathcal{F}^{A_i}$

(ii) $\mathcal{F}_{\bigcap_{i \in I} A_i} \subseteq \bigcap_{i \in I} \mathcal{F}_{A_i}$; $\bigcup_{i \in I} \mathcal{F}^{A_i} \subseteq \mathcal{F}^{\bigcap_{i \in I} A_i}$

5.2 (continued) Show that

(i) the family $\{\mathcal{F}^K : K \in \mathcal{K}\}$ is closed under finite intersections;

(ii) the family $\{\mathcal{F}^K_{G_1,\ldots,G_n} : K \in \mathcal{K}, G_i \in \mathcal{G}, n \geq 0\}$ is a base for a topology of $\mathcal{F}$.

5.3 Let U be a finite set. Let $T : 2^U \to [0, 1]$ such that

(i) $T(\emptyset) = 0$, $T(U) = 1$

(ii) $\forall n \geq 2$ and $A_1, A_2, \ldots, A_n \subseteq U$,

$$T\left(\bigcap_{i=1}^n A_i\right) \leq \sum_{\emptyset \neq I \subseteq \{1,2,\ldots,n\}} (-1)^{|I|+1} T\left(\bigcup_{i \in I} A_i\right).$$

Show that T determines uniquely a probability measure Q on the power set of 2^U such that $Q(\mathcal{F}_A) = T(A)$, $\forall A \subseteq U$, where

$$\mathcal{F}_A = \{B \subseteq U : A \cap B \neq \emptyset\}.$$

5.4 Let $X : (\Omega, \mathcal{A}, P) \to (\mathbb{R}^d, \mathcal{B}(\mathbb{R}^d))$ be a random vector. Let $T : \mathcal{K}(\mathbb{R}^d) \to [0, 1]$, $T(K) = P(X \in K)$. Show that T satisfies the following properties:

(i) If $K_n \searrow K$ then $T(K_n) \searrow T(K)$.

(ii) T is monotonic: $K_1 \subseteq K_2 \Rightarrow T(K_1) \leq T(K_2)$.

(iii) For $n \geq 2$, and $K_1, K_2, \ldots, K_n$ in $\mathcal{K}$,

$$T\left(\bigcap_{i=1}^n K_i\right) \leq \sum_{\emptyset \neq I \subseteq \{1,2,\ldots,n\}} (-1)^{|I|+1} T\left(\bigcup_{i \in I} K_i\right).$$

5.5 Let $f : \mathbb{R} \to [0,1]$ be semi-upper-continuous. Let $(\Omega, \mathcal{A}, P)$ be a probability space, and $\alpha : \Omega \to [0,1]$ be a random variable, uniformly distributed.
Compute the capacity functional T of the random closed set on $\mathbb{R}$,

$$S(\omega) = \{x \in \mathbb{R} : f(x) \geq \alpha(\omega)\}$$

to show that the set function defined on $\mathcal{K}(\mathbb{R})$ by $K \to \sup\{f(x) : x \in K\}$ is
a capacity functional.

5.6 Let $\mathcal{F}(\mathbb{R}^d)$ be topologized by hit-or-miss topology. Verify that its Borel
σ-field $\mathcal{B}(\mathcal{F})$ is also generated by

   i) $\{\mathcal{F}_G : G \in \mathcal{G}\}$

or by

   ii) $\{\mathcal{F}^K : K \in \mathcal{K}\}$.

5.7 Let $S : (\Omega, \mathcal{A}) \to (\mathcal{F}(\mathbb{R}^d), \mathcal{B}(\mathcal{F}))$. Show that S is a random closed set if
it satisfies any of the following:

   i) $\forall F \in \mathcal{F}, S^{-1}(\mathcal{F}_F) \in \mathcal{A}$

   ii) $\forall K \in \mathcal{K}, S^{-1}(\mathcal{F}_K) \in \mathcal{A}$

   iii) $\forall G \in \mathcal{G}, S^{-1}(\mathcal{F}_G) \in \mathcal{A}$.

5.8 Let $X, Y : (\Omega, \mathcal{A}) \to (\mathcal{F}(\mathbb{R}^d), \mathcal{B}(\mathcal{F}))$. Show that $X \cap Y$ is $\mathcal{A}$-$\mathcal{B}(\mathcal{F})$-
measurable.

5.9 Let $T : \mathcal{K}(\mathbb{R}^d) \to [0,1]$ be a capacity functional. Extend T to $\mathcal{B}(\mathbb{R}^d)$ as

$$T(A) = \sup\{T(K) : K \in \mathcal{K}, K \subseteq A\}.$$

Show that T is alternating of infinite order on $\mathcal{B}(\mathbb{R}^d)$.

5.10 Let T be the capacity functional of the random closed set S on $\mathbb{R}^d$. Show
that, $\forall A \in \mathcal{B}(\mathbb{R}^d)$, $T^*(A) = P(S \cap A \neq \emptyset)$, where T^* denotes the extension of
T on $\mathcal{K}(\mathbb{R}^d)$ to the power set of $\mathbb{R}^d$) (Section 5.3).

Chapter 6

The Choquet Integral

This chapter is devoted to the concept of the integral with respect to *nonadditive set functions*, known as the *Choquet integral*. This concept of integral will be used in Chapter 7 to study the convergence in distribution of random closed sets when the nonadditive set functions are capacity functionals. However the Choquet integral is also useful in many other fields such as *mathematical economics* (e.g., Marinacci and Montrucchio [72]), *multicriteria decision making* (e.g., Grabisch et al. [42]). As such we are going to present the material in a fairly general context as well as providing some motivations.

6.1 Some Motivations

Let X be a nonnegative random variable defined on $(\Omega, \mathcal{A}, P)$. Then it is well known that

$$EX = \int_\Omega X(\omega)dP(\omega) = \int_0^{+\infty} P(X > t)dt.$$

It is interesting to note that the above equality can be proved by using Robbins' formula (Chapter 2) applied to the random closed set $\omega \to [0, X(\omega)]$ of $\mathbb{R}^+$, see Exercise 6.1, by noting that

$$\int_0^{+\infty} P(X > t)dt = \int_0^{+\infty} P(X \geq t)dt. \qquad (6.1)$$

The equality (6.1) can be proved as follows. For $n \geq 1$ and $t \in \mathbb{R}$, we have

$$\left\{\omega : X(\omega) \geq t + \frac{1}{n}\right\} \subseteq \{\omega : X(\omega) > t\} \subseteq \{\omega : X(\omega) \geq t\}.$$

Since P is a monotone increasing set function, it follows that

$$P\left(X \geq t + \frac{1}{n}\right) \leq P(X > t) \leq P(X \geq t).$$

The function $t \rightarrow P(X \geq t)$ is monotonic nonincreasing on $\mathbb{R}$, it is continuous except in some countable subset $D \subseteq \mathbb{R}$. Thus, for $t \notin D$,

$$\lim_{n \to \infty} P\left(X \geq t + \frac{1}{n}\right) = P(X \geq t),$$

and hence

$$P(X \geq t) = P(X > t), \quad \forall t \notin D,$$

which implies that

$$\int_0^{+\infty} P(X > t)dt = \int_0^{+\infty} P(X \geq t)dt.$$

Remark. The above proof depends only on the measurability of X and the monotonicity of P. As such, the result still holds when the probability measure P is replaced by any monotone set function on $\mathcal{A}$.

We have seen in Section 3.4 that in the case of incomplete statistical information, the lower bound on expected values is written in terms of the distribution F (of a random set), which is a monotone set function, namely,

$$\int_0^{+\infty} F(X \geq t)dt.$$

This expression is referred to as the *Choquet integral* of the function X with respect to the monotone (not necessarily additive) set function F, in view of the pioneering work of Choquet [15].

Another similar situation in *coarse data* analysis is this. Let X be a random variable of interest, defined on $(\Omega, \mathcal{A}, P)$, and $g : \mathbb{R} \rightarrow \mathbb{R}^+$, measurable. Suppose that X is not observable and S is a coarsening of X, i.e., $S : \Omega \rightarrow \mathcal{F}(\mathbb{R})$ is a random closed set on $\mathbb{R}$ such that $P(X \in S) = 1$. Let

$$g_*(\omega) = \inf\{g(x) : x \in S(\omega)\} \leq g(X(\omega)) \leq \inf\{g(x) : x \in S(\omega)\} \leq g^*(\omega).$$

Thus, $g_* \leq g(X) \leq g^*$, a.s., and hence

$$E(g_*) \leq E(g(X)) \leq E(g^*).$$

Remark. Of course, we need to verify that g_* and g^* are measurable! See Exercise 6.5.

Let $F_*, F^* : \mathcal{B}(\mathbb{R}) \rightarrow [0,1]$ be

$$F_*(B) = P(S \subseteq B), \quad F^*(B) = P(S \cap B \neq \emptyset).$$

We have

$$E(g_*) = \int_\Omega g_*(\omega)dP(\omega) = \int_0^{+\infty} P(g_* > t)dt = \int_0^{+\infty} P[g_*^{-1}(t, +\infty)]dt =$$

$$\int_0^{+\infty} P[S \subseteq g^{-1}(t, +\infty)]dt =$$

$$\int_0^{+\infty} F_*[g^{-1}(t, +\infty)]dt = \int_0^{+\infty} F_*(X : g(x) > t)dt.$$

Similarly,

$$E(g^*) = \int_0^{+\infty} F^*(x : g(x) > t)dt.$$

Although both F_* and F^* are not probability measures on $\mathcal{B}(\mathbb{R})$, they can be used for approximate inference process. Thus, Choquet integrals represent some practical quantities of interest.

Another important motivation for using the Choquet integral, in the finite case, is in the theory of multicriteria decision making.

The problem of ranking of alternatives with respect to a set of criteria is this. Let $N = \{1, 2, \ldots, n\}$ be the set of criteria under consideration in some decision problem. An evaluation is a map, say, $X : N \to \mathbb{R}^+$. Given several $X^{(i)} : N \to \mathbb{R}^+$, $i = 1, 2, \ldots, n$, we wish to linearly rank them according to N. For this to be possible, we need to map each X, viewed as a vector in $(\mathbb{R}^+)^n$, to $\mathbb{R}$. Such a mapping is called an *aggregation operator*.

Let $\alpha = (\alpha_1, \alpha_2, \ldots, \alpha_n)$ with $\alpha_i \geq 0$ and $\sum_{i=1}^n \alpha_i = 1$. The value α_i is intended to denote the degree of importance of the criterion i. Then the weighted average (linear) aggregation operator is

$$M_\alpha : \mathbb{R}^n \to \mathbb{R}, \quad M_\alpha(X) = \sum_{i=1}^n \alpha_i x_i$$

where $x_i = X(i)$, the value of X at i.

If we view α as a probability measure on 2^N, i.e., for $A \subseteq N$,

$$\alpha(A) = \sum_{i \in A} \alpha_i,$$

then $M_\alpha(X) = E_\alpha(X)$, the expected value of X with law α.

Consider the "order statistic" $x_{(1)} \leq x_{(2)} \leq \ldots \leq x_{(n)}$ associated with the values of X, and let

$$A_{(i)} = \{(i), (i+1), \ldots, (n)\} \subseteq N, \quad i = 1, 2, \ldots, n.$$

Then

$$M_\alpha(X) = \sum_{i=1}^n \alpha_i x_i = \sum_{i=1}^n \alpha_{(i)} x_{(i)} = \sum_{i=1}^n (x_{(i)} - x_{(i-1)})\alpha(A_{(i)}),$$

where $x_{(0)} = 0$.

It was argued in the theory of multicriteria decision making that when the criteria are interactive (i.e., not independent), nonlinear aggregation operators are more suitable for ranking. In some situations where degrees of important criteria can be determined by pairwise comparisons, such as in the *Analytical Hierarchy Process* (Saaty [107]), linear aggregation operators are still in use.

As we will see in the next section, the Choquet integral is nonlinear. On the other hand, the Choquet integral generalizes the Lebesgue integral and hence generalizes weighted average operators. As such, it is a plausible candidate for a nonlinear aggregation operator in decision making.

From its definition, or simply replacing the probability measure α in the above $M_\alpha(X)$, by the aggregation operator $M_\mu(\cdot)$, where μ is a map from $2^N \to [0,1]$ such that $\mu(N) = 1$, $\mu(\emptyset) = 0$, and for $A \subseteq B$, $\mu(A) \leq \mu(B)$, we have

$$M_\mu(X) = \sum_{i=1}^{n} (x_{(i)} - x_{(i-1)})\mu(A_{(i)}),$$

which is the *discrete version* of the Choquet integral of X with respect to μ.

Note that for $A \subseteq N$, the value $\mu(A)$ is intended to represent the degree of importance of the collection of criteria in A, whereas for each $i \in N$, $\mu(\{i\})$ is the degree of importance of the individual criterion i. In the context of coalitional games (or transferable utility games), where N is the set of players, each subset A of N is a coalition, and $\mu(A)$ is interpreted as the amount the members in A can achieve by teaming up. Note, however, the subtle distinction with *scaling weights* in decision with multiple objectives (see Keeney and Raifa [59]), where (additive) aggregation is used to obtain a utility function for optimization purpose.

6.2   The Choquet Integral

In view of the discussion in Section 6.1, we are going to formulate a fairly general concept of the Choquet integral of numerical functions with respect to set functions that need not be σ-additive. This concept of the Choquet integral will generalize the Lebesgue integral. Note that, in Chapter 7, we will consider a specific setting for studying the convergence in distribution of random closed sets, namely, the Choquet integral for continuous and bounded functions defined on $\mathbb{R}^d$ (or, more generally, on a Hausdorff, locally compact, second countable space) with respect to set functions on $\mathcal{B}(\mathbb{R}^d)$.

In the following, Ω is a set, and μ is real-valued set function, defined on some class $\mathcal{A}$ of subsets of Ω (containing the empty set $\emptyset$), such that $\mu(\emptyset) = 0$, and for $A, B \in \mathcal{A}$ with $A \subseteq B$, we have $\mu(A) \leq \mu(B)$, i.e., μ is a *monotone nondecreasing* set function.

Standard measure theory is a theory of integrals motivated by the need to compute important quantities such as volumes, expected values, and so on. Similarly, here we are going to consider "integrals" of real-valued functions f defined on Ω. As in analysis, it is convenient to extend the possible values to $\pm\infty$, so that the range of functions is the interval $[-\infty, +\infty]$, and all suprema, such as $\sup\{f(\omega) : \omega \in A \subseteq \Omega\}$, exist. Arithmetic operations are extended from $\mathbb{R} = (-\infty, +\infty)$ to $\overline{\mathbb{R}} = [-\infty, +\infty]$ in the usual way.

For practical purposes, $\mathcal{A}$ will be taken as a σ-field of subsets of Ω. Let $f : \Omega \to \overline{\mathbb{R}}$ be $\mathcal{A}$-$\mathcal{B}(\overline{\mathbb{R}})$-measurable, where $\mathcal{B}(\overline{\mathbb{R}})$ is the Borel σ-field on $\overline{\mathbb{R}}$ generated by $\mathcal{B}(\mathbb{R})$ and $\{-\infty, +\infty\}$. Note that

$$\mathcal{B}(\overline{\mathbb{R}^+}) = \mathcal{B}([0, +\infty]) = \{A \cap [0, +\infty] : A \in \mathcal{B}(\overline{\mathbb{R}})\} = \{A : A \subseteq [0, +\infty] \cap \mathcal{B}(\overline{\mathbb{R}})\}.$$

To avoid meaningless expressions like $-\infty + \infty$, we start out by considering nonnegative and nonpositive functions separately. Thus, let $f : \Omega \to \overline{\mathbb{R}^+}$ be $\mathcal{A}$-$\mathcal{B}(\overline{\mathbb{R}^+})$-measurable. Then there exists an increasing sequence $f_1, f_2, \ldots$ of simple functions $f_n : \Omega \to [0, +\infty]$, that is, functions of the form $\sum_{j=1}^{n} a_j 1_{A_j}(\omega)$ with $a_j \in \mathbb{R}^+$ and the A_j pairwise disjoint elements of $\mathcal{A}$, such that for all $\omega \in \Omega$, $f(\omega) = \lim_{n\to\infty} f_n(\omega)$. If $f : \Omega \to [-\infty, \infty]$, then we write

$$f(\omega) = f(\omega)1_{f\geq 0}(\omega) + f(\omega)1_{f<0}(\omega) =$$

$$f(\omega)1_{f\geq 0}(\omega) - (-f(\omega)1_{f<0}(\omega)) = f^+(\omega) - f^-(\omega).$$

Then f^+ and f^- are maps $\Omega \to [0, \infty]$, and f is measurable if and only if both f^+ and f^- are measurable. When μ is a nonnegative σ-additive measure on $(\Omega, \mathcal{A})$, the Lebesgue integral of a nonnegative f with respect to μ is defined to be

$$\int_\Omega f(\omega)d\mu(\omega) = \lim_{n\to\infty} f_n(\omega)d\mu(\omega),$$

where f_n is simple and converges from below to f, and

$$\int f_n(\omega)d\mu(\omega) = \sum_{j=1}^{k_n} a_j \mu(A_j)$$

when the A_j's form a $\mathcal{A}$-partition of Ω. Because of the additivity of μ, the quantity $\int_\Omega f(\omega)d\mu(\omega)$ is well defined. It is independent on the particular choices of the f_n.

For $f : \Omega \to \mathbb{R}$ measurable, we define

$$\int_\Omega f(\omega)d\mu(\omega) = \int_\Omega f^+(\omega)d\mu(\omega) - \int_\Omega f^-(\omega)d\mu(\omega)$$

provided that both terms in the right-hand side are not ∞. When $\mu(\Omega) < \infty$,

$$\int_\Omega f(\omega)d\mu(\omega) = \int_0^\infty \mu(f > t)dt + \int_{-\infty}^0 [\mu(f > t) - \mu(\Omega)]dt. \qquad (6.2)$$

Indeed, since $f^+ \geq 0$, we have

$$\int_\Omega f^+(\omega)d\mu(\omega) = \int_0^\infty \mu(f^+ > t)dt = \int_0^\infty \mu(f > t)dt.$$

Similarly, $\int_\Omega f^-(\omega)d\mu(\omega) = \int_0^\infty \mu(f^- > t)dt$. For each $t > 0$, $(f^- > t)$ and $(f \geq -t)$ form a partition of Ω. By the additivity of μ,

$$\mu(f^- > t) = \mu(\Omega) - \mu(f \geq -t),$$

and (6.2) follows.

Still in the case $\mu(\Omega) < \infty$, for $A \in \mathcal{A}$, we get in a similar way that

$$\int_A f d\mu = \int_\Omega (1_A f)d\mu =$$

$$\int_0^\infty \mu((f > t) \cap A)dt + \int_{-\infty}^0 [\mu((f > t) \cap A) - \mu(A)]dt.$$

For $f \geq 0$, we have $\int_\Omega f d\mu = \int_\Omega \mu(f > t)dt$. The right-hand side is an integral of the numerical function $t \to \mu(f > t)$ with respect to the Lebesgue measure dt on $[0, \infty]$. It is well defined as long as the function is measurable. Thus if μ is a monotone set function on $(\Omega, \mathcal{A})$ and f is $\mathcal{A}$-$[0, \infty]$-measurable, then $(f > t) \in \mathcal{A}$ for all $t \in [0, \infty]$. Since μ is increasing, $t \to \mu(f > t)$ is $\mathcal{B}([0, \infty])$-$\mathcal{B}([0, \infty])$-measurable since it is a decreasing function. This is proved as follows: for $t \in [0, \infty]$, let $\varphi(t) = \mu(f > t)$. Then

$$(\varphi > t) = \begin{cases} [a, \infty] & \text{if } a = \inf\{\varphi < t\} \text{ is attained} \\ (a, \infty] & \text{if not} \end{cases}$$

In either case, $(\varphi < t) \in \mathcal{B}([0, \infty])$. Indeed if $\inf\{\varphi < t\}$ is attained at a, then for $b \geq a$, we have $\varphi(b) \leq \varphi(a) < x$ so that $b \in (\varphi < x)$. Conversely, if $\varphi(c) < x$, then $a \leq x$ by the definition of a.

If $a = \inf\{\varphi < t\}$ is not attained, then if $\varphi(b) < x$, we must have $a < b$. Conversely, if $c > a$, then for $\varepsilon = c - a > 0$, there exists y such that $\varphi(y) < x$ and $a < y < c$. But φ is decreasing, so $\varphi(c) \leq \varphi(y) < x$. Thus $c \in (\varphi < x)$.

Thus for $f : \Omega \to [0, \infty]$, one can define the Choquet integral of f with respect to μ as $C_\mu(f) = \int_0^\infty \mu(f > t)dt$. A general definition of the Choquet integral is as follows.

DEFINITION 6.1 *When $f : \Omega \to [-\infty, \infty]$ and $\mu(\Omega) < \infty$, the Choquet integral of f with respect to μ is defined as*

$$C_\mu(f) = \int_0^\infty \mu(f > t)dt + \int_{-\infty}^0 [\mu(f > t) - \mu(\Omega)]dt.$$

We also write $C_\mu(f) = \int_\Omega f d\mu$. And for $A \in \mathcal{A}$,

$$C_\mu(1_A f) = \int_A f d\mu =$$

$$\int_0^\infty \mu((f > t) \cap A)dt + \int_{-\infty}^0 [\mu((f > t) \cap A) - \mu(A)]dt.$$

Remarks.

(i) In view of Exercise 6.3, when $\mu(\Omega) < \infty$,

$$C_\mu(f) = \int_0^\infty \mu(f > t)dt + \int_{-\infty}^0 [\mu(f > t) - \mu(\Omega)]dt =$$

$$\int_0^\infty \mu(f \geq t)dt + \int_{-\infty}^0 [\mu(f \geq t) - \mu(\Omega)]dt.$$

(ii) As shown in Marinacci and Montrucchio [72], the above definition of the Choquet integral is the *unique translation invariant extension* from non-negative functions to real-valued functions, i.e., for $\alpha \in \mathbb{R}$, $C_\mu(f + \alpha 1) = C_\mu(f)) + \alpha C_\mu(1)$, where 1 stands for the function $1(\omega) = 1$, $\forall \omega \in \Omega$. See Exercise 6.9.

We say that f is μ-integrable on A when the above right side is finite. When $\mu(\Omega) < \infty$, say $\mu(\Omega) = 1$, and μ represents subjective evaluations concerning variables, then the "variable" f has the "distribution"

$$\varphi_f(t) = \begin{cases} \mu(f > t) & \text{if } t \geq 0 \\ \mu(f \geq t) - \mu(\Omega) & \text{if } t < 0. \end{cases}$$

The "expected value" of f under μ is the Lebesgue integral of the distribution $\varphi_f(t)$, that is $\int_{-\infty}^\infty \varphi_f(t)dt$. Some remarks are in order.

When $f = 1_A$ with $A \in \mathcal{A}$, we have $\mu(f > t) = \mu(A)1_{[0,1)}(t)$, so that

$$C_\mu(1_A) = \mu(A).$$

More generally, for $f(\omega) = \sum_{i=1}^n a_i 1_{A_i}(\omega)$ with the A_i's pairwise disjoint subsets of Ω and $a_0 = 0 < a_1 < \ldots < a_n$, we have

$$\mu(f > t) = \sum_{i=1}^n \mu\left(\bigcup_{j=i}^n A_j\right) 1_{[a_{i-1}, a_i)}$$

so that

$$C_\mu(f) = \sum_{i=1}^{n} (a_i - a_{i_1})\mu \left(\bigcup_{j=i}^{n} A_j \right).$$

For an arbitrary simple function of the form $f = \sum_{i=1}^{n} a_i 1_{A_i}$, and with the A_i's forming a measurable partition of Ω and

$$a_1 < \ldots < a_k < 0 < a_{k+1} < \ldots < a_n$$

we have for $t \geq 0$

$$(f > t) = \begin{cases} \emptyset & \text{if } t \in [a_n, \infty) \\ \bigcup_{i=j+1}^{n} A_i & \text{if } t \in [a_j, a_{j+1}) \\ \bigcup_{i=k+1}^{n} A_i & \text{if } t \in [0, a_{k+1}) \end{cases}$$

and for $t < 0$

$$(f \geq t) = \begin{cases} \bigcup_{i=k+1}^{n} A_i & \text{if } t \in (a_k, 0) \\ \bigcup_{i=j+1}^{n} A_i & \text{if } t \in (a_j, a_{j+1}] \\ \Omega = \bigcup_{i=1}^{n} A_i & \text{if } t \in (-\infty, a_1]. \end{cases}$$

Thus

$$\int_0^\infty \mu(f > t)dt = a_{k+1}\mu \left(\bigcup_{i=k+1}^{n} A_i \right) + \ldots +$$

$$(a_{j+1} - a_j)\mu \left(\bigcup_{i=j+1}^{n} A_i \right) + \ldots + (a_n - a_{n-1})\mu(A_n).$$

Also

$$\int_{-\infty}^0 [\mu(f \geq t) - \mu(\Omega)]dt = (a_2 - a_1) \left[\mu \left(\bigcup_{i=2}^{n} A_i \right) - \mu \left(\bigcup_{i=1}^{n} A_i \right) \right] + \ldots +$$

$$(a_{j+1} - a_j) \left[\mu \left(\bigcup_{i=j+1}^{n} A_i \right) - \mu \left(\bigcup_{i=j}^{n} A_i \right) \right] + \ldots +$$

$$(-a_k) \left[\mu \left(\bigcup_{i=k+1}^{n} A_i \right) - \mu \left(\bigcup_{i=1}^{n} A_i \right) \right],$$

so that

$$C_\mu(f) = \sum_{j=1}^n a_j \left[\mu\left(\bigcup_{i=j+1}^n A_i \right) - \mu\left(\bigcup_{i=j}^n A_i \right) \right].$$

Example 1. Let T be the capacity functional of a random closed st S on $\mathbb{R}^d$. Let $g : \mathbb{R}^d \to \mathbb{R}^+$ measurable. Since T is extended to the power set of $\mathbb{R}^d$, in particular to $\mathcal{B}(\mathbb{R}^d)$, we can consider

$$\int_0^{+\infty} T(x \in \mathbb{R}^d : g(x) \geq t)dt.$$

By Exercise 5.10,

$$T(g \geq t) = P(S \cap (g \geq t) \neq \emptyset).$$

Obviously,

$$S(\omega) \cap (g \geq t) \neq \emptyset \Leftrightarrow \sup\{g(x) : x \in S(\omega)\} \geq t.$$

Thus,

$$\int_0^{+\infty} T(g \geq t)dt =$$

$$\int_0^{+\infty} P[\sup\{g(x) : x \in S\} \geq t]dt =$$

$$E[\sup\{x : x \in S\}].$$

Example 2. Let $(\Omega, \mathcal{A})$ be a measurable space. For $\emptyset \neq A \in \mathcal{A}$, let $u_A : \mathcal{A} \to \{0, 1\}$ be

$$u_A(B) = \begin{cases} 0 \text{ if } A \subseteq B, \\ 1 \text{ if not.} \end{cases}$$

Let $f : \Omega \to \mathbb{R}^+$, $\mathcal{A}$-$\mathcal{B}(\mathbb{R})$-measurable and bounded. Then

$$C_{u_A}(f) = \int_0^{+\infty} u_A(x : f(x) \geq t)dt = \int_0^{\inf\{f(x): x \in A\}} dt = \inf\{f(x) : x \in A\}$$

since

$$A \subseteq (f \geq t) \Leftrightarrow t \leq \inf\{f(x) : x \in A\}$$

so that

$$u_A(f \geq t) = \begin{cases} 0 \text{ if } t \leq \inf\{f(x) : x \in A\} \\ 1 \text{ otherwise.} \end{cases}$$

While the Choquet integral is monotone and positive homogeneous of degree one, that is, $f \leq g$ implies that $C_\mu(f) \leq C_\mu(g)$ and for $\lambda > 0$, $C_\mu(\lambda f) = \lambda C_\mu(f)$, its additivity fails. For example, if $f = (1/4)1_A$ and $g = (1/2)1_B$, then

$$C_\mu(f + g) \neq C_\mu(f) + C_\mu(g)$$

as an easy calculation shows. However, if we consider two simple functions f and g of the form

$$f = a1_A + b1_B \text{ with } A \cap B = \emptyset, \ 0 \leq a \leq b,$$

$$g = \alpha 1_A + \beta 1_B \text{ with } 0 \leq \alpha \leq \beta$$

then

$$C_\mu(f + g) = C_\mu(f) + C_\mu(g).$$

More generally, this equality holds for $f = \sum_{j=1}^{n} a_j 1_{A_j}$ and $g = \sum_{j=1}^{n} b_j 1_{A_j}$, with the A_j pairwise disjoint and the a_i and b_i increasing and nonnegative. Such pairs of functions satisfy the inequality

$$(f(\omega) - f(\omega'))(g(\omega) - g(\omega')) \geq 0.$$

That is, the pair is *comonotonic*, or *similarly ordered*. It turns out that the concept of comonotonicity of real-valued bounded measurable functions is essential for the characterization of the Choquet integrals.

DEFINITION 6.2   *Two real-valued functions f and g defined on Ω are comonotonic if for all ω and ω' in Ω,*

$$(f(\omega) - f(\omega'))(g(\omega) - g(\omega')) \geq 0.$$

Roughly speaking, this means that f and g have the same "tableau of variation." Here are a few elementary facts about comonotonic functions.

- The comonotonicity relation is symmetric and reflexive, but not transitive.

- Any function is comonotonic with a constant function.

- If f and g are comonotonic and r and s are positive numbers, then rf and sg are comonotonic.

- As we saw above, two functions $f = \sum_{j=1}^{n} a_j 1_{A_j}$ and $g = \sum_{j=1}^{n} b_j 1_{A_j}$, with the A_j pairwise disjoint and the a_i and b_i increasing and nonnegative, are comonotonic.

DEFINITION 6.3 *A functional H from the space $\mathbb{B}$ of bounded real-valued measurable functions on $(\Omega, \mathcal{A})$ to $\mathbb{R}$ is* comonotonic additive *if whenever f and g are comonotonic, $H(f + g) = H(f) + H(g)$.*

If $H(f) = \int_{\Omega} f d\mu$ with μ being the Lebesgue measure, then H is additive, and in particular, comonotonic additive. Choquet integrals are comonotonic additive.

Here are some facts about comonotonic additivity.

- If H is comonotonic additive, then $H(0) = 0$. This follows since 0 is comonotonic with itself, whence $H(0) = H(0 + 0) = H(0) + H(0)$.

- If H is comonotonic additive and $f \in \mathbb{B}$, then for positive integers n, $H(nf) = nH(f)$. This is an easy induction. It is clearly true for $n = 1$, and for $n > 1$ and using the induction hypothesis,

$$H(nf) = H(f + (n - 1)f) = H(f) + H((n - 1)f) =$$

$$H(f) + (n - 1)H(f) = nH(f).$$

- If H is comonotonic additive and $f \in \mathbb{B}$, then for positive integers m and n, $H((m/n)f) = (m/n)H(f)$. Indeed,

$$(m/n)H(f) = (m/n)H(n(f/n)) = mH(f/n) = H((m/n)f).$$

- If H is comonotonic additive and monotonic increasing, then $H(rf) = rH(f)$ for positive r and $f \in \mathbb{B}$. Just take an increasing sequence of positive rational numbers r_i converging to r. Then $H(r_i f) = r_i H(f)$ converges to $rH(f)$ and $r_i f$ converges to rf. Thus $H(r_i f)$ converges also to $H(rf)$.

We now state the result concerning the characterization of Choquet integrals. Consider the case when $\mu(\Omega) = 1$. Let

$$C_\mu(f) = \int_0^\infty \mu(f > t)dt + \int_{-\infty}^0 [\mu(f \geq t) - 1]dt.$$

THEOREM *The function C_μ on $\mathbb{B}$ satisfies the following:*

(i) $C_\mu(1_\Omega) = 1$;

(ii) C_μ is monotone increasing;

(iii) C_μ is comonotone additive.

Conversely, if H is a functional on $\mathbb{B}$ satisfying these three conditions, then H of the form C_μ for the monotone set function μ defined on $(\Omega, \mathcal{A})$ by $\mu(A) = H(1_A)$.

The first two parts are trivial. To get the third, it is sufficient to show that C_μ is comonotonic additive on simple functions. The proof is cumbersome and we omit the details.

For the converse, if $\mu(A) = H(1_A)$, then by the comonotonic additivity of H, $H(0) = 0$, so $\mu(\emptyset) = 0$. By the second condition, μ is a monotone set function. For the rest it is sufficient to consider nonnegative simple functions. We refer to Schneidler [112] for the details.

6.3 Radon-Nikodym Derivatives

Let μ and ν be two σ-additive set functions (measures) defined on a σ-algebra $\mathcal{U}$ of subsets of a set U. If there is a $\mathcal{U}$-measurable function $f : U \to [0, \infty)$ such that $\mu(A) = \int_A f d\nu$ for all $A \in \mathcal{U}$, then f is called (a version of) the *Radon-Nikodym derivative* of μ with respect to ν, and is written as $f = d\mu/d\nu$.

It is well known that the situation above happens if and only if μ is absolutely continuous with respect to ν, in symbols $\mu \ll \nu$, that is, if $\nu(A) = 0$, then $\mu(A) = 0$. When μ and ν are no longer additive, a similar situation still exists. Here is the motivating example. Let $f : U \to [0, \infty)$ be μ-measurable. Consider the maxitive measure $\mu(A) = \sup_{u \in A} f(u)$. Also let $\nu_0 : \mathcal{U} \to [0, \infty)$ be defined by

$$\nu_0 = \begin{cases} 0 \text{ if } A = \emptyset \\ 1 \text{ if } A \neq \emptyset \end{cases}$$

Then μ can be written as the Choquet integral of f with respect to the ν_0. Indeed

$$\nu_0(\{u : f(u) \geq t\} \cap A) = \begin{cases} 0 \text{ if } f(u) < t \text{ for } u \in A \\ 1 \text{ if } f(u) \geq t \text{ for some } u \in A \end{cases}$$

Thus

$$\int_0^\infty \nu_0(\{u : f(u) \geq t\} \cap A) dt = \int_0^{\sup_{u \in A} f(u)} dt = \mu(A).$$

By analogy with ordinary measure theory, we say that f is the Radon-Nikodym derivative of μ with respect to ν_0.

DEFINITION 6.4 *Let $\mu, \nu : U \to [0, \infty]$ be two increasing set-functions. If there exists a $\mathcal{U}$-measurable function $f : U \to [0, \infty)$ such that for $A \in \mathcal{U}$*

$$\mu(A) = \int_0^\infty \nu(\{u : f(u) \geq t\} \cap A)dt,$$

then μ is said to have f as its Radon-Nikodym derivative with respect to ν, and we write $f = \dfrac{d\mu}{d\nu}$, or $d\mu = fd\nu$.

Here is another example. Let $U = \mathbb{R}$ and $\mathcal{U} = \mathcal{B}(\mathbb{R})$, the Borel σ-field of $\mathbb{R}$. Let $B = [0, 1)$, $A \in \mathcal{B}(\mathbb{R})$, $d(A, B) = \inf\{|x - y| : x \in A, \ y \in B\}$, and A' be the complement of A in $\mathcal{U}$. Define μ and ν by

$$\mu(A) = \begin{cases} 0 & \text{if} & A = \emptyset \\ \dfrac{1}{2}\sup\{|x| : x \in A\} & \text{if} \ A \neq \emptyset \text{ and } d(A, B') > 0 \\ 1 & & A \neq \emptyset \text{ and } d(A, B') = 0 \end{cases}$$

$$\nu(A) = \begin{cases} 0 \text{ if } A = \emptyset \\ \dfrac{1}{2} \text{ if } A \neq \emptyset \text{ and } d(A, B') > 0 \\ 1 \text{ if } A \neq \emptyset \text{ and } d(A, B') = 0. \end{cases}$$

Then $f(x) = x$ if $x \in B$ and 1 otherwise is $\dfrac{d\mu}{d\nu}$.

Obviously, if $d\mu = fd\nu$, then $\mu \ll \nu$. But unlike the situation for ordinary measures, the absolute continuity is only a necessary condition for μ to admit a Radon-Nikodym derivative with respect to ν. For example, with $U = \mathbb{R}$ and $\mathcal{U} = \mathcal{B}(\mathbb{R})$ as before, and $\mathbb{N}$ the positive integers, let μ and ν be defined by

$$\mu(A) = \begin{cases} 0 \text{ if } A \cap \mathbb{N} = \emptyset \\ 1 \text{ if } A \cap \mathbb{N} \neq \emptyset \end{cases}$$

$$\nu(A) = \begin{cases} 0 & \text{if } A \cap \mathbb{N} = \emptyset \\ \sup\left\{\dfrac{1}{x} : x \in A \cap \mathbb{N}\right\} & \text{if } A \cap \mathbb{N} \neq \emptyset. \end{cases}$$

By construction, $\mu \ll \nu$. Suppose that $d\mu = fd\nu$. Then $f : \mathbb{R} \to [0, \infty)$ is a $\mathcal{B}(\mathbb{R})$-measurable function such that for $A \in \mathcal{B}(\mathbb{R})$, $\mu(A) = \int_A fd\nu$ (in Choquet's sense). We then have for $n \in \mathbb{N}$,

$$\mu(\{n\}) = \int_0^\infty \nu((f \geq t) \cap \{n\})dt$$

$$= \int_0^{f(n)} \nu(\{n\})dt = \int_0^{f(n)} \frac{1}{n}dt = \frac{1}{n}f(n).$$

But by construction of μ, $\mu(\{n\}) = 1$ so that $f(n) = n$. Now

$$\mu(\mathbb{N}) = \int_{\mathbb{N}} f d\nu = \int_0^\infty \nu((f \geq t) \cap \mathbb{N}) dt = \sum_{n=1}^\infty \int_{n-1}^n \nu((f \geq t) \cap \mathbb{N}) dt$$

$$\geq \sum_{n=1}^\infty \int_{n-1}^n \nu((f \geq n) \cap \mathbb{N}) dt = \sum_{n=1}^\infty \int_{n-1}^n \nu(\{k : f(k) \geq n\}) dt$$

$$= \sum_{n=1}^\infty \int_{n-1}^n \nu(\{k : k \geq n\}) dt = \sum_{n=1}^\infty \sup \left\{ \frac{1}{k} : k \geq n \right\}$$

$$= \sum_{n=1}^\infty \frac{1}{n} = \infty.$$

This contradicts the fact that $\mu(\mathbb{N}) = 1$. Thus, even though $\mu \ll \nu$, μ does not admit a Radon-Nikodym derivative with respect to ν. Depending on additional properties of μ and ν, sufficient conditions for μ to admit a Radon-Nikodym derivative with respect to ν can be found.

It is well known that probability measures (distributions of random variables) can be defined in terms of probability density functions via Lebesgue integration. For distributions of random sets, namely, capacities alternating of infinite order, a similar situation occurs when the Radon-Nikodym property is satisfied. Specifically, suppose that $d\mu = f d\nu$. Then μ is alternating of infinite order whenever ν is. We restrict the proof of this fact to the case where $\mathcal{U}$ is finite, in which case f can take only finitely many values, say, $x_1 \leq x_2 \leq \ldots \leq x_n$ with $x_0 = 0$, we can write

$$\mu(A) = \int_A f d\nu$$

$$= \sum_{k=1}^n \int_{x_{k-1}}^{x_k} \nu(\{x \in A : f(x) \geq t\}) dt$$

$$= \sum_{k=1}^n (x_k - x_{k-1}) \nu(B_k \cap A)$$

where $B_k = \{u \in U : f(u) \geq x_k\}$ for $k \leq n$. Take $A = \bigcap_{i=1}^m A_i$. Then we have

$$\mu \left(\bigcap_{i=1}^m A_i \right) = \sum_{k=1}^n (x_k - x_{k-1}) \nu \left(B_k \cap \left(\bigcap_{i=1}^m A_i \right) \right).$$

Since $B_k \cap \left(\bigcap_{i \in I} A_i \right) = \bigcap_{i \in I} (B_k \cap A_i)$ and ν is alternating of infinite order, we have

$$\nu\left(B_k \cap \left(\bigcap_{i=1}^{m} A_i\right)\right) \leq \sum_{\emptyset \neq I \subseteq \{1,2,\dots,m\}} (-1)^{|I|+1}\nu\left(\bigcup_{i \in I}(B_k \cap A_i)\right)$$

$$= \sum_{\emptyset \neq I \subseteq \{1,2,\dots,m\}} (-1)^{|I|+1}\nu\left(B_k \cap \left(\bigcup_{i \in I} A_i\right)\right).$$

Thus

$$\mu\left(\bigcap_{i=1}^{m} A_i\right) \leq \sum_{k=1}^{n}(x_k - x_{k-1}) \sum_{\emptyset \neq I \subseteq \{1,2,\dots,m\}} (-1)^{|I|+1}\nu\left(B_k \cap \left(\bigcap_{i \in I} A_i\right)\right)$$

$$= \sum_{\emptyset \neq I \subseteq \{1,2,\dots,m\}} (-1)^{|I|+1} \sum_{k=1}^{n}(x_k - x_{k-1})\nu\left(B_k \cap \left(\bigcap_{i \in I} A_i\right)\right)$$

$$= \sum_{\emptyset \neq I \subseteq \{1,2,\dots,m\}} (-1)^{|I|+1}\mu\left(\bigcap_{i \in I} A_i\right).$$

Now, we establish a new version of Graf's theorem [43] for submeasures.

DEFINITION 6.5   *Let $(U,\mathcal{U})$ be a measurable space. A set-function $\mu : \mathcal{U} \to [0,\infty)$ is called a submeasure if $\mu(\emptyset) = 0$, μ is monotone increasing, and σ-subadditive, i.e.,*

$$\mu\left(\bigcup_{n\geq 1} A_n\right) \leq \sum_{n\geq 1}\mu(A_n) \quad \text{for any } \{A_n, n \geq 1\} \subseteq \mathcal{U}.$$

Note that any capacity in Graf's sense is a submeasure, but the class of submeasures is strictly larger than the class of capacities in the sense of Graf [43], see an example below.

Let us recall the following fact of Graf [43]. Let $(U,\mathcal{U})$ be a measurable space and let $\{A(t) : t \in [0,\infty)\} \subset \mathcal{U}$ be a family of measurable sets. A decreasing family $\{B(t) : t \in [0,\infty)\} \subset \mathcal{U}$ is defined as follows: first let $B(0) = U$ and for every $t > 0$ let $Q(t) = Q \cap [0,t)$ and $Q^+ = Q \cap [0,\infty)$, where Q denotes the set of rational numbers, let $B(t) = \bigcap_{q \in Q(t)} A(q)$. Since $Q(t)$ is countable, it follows that $B(t)$ is $\mathcal{U}$-measurable for every $t \in [0,\infty)$. We define a function $f : U \to [0,\infty)$ by Graf's formula:

$$f(x) = \begin{cases} 0 & \text{if } x \in \bigcap_{t\in Q^+} B(t) \\ \sup\{t : x \in B(t)\} & \text{otherwise.} \end{cases} \tag{6.3}$$

Observe that f is $\mathcal{U}$-measurable.

Now we prove the following result which is our version of Graf's theorem for submeasures.

THEOREM   Let $\mu, \nu : \mathcal{U} \to [0, \infty)$ be submeasures. Then (μ, ν) has the RNP if and only if there exists a family $\{A(t) : t \in [0, \infty)\} \subset \mathcal{U}$, satisfying the following conditions:

(i) $\mu\left(\bigcap_{q\in Q^+} A(q)\right) = \nu\left(\bigcap_{q\in Q^+} A(q)\right) = 0;$

(ii) $\mu(B(t) \backslash A(t)) = 0$ for every $t \in [0, \infty);$

(iii) For any $s, t \in [0, \infty)$ with $s < t$ and $A \in \mathcal{U}$,

$$s[\nu(A \cap A(s)) - \nu(A \cap A(t))] \le \mu(A \cap A(s)) - \mu(A \cap A(t))$$
$$\le t[\nu(A \cap A(s)) - \nu(A \cap A(t))].$$

Remark. Conditions (i)–(iii) are natural. In fact, we want to find a measurable function $f : U \to [0, \infty)$ such that

$$\mu(A) = \int_A f d\nu = \int_0^\infty \nu(A \cap \{f \ge t\}) dt \quad \forall A \in \mathcal{U}.$$

If $A \subseteq \{x : f(x) < \alpha\}$, then $\mu(A) = \int_0^\alpha \nu(A \cap \{f \ge t\}) dt$. To compute this definite integral, we divide $[0, \alpha]$ into small intervals and then take the Riemann sum. The first inequality of (iii) is used to show that $\mu(A) \le \int_0^\alpha \nu(A \cap \{f \ge t\}) dt$ and the second inequality of (iii) is used to show that $\mu(A) \ge \int_0^\alpha \nu(A \cap \{f \ge t\}) dt$.

Proof.   Assume that (μ, ν) has the RNP. Let $f = d\mu/d\nu$. For every $t \in [0, \infty)$, denote $A(t) = \{f \ge t\}$. Then it suffices to show that the family $\{A(t) : t \in [0, \infty)\}$ satisfies the conditions (i)–(iii). Since $f(x) < \infty$ for every $x \in U$, $\bigcap_{q\in Q^+} A(q) = \emptyset$. Thus

$$\mu\left(\bigcap_{q\in Q^+} A(q)\right) = \nu\left(\bigcap_{q\in Q^+} A(q)\right) = 0,$$

proving (i). It is easy to see that $B(t) = A(t)$ for every $t \in [0, \infty)$ and hence (ii) follows. (iii) can be obtained by noting that

$$\mu(A \cap A(s)) - \mu(A \cap A(t)) = \int_{A \cap A(s)} f d\nu - \int_{A \cap A(t)} f d\nu =$$

$$\int_0^\infty [\nu(A \cap A(s) \cap \{f \ge a\}) - \nu(A \cap A(t) \cap \{f \ge a\})] d\alpha \le$$

$$\int_0^s [v(A \cap A(s) \cap \{f \geq \alpha\}) - \nu(A \cap A(t) \cap \{f \geq \alpha\})]d\alpha =$$

$$\int_0^s [\nu(A \cap A(s)) - \nu(A \cap A(t)]d\alpha = s[\nu(A \cap A(s)) - \nu(A \cap A(t))]$$

and

$$\mu(A \cap A(s)) - \mu(A \cap A(t)) =$$

$$\int_0^\infty [\nu(A \cap A(s) \cap \{f \geq \alpha\}) - \nu(A \cap A(t) \cap \{f \geq \alpha\})]d\alpha =$$

$$\int_0^t [\nu(A \cap A(s) \cap \{f \geq \alpha\}) - \nu(A \cap A(t))]d\alpha \leq$$

$$\int_0^t [\nu(A \cap A(s)) - \nu(A \cap A(t))]d\alpha = t[\nu(A \cap A(s)) - \nu(A \cap A(t))].$$

Now we prove that the existence of a family $\{A(t) : t \in [0, \infty)\}$, satisfying the conditions (i)–(iii), will imply the RNP for (μ, ν). First we establish some relations between the two families $\{A(t)\}$ and $\{B(t)\}$.

CLAIM 1. $\mu \left(\bigcap_{k=1}^\infty B(i) \right) = \nu \left(\bigcap_{k=1}^\infty B(k) \right) = 0.$

Proof. Observe that $\bigcap_{q \in Q^+} B(q) = \bigcap_{q \in Q^+} A(q)$. Thus by (i), we have

$$\mu \left(\bigcap_{q \in Q^+} B(q) \right) = \nu \left(\bigcap_{q \in Q^+} B(q) \right) = 0,$$

which implies that $\mu \left(\bigcap_{n \in \mathbb{N}} B(n) \right) = \nu \left(\bigcap_{n \in \mathbb{N}} B(n) \right) = 0$ as $B(n)$ is decreasing in n.

CLAIM 2. $\lim_{n \to \infty} \mu(B(n)) = \lim_{n \to \infty} \nu(B(n)) = 0.$

Proof. Since

$$B(n) = \left[B(n) \cap \left(\bigcap_{k=1}^\infty B(k) \right) \right] \cup \left[B(n) \cap \left(\bigcap_{k=1}^\infty B(k) \right)^c \right],$$

$$\mu(B(n)) \leq \mu \left(\bigcap_{k=1}^\infty B(k) \right) + \mu \left(B(n) \cap \left(\bigcup_{k=1}^\infty B^c(k) \right) \right)$$

$$\leq \sum_{k=1}^\infty \mu[B(n) \backslash B(k)] \quad \text{(by Claim 1)}.$$

Since $\{B(n)\}$ is decreasing,

$$\lim_{n\to\infty} \mu(B(n)\backslash B(k)) = 0 \text{ for each } k \in \mathbb{N},$$

and the result follows. A similar reasoning applies to ν.

CLAIM 3.   *For every $A \in \mathcal{U}$ we have*

$$\mu(A) = \lim_{n\to\infty} \mu(A \cap B^c(n)), \quad \nu(A) = \lim_{n\to\infty} \nu(A \cap B^c(n)).$$

Proof. For any $n \in \mathbb{N}$, we have

$$\mu(A \cap B^c(n)) \le \mu(A) \le \mu(A \cap B(n)) + \mu(A \cap B^c(n))$$
$$\le \mu(B(n)) + \mu(A \cap B^c(n)),$$

and the result follows from Claim 2. Similarly

$$\nu(A) = \lim_{n\to\infty} \nu(A \cap B^c(n)).$$

CLAIM 4. $\mu(A(t)\backslash B(t)) = \nu(A(t)\backslash B(t)) = 0$ *for every $t \in [0, \infty)$.*

Proof. Let $A = A(t)\backslash A(s)$ in (iii), then we obtain

$$-s\nu[A(t)\backslash A(s)] \le -\mu[A(t))\backslash A(s)] \le -t\nu[A(t))\backslash A(s))]$$

for any $s, t \in [0, \infty)$ with $s < t$. It follows that

$$\nu[A(t)\backslash A(s)] = \mu[A(t))\backslash A(s)] = 0 \tag{6.4}$$

for any $s, t \in [0, \infty)$ with $s < t$. Observe that

$$A(t)\backslash B(t) = A(t)\backslash \left(\bigcap_{q\in Q(t)} A(q) \right) = \bigcup_{q\in Q(t)} (A(t)\backslash A(q)).$$

Since $q < t$, we obtain from (6.4),

$$\mu(A(t)\backslash B(t)) \le \sum_{q\in Q(t)} \mu(A(t)\backslash A(q)) = 0 \text{ for every } t \in [0, \infty)$$

and

$$\nu(A(t)\backslash B(t)) \le \sum_{q\in Q(t)} \nu(A(t)\backslash A(q)) = 0 \text{ for every } t \in [0, \infty).$$

Note that Claim 4 and (ii) imply that

$$\mu(A(t)) = \mu(B(t)), \quad \nu(A(t)) = \nu(B(t)) \;\forall t \in [0, \infty).$$

CLAIM 5. For any $t \in [0, \infty)$ and $A \in \mathcal{U}$,

$$\mu(A \cap A(t)) = \mu(A \cap B(t)) \quad \text{and} \quad \nu(A \cap A(t)) = \nu(A \cap B(t)).$$

Proof. Note that

$$A \cap A(t) = [A \cap (A(t) \backslash B(t))] \cup [A \cap A(t) \cap B(t)].$$

By Claim 4, we obtain $\mu(A \cap A(t)) = \mu(A \cap A(t) \cap B(t))$. Similarly we can obtain $\mu(A \cap B(t)) = \mu(A \cap B(t) \cap A(t))$. Thus $\mu(A \cap A(t)) = \mu(A \cap B(t))$. A similar reasoning applies to ν.

CLAIM 6. For any $s, t \in [0, \infty)$ with $s < t$ and $A \in \mathcal{U}$,

$$s[\nu(A \cap B(s)) - \nu(A \cap B(t))] \leq \mu(A \cap B(s)) - \mu(A \cap B(t))$$
$$\leq t[\nu(A \cap B(s)) - \nu(A \cap B(t))].$$

Proof. By (iii) and Claim 5,

$$s[\nu(A \cap B(s)) - \nu(A \cap B(t))] = s[\nu(A \cap A(s)) - \nu(A \cap A(t))]$$
$$\leq \mu(A \cap A(s)) - \mu(A \cap A(t))$$
$$= \mu(A \cap B(s)) - \mu(A \cap B(t))$$

and

$$\mu(A \cap B(s)) - \mu(A \cap B(t)) = \mu(A \cap A(s)) - \mu(A \cap A(t))$$
$$\leq t[\nu(A \cap A(s)) - \nu(A \cap A(t))]$$
$$= t[\nu(A \cap B(s)) - \nu(A \cap B(t))].$$

Remark. Observe that conditions (i)–(iii) imply the absolute continuity of μ with respect to ν. In fact, assume that $\nu(A) = 0$. Then from Claim 6,

$$\mu(A \cap B(s)) - \mu(A \cap B(t)) = 0 \text{ for every } s, t \in [0, \infty) \text{ with } s < t.$$

Letting $t = n \to \infty$, we obtain from Claim 2, $\mu(A \cap A(s)) = 0$ for every $s \in [0, \infty)$. It follows that $\mu(A \cap B(0)) = \mu(A \cap U) = \mu(A) = 0$.

Now we are able to complete the proof of our result.

CLAIM 7. $\mu(A) = \int_A f d\nu$ for every $A \in \mathcal{U}$.

Proof. Let $f : U \to [0, 1]$ be given in (6.3) and $A_n = A \cap \{f < n\} = A \cap B^c(n)$ for every $A \in \mathcal{U}$ and $n \in \mathbb{N}$. Then as $A_n \subset \{f < n\}$,

$$\int_0^n \nu(A_n \cap \{f \geq t\})dt = \int_0^\infty \nu(A_n \cap \{f \geq t\})dt. \tag{6.5}$$

We first prove that

$$\mu(A_n) = \int_0^\infty \nu(A_n \cap \{f \geq t\}dt \text{ for every } n \in \mathbb{N}.$$

Let $0 = t_0 < t_1 < \ldots < t_k = n$,

$$s_k = \sum_{i=1}^k (t_i - t_{i-1})\nu(A_n \cap \{f \geq t_i\}),$$

and

$$S_k = \sum_{i=1}^k (t_i - t_{i-1})\nu(A_n \cap \{f \geq t_{i-1}\}).$$

Then we have

$$s_k \leq \int_0^n \nu(A_n \cap \{f \geq t\})dt \leq S_k.$$

Note that as $\max\{t_i - t_{i-1} : i = 1, \ldots, k\} \to 0$,

$$s_k \to \int_0^n \nu(A_n \cap \{f \geq t\})dt, \quad S_k \to \int_0^n \nu(A_n \cap \{f \geq t\})dt. \tag{6.6}$$

By using the first inequality in Claim 6 and the fact that $\mu(A_n \cap \{f \geq t_k\}) =$

$\nu(A_n \cap \{f \geq t_k\}) = 0$, we obtain

$$s_k = \sum_{i=1}^{k}(t_i - t_{i-1})\nu(A_n \cap \{f \geq t_i\})$$

$$= \sum_{i=1}^{k} t_i\nu(A_n \cap \{f \geq t_i\}) - \sum_{i=1}^{k} t_{i-1}\nu(A_n \cap \{f \geq t_i\})$$

$$= \sum_{i=1}^{k-1} t_i[\nu(A_n \cap \{f \geq t_i\}) - \nu(A_n \cap \{f \geq t_{i+1}\})]$$

$$\leq \sum_{i=1}^{k-1}[\mu(A_n \cap \{f \geq t_i\}) - \mu(A_n \cap \{f \geq t_{i+1}\})]$$

$$= \mu(A_n \cap \{f \geq t_1\}) - \mu(A_n \cap \{f \geq t_k\})$$

$$= \mu(A_n \cap \{f \geq t_1\}) \leq \mu(A_n).$$

Hence $s_k \leq \mu(A_n)$. Similarly, by using the second inequality in Claim 6, $S_k \geq \mu(A_n)$. Therefore $s_k \leq \mu(A_n) \leq S_k$. Combining (6.5) and (6.6), we obtain

$$\mu(A_n) = \int_0^n \nu(A_n \cap \{f \geq t\})dt = \int_{A_n} f d\nu.$$

By Claim 3 we have

$$\mu(A) = \lim_{n\to\infty} \mu(A_n) = \int_0^\infty \nu(A \cap \{f \geq t\})dt = \int_A f d\nu.$$

Consequently the theorem is proved. □

The following example is a pair of set-functions that are not capacities in Graf's sense, but are submeasures. Moreover, they satisfy the conditions of the Theorem and hence (μ, ν) has the RNP.

Example. Let U be a Banach space, $\mathcal{U} = \mathcal{B}(U)$, and $S = \{x : ||x|| < 1\}$. Define

$$\mu(A) = \begin{cases} 0 & \text{if } A = \emptyset \\ \frac{1}{2}\sup_{x\in A}||x|| & \text{if } A \neq \emptyset \text{ and } d(A, S^c) > 0 \\ 1 & \text{if } A \neq \emptyset \text{ and } d(A, S^c) = 0 \end{cases}$$

where S^c is the set complement of S and

$$d(A, B) = \inf\{||x - y|| : x \in A, \ y \in B\}.$$

Proof. First observe that μ and ν are not capacities in the sense of Graf. In fact, for $S_n = \{x : ||x|| \leq 1 - 1/n\}$, we have $S_n \nearrow S = \bigcup_{n=1}^{\infty} S_n$, but

$$\mu(S) = \nu(S) = 1 > \frac{1}{2} = \lim_{n\to\infty} \mu(S_n) = \lim_{n\to\infty} \nu(S_n).$$

Therefore the lower continuity condition of capacities is violated.

It is easy to see that μ and ν are submeasures. Moreover, applying the Theorem we will show that (μ, ν) has the RNP.

To verify the conditions (i)–(iii) of the theorem let

$$A(t) = \begin{cases} \{x : \|x\| \geq t\} & \text{for } t \in [0, 1] \\ \emptyset & \text{for } t > 1 \end{cases}$$

Then $\bigcap_{q \in Q^+} A(q) = \emptyset$, and hence (i) holds. Clearly, $B(t) = \bigcap_{q \in Q(t)} A(t)$, which implies (ii). Thus it suffices to verify that (iii) holds. Note that $A(t) \subset A(s)$ for all $s < t$, and (iii) holds if $A \cap A(s) = \emptyset$. So we need only consider the case where $A \cap A(s) \neq \emptyset$ (in this case, $s \leq 1$ as $A(s) = \emptyset$ for $s > 1$).

Case 1. $d(A \cap A(s), S^c) > 0$ and $A \cap A(t) = \emptyset$. We have

$$s \leq \sup_{A \cap A(s)} \|x\| \leq t$$

and $\mu(A \cap A(t)) = \nu(A \cap A(t)) = 0$. Therefore

$$\mu(A \cap A(s)) = \frac{1}{2} \sup_{A \cap A(s)} \|x\| \geq \frac{s}{2} = s[\nu(A \cap A(s))],$$

as $\nu(A \cap A(s)) = 1/2$. Also

$$\mu(A \cap A(s)) = \frac{1}{2} \sup_{A \cap A(s)} \|x\| \leq \frac{t}{2} = t[\nu(A \cap A(s))],$$

hence (iii) follows.

Case 2. $d(A \cap A(s), S^c) > 0$ and $A \cap A(t) \neq \emptyset$. We have $d(A \cap A(t), S^c) \geq d(A \cap A(s), S^c) > 0$, $\nu(A \cap A(s)) = \nu(A \cap A(t)) = 1/2$, and

$$\nu(A \cap A(s)) = \frac{1}{2} \sup_{A \cap A(s)} \|x\| = \frac{1}{2} \sup_{A \cap A(t)} \|x\| = \mu(A \cap A(t)).$$

Thus (iii) holds.

Case 3. $d(A \cap A(s), S^c) = 0$ and $A \cap A(t) = \emptyset$. We have $t \geq 1$ and $\mu(A \cap A(s)) = \nu(A \cap A(s)) = 1$. Therefore

$$s[\nu(A \cap A(s))] = s \leq 1 = \mu(A \cap A(s)) \leq t = t\nu(A \cap A(s)),$$

i.e., (iii) holds.

Case 4. $d(A \cap A(s), S^c) = 0$ and $A \cap A(t) \neq \emptyset$. We have $s \leq 1$, $t \leq 1$, $\mu(A \cap A(s)) = \nu(A \cap A(s)) = 1$, and $d(A \cap A(t), S^c) = 0$. Thus $\mu(A \cap A(t)) = \nu(A \cap A(t)) = 1$ and hence (iii) follows.

Since (μ, ν) satisfies the conditions (i)–(iii) of the theorem, (μ, ν) has the RNP and hence $d\mu/d\nu$ exists. Moreover, the Radon-Nikodym derivative $d\mu/d\nu$ can be obtained by:

$$\frac{d\mu}{d\nu} = f(x) = \begin{cases} \|x\| & \text{if } x \in S \\ 1 & \text{if } x \in S^c \end{cases}$$

□

6.4 Exercises

6.1 Let X be a nonnegative random variable, defined on a probability space $(\Omega, \mathcal{A}, P)$. Consider the random closed set S on $\mathbb{R}^+$ defines by $S(\omega) = [0, X(\omega)]$. Verify that the covering function of S on $\mathbb{R}^+$ is

$$\pi(x) = P(x \in S) = P(X \geq x).$$

Use the Robbins' formula (Section 2.2) to show that $EX = \int_0^{+\infty} P(X \geq x)dx$.

6.2 Let X be an integrable random variable, defined on $(\Omega, \mathcal{A}, P)$. Show that

$$EX = \int_0^{+\infty} P(X > t)dt + \int_{-\infty}^0 [P(X > t) - 1]dt.$$

6.3 Let μ be a set function, defined on $\mathcal{B}(\mathbb{R}^d)$, with values in $[0, 1]$, such that $\mu(\emptyset) = 0$ and μ is monotone increasing. Let $f : \mathbb{R}^d \to \mathbb{R}$, continuous and bounded. Show that

$$\int_0^{+\infty} \mu(f > t)dt + \int_{-\infty}^0 [\mu(f > t) - \mu(\mathbb{R}^d)]dt =$$

$$\int_0^{+\infty} \mu(f \geq t)dt + \int_{-\infty}^0 [\mu(f \geq t) - \mu(\mathbb{R}^d)]dt.$$

6.4 Let $N = \{1, 2, \ldots, n\}$, and $\mu : 2^N \to [0, 1]$ such that $\mu(N) = 1$, $\mu(\emptyset) = 0$ and for $A \subseteq B$, $\mu(A) \leq \mu(B)$. For $X : N \to \mathbb{R}^+$, let

$$C_\mu(X) = \sum_{i=1}^n (x_{(i)} - x_{(i-1)})\mu(A_{(i)}),$$

where $x_i = X(i)$, $x_{(1)} \leq x_{(2)} \leq \ldots \leq x_{(n)}$ is an arrangement of x_i's in an increasing order, with $x_{(0)} = 0$, and $A_{(i)} = \{(i), (i+1), \ldots, (n)\}$.

a) Verify that $C_\mu(X) = \int_0^{+\infty} \mu(X > t)dt$.

b) If $Y : N \to \mathbb{R}^+$ is such that $X \le Y$ in $\mathbb{R}^n$ (i.e., $x_i \le y_i$, $\forall I = 1, 2, \ldots, n$), then

$$C_\mu(X) \le C_\mu(Y).$$

c) Show that if $x \in \mathbb{R}^n$, $r > 0$, and $s \in \mathbb{R}$, then

$$C_\mu(rX + s) = rC_\mu(X) + s,$$

where

$$rX + s = (rx_1 + s, rx_2 + s, \ldots, rx_n + s).$$

d) For $A \subseteq N$, verify that
$$C_\mu(1_A) = \mu(A).$$

e) Show that there exist 2^n functions $f_A : \mathbb{R}^n \to \mathbb{R}$, $A \subseteq N$, such that if $X : N \to \mathbb{R}^+$, then

$$C_\mu(X) = \sum_{A \subseteq N} \mu(A)f_A(X)$$

(where X is viewed as a vector in $\mathbb{R}^n$).

Hint: consider the Möbius inverse of μ.

6.5 Let S be a random closed set on $\mathbb{R}$, defined on $(\Omega, \mathcal{A}, P)$ and $g : \mathbb{R} \to \mathbb{R}^+$ measurable. We say that S is *strongly measurable* if $\forall B \in \mathcal{B}(\mathbb{R})$,

$$B_* = \{\omega : S(\omega) \subseteq B\} \in \mathcal{A}, \quad B^* = \{\omega : S(\omega) \cap B \ne \emptyset\} \in \mathcal{A}.$$

Let
$$g_*(\omega) = \inf\{g(x) : x \in S(\omega)\}, \quad g^*(\omega) = \sup\{g(x) : x \in S(\omega)\}.$$
Show that if S is strongly measurable, then g_* and g^* are measurable.

6.6 Let T be the capacity functional of a random closed set S on $\mathbb{R}^d$, and $g : \mathbb{R}^d \to \mathbb{R}^+$ measurable. Let F be the dual of T, i.e., $F(A) = 1 - T(A^c)$, $A \in \mathcal{B}(\mathbb{R}^d)$. Show that

$$\int_0^{+\infty} F(g \ge t)dt = \int_0^{+\infty} P[\inf\{g(x) : x \in S\} \ge t]dt.$$

6.7* Let $(\Omega, \mathcal{A})$ be a measurable space, and $f : \Omega \to \mathbb{R}^+$, $\mathcal{A}$-$\mathcal{B}(\mathbb{R}^+)$-measurable. Let $\mu, \nu : \mathcal{A} \to \mathbb{R}^+$ be monotone set functions such that $\mu(\emptyset) = \nu(\emptyset) = 0$.

Suppose $\mu(A) = \int_0^{+\infty} \nu(\{x \in A, f(x) \geq t\})dt$, $\forall A \in \mathcal{A}$ (i.e., the pair (μ, ν) has the Radon-Nikodym property, RNP). Show that μ is alternating of infinite order whenever ν is alternating of infinite order.

6.8 Let $(\Omega, \mathcal{A})$ be a measurable space. Let $\Gamma : \mathcal{A} \to \{0, 1\}$ be defined by

$$\Gamma(A) = \begin{cases} 0 & \text{if } A \neq \emptyset \\ 1 & \text{if } A = \emptyset \end{cases}$$

Recall that $\mu : \mathcal{A} \to \mathbb{R}^+$ is maxitive if $\mu(A \cup B) = \max(\mu(A), \mu(B))$.
Let $\mu : \mathcal{A} \to \mathbb{R}^+$ be such that $\mu(\emptyset)) = 0$, for $A \subseteq B$, $\mu(A) \leq \mu(B)$. Show that μ is maxitive if and only if (μ, Γ) has the RNP on every *finite* subalgebra of $\mathcal{A}$.

6.9 Let $(\Omega, \mathcal{A})$ be a measurable space, and $\mathbb{B}(\Omega)$, $\mathbb{B}^+(\Omega)$ denote the classes of measurable, bounded and of nonnegative measurable bounded (real-valued) functions, respectively. Let $\mu : \mathcal{A} \to \mathbb{R}$ such that $\mu(\emptyset) = 0$, and μ is monotone. For $f \in \mathbb{B}^+(\Omega)$, let

$$C_\mu(f) = \int_0^{+\infty} \mu(f \geq t)dt$$

and for $f \in \mathbb{B}(\Omega)$, let

$$C_\mu^*(f) = \int_0^{+\infty} \mu(f \geq t)dt + \int_{-\infty}^0 [\mu(f \geq t) - \mu(\Omega)]dt.$$

Show that $C_\mu^*(\cdot)$ is the unique translation invariant extension of $C_\mu(\cdot)$ in the sense that for each $\alpha \in \mathbb{R}$, and $f \in \mathbb{B}(\Omega)$, $C_\mu^*(f + \alpha 1) = C_\mu^*(f) + \alpha C_\mu^*(1)$.

6.10 Let X be a nonnegative random variable defined on $(\Omega, \mathcal{A}, P)$. Use the Fubini's Theorem (see Appendix) to show that

$$EX = \int_0^{+\infty} P(X > t)dt.$$

Chapter 7

Choquet Weak Convergence

This chapter is about various concepts of convergence of sequences of random closed sets on $\mathbb{R}^d$. They are generalizations of familiar concepts for random vectors. Our focus is on the convergence in distribution in which the concept of the Choquet integral with respect to capacity functionals is used to generalize weak convergence of probability measures on metric spaces.

7.1 Stochastic Convergence of Random Sets

Let $(\Omega, \mathcal{A}, P)$ be a probability space on which all considered random elements are defined. Recall that the space of closed sets $\mathcal{F}(\mathbb{R}^d)$ is topologized by the hit-or-miss topology $\mathcal{T}$, and $\mathcal{B}(\mathcal{F})$ denotes the Borel σ-filed of $\mathcal{F}(\mathbb{R}^d)$. Let S, S_n be random closed sets, i.e., $S, S_n : \Omega \to \mathcal{F}(\mathbb{R}^d)$, $\mathcal{A}$-$\mathcal{B}(\mathcal{F})$-measurable.

DEFINITION 7.1   *The sequence* $(S_n, n \geq 1)$ *is said to converge* almost surely (a.s.) *to* S *if* $P(\omega : S_n(\omega) \to S(\omega)) = 1$. *In other words,* $S_n \to S$ *a.s. if there exists* $\Omega_0 \in \mathcal{A}$ *with* $P(\Omega_0) = 0$ *such that for* $\omega \in \Omega_0^c$, $S_n(\omega) \to S(\omega)$ *as* $n \to \infty$.

Here S and S_n are random closed sets, the convergence $S_n(\omega) \to S(\omega)$ is of course taken in the sense of topology $\mathcal{T}$. Specifically, $S_n(\omega) \to S(\omega)$ in $\mathcal{T}$ if for any neighborhood V of the set $S(\omega)$, in the topology $\mathcal{T}$, there exists $n(V)$ such that $S_n(\omega) \in V$ for all $n \geq n(V)$.

Let us elaborate a little bit on this convergence. Details for proofs can be found, e.g., in Mathéron [73], Salinetti and Wets [108]. Let $F, F_n \in \mathcal{F}(\mathbb{R}^d)$, we write $F_n \xrightarrow{\mathcal{T}} F$ to mean that F_n converges to F in the topology $\mathcal{T}$. The following are equivalent:

a) $F_n \xrightarrow{\mathcal{T}} F$

b) If $G \in \mathcal{G}$ and $G \cap F \neq \emptyset$, then $G \cap F_n \neq \emptyset$ for all n sufficiently large (i.e., there exists $n(G)$ such that $\forall n \geq n(G)$, $G \cap F_n \neq \emptyset$); *and if* $K \in \mathcal{K}$ and $K \cap F \neq \emptyset$, then $K \cap F_n \neq \emptyset$ for all n sufficiently large.

Remark. The above tractable criterion for the convergence in $\mathcal{T}$ is valid for general Hausdorff, locally compact, second countable spaces. For $\mathbb{R}^d$ or metric spaces, the situation is simpler. Let ρ denote the Euclidean metric on $\mathbb{R}^d$, then $F_n \xrightarrow{\mathcal{T}} F$ if and only if $\forall x \in \mathbb{R}^d$, the sequence $\{\rho(x, F_n), n \geq 1\}$ converges in $\overline{\mathbb{R}^+}$ to $\rho(x, F)$, where $\rho(x, F) = \inf\{\rho(x, y) : y \in F\}$.

The hit-or-miss topology $\mathcal{T}$ turns out to be the topology that corresponds to a notion of convergence of closed sets in topological spaces. Specifically, the following *definition of convergence* of closed sets corresponds precisely to the convergence in the topology $\mathcal{T}$ (see Beer [9] for details).

A sequence of subsets A_n in $\mathbb{R}^d$ is said to converge in the *Painlevé sense* to the set A if

$$\liminf A_n = \limsup A_n = A,$$

where

$$\liminf A_n = \{x \in \mathbb{R}^d : x_n \to x, x_n \in A_n, n \geq 1\}$$

$$\limsup A_n = \{x \in \mathbb{R}^d : x_{n(k)} \to x, x_{n(k)} \in A_{n(k)},$$

$$\text{with } n(k), k \geq 1, \text{ being a subsequence}\}.$$

Remarks.

(i) Clearly $\liminf A_n \subseteq \limsup A_n$. On $\mathbb{R}^d$, $F, F_n \in \mathcal{F}(\mathbb{R}^d)$, we have $F_n \xrightarrow{\mathcal{T}} F$ if and only if $F_n \to F$ in the above Painlevé sense.

(ii) Do not confuse the above concepts of $\liminf A_n$ and $\limsup A_n$ with are defined in a topological context with set-theoretic concepts! Recall that in discussing "continuity property" of probability measures, the following set-theoretic concept of convergence of sequences of events is considered. A sequence of subsets A_n is said to *converge* to a set A if $\underline{\lim} A_n = \overline{\lim} A_n = A$, where

$$\underline{\lim} A_n = \bigcup_{n \geq 1} \bigcap_{n \geq k} A_k, \qquad \overline{\lim} A_n = \bigcap_{n \geq 1} \bigcup_{n \geq k} A_k.$$

Note that in a topological context, $\underline{\lim} A_n \subseteq \liminf A_n$ and $\overline{\lim} A_n \subseteq \limsup A_n$.

(iii) A natural question to ask when a definition of convergence of sequences of "points" in a space is given is this. Is there a topology on the space such that the given concept of convergence corresponds to the convergence in that topology? For this topic, see e.g., Kelley [60], Klein and Thompson [62].

(iv) The space $\mathcal{F}$ is metrizable. Let Δ be any metric on $\mathcal{F}$ compatible with the topology $\mathcal{T}$. Then $S_n \xrightarrow{a.s.} S$ iff $\Delta(S_n, S) \to 0$, a.s., as $n \to \infty$.

In the context of coarsening schemes for coarse data, we deal with *non-empty* random closed sets, say, on metric spaces such as $\mathbb{R}^d$. Let $\mathcal{F}_0(\mathbb{R}^d)$ denote $\mathcal{F}(\mathbb{R}^d) \setminus \{\emptyset\}$. A metric on $\mathcal{F}_0$ is given by

$$W_\rho(A, B) = \sum_{n=1}^{\infty} \frac{1}{2^n} \min\{1, |\rho(x_n, A) - \rho(x_n, B)|\},$$

where ρ is the euclidean metric on $\mathbb{R}^d$, $D = \{x_n : n \geq 1\}$ is a countable dense subset of $\mathbb{R}^d$, and $\rho(x, A) = \inf\{\rho(x, y) : y \in A\}$. Note that this metric is compatible with the Wijsman topology on $\mathcal{F}_0$, i.e., the weakest topology on $\mathcal{F}_0$ making all the maps $\rho(x, \cdot) : \mathcal{F}_0 \to \mathbb{R}^+$, $x \in \mathbb{R}^d$, continuous. In this topology, $F_n, F \in \mathcal{F}_0$, $F_n \to F$ is equivalent to $\rho(x, F_n) \to \rho(x, F)$, as $n \to \infty$, $\forall x \in \mathbb{R}^d$.

In the special case where our base space U is a *compact* subspace of $\mathbb{R}^d$, the a.s. convergence of *random compact sets* can be expressed in terms of the Hausdorff distance. Specifically, the sequence of nonempty random compact sets S_n converges a.s. to a nonempty compact set S iff $S_n \to S$ in the Hausdorff distance H_ρ, a.s., as $n \to \infty$, where the *Hausdorff distance H_ρ* on $\mathcal{K}_0(U)$ is defined as

$$H_\rho(A, B) = \max \left\{ \sup_{x \in A} \rho(x, B), \sup_{x \in B} \rho(x, A) \right\}$$

$$(= \inf\{\varepsilon > 0 : A \subseteq B^\varepsilon, B \subseteq A^\varepsilon\}, \quad \text{where } A^\varepsilon = \{x \in U : \rho(x, A) \leq \varepsilon\}).$$

Students who are interested in various topologies on the spaces of closed sets of topological spaces should read a text like Beer [9].

We turn now to the concept of *convergence in probability* of random closed sets. Random vectors are random elements with values in the metric space $(\mathbb{R}^d, \rho)$. While the metric property of $\mathbb{R}^d$ is essential for the concept of a.s. convergence, e.g., if $X_n \to X$, a.s., as $n \to \infty$, then X is a random vector (i.e., measurable), see Dudley [24].

The formulation of convergence in probability requires a little more, namely, the *separability* of $\mathbb{R}^d$. Indeed, to define $X_n \xrightarrow{P} X$ as

$$\forall \varepsilon > 0, \quad \lim_{n \to \infty} P\{\omega : \rho(X_n(\omega), X(\omega)) > \varepsilon\} = 0,$$

we need $\rho(X_n, X)$ to be a random variable! Now $\rho(X_n(\omega), X(\omega)) = (\rho \circ f)(\omega)$, where

$$f(\omega) = (X_n(\omega), X(\omega)).$$

On the other hand, $\rho : \mathbb{R}^d \times \mathbb{R}^d \to \mathbb{R}^+$ is continuous in the product topology of $\mathbb{R}^d \times \mathbb{R}^d$ (i.e., topology on $\mathbb{R}^d \times \mathbb{R}^d$ generated by the base $\{A \times B : A, B \text{ opens of } \mathbb{R}^d\}$), and hence measurable with respect to the Borel σ-field $\mathcal{B}(\mathbb{R}^d \times \mathbb{R}^d)$ in the metric product topology.

Now $\mathcal{B}(\mathbb{R}^d \times \mathbb{R}^d) = \mathcal{B}(\mathbb{R}^d) \otimes \mathcal{B}(\mathbb{R}^d)$, the product σ-field (see Appendix), since $\mathbb{R}^d$ is second countable. Thus $\rho(X_n, X)$ is indeed a random variable.

Remark. The fact that $\mathcal{B}(\mathbb{R}^d \times \mathbb{R}^d) = \mathcal{B}(\mathbb{R}^d) \otimes \mathcal{B}(\mathbb{R}^d)$ can be seen as follows, in the general context. Let (U, τ) and (V, ν) be two topological spaces. The Borel σ-field $\mathcal{B}(U \times V)$ is the smallest σ-field making the projections $U \times V \to U$ and $U \times V \to V$ measurable. As such, $\mathcal{B}(U) \otimes \mathcal{B}(V) \subseteq \mathcal{B}(U \times V)$. Suppose that both U and V are second countable. Let $\mathbb{B}(U)$ and $\mathbb{B}(V)$ be countable bases of U, V, respectively. Then, $\Theta = \{A \times B : A \in \mathbb{B}(U), B \in \mathbb{B}(V)\}$ is a countable base for the product topology $\mathcal{T}$ of $U \times V$, and we have $\Theta \subseteq \mathcal{B}(U) \otimes \mathcal{B}(V)$. On the other hand, each open set in $\mathcal{T}$ is a countable union of open sets in Θ, and hence $\mathcal{T} \subseteq \mathcal{B}(U) \otimes \mathcal{B}(V)$ implying that $\mathcal{B}(U \times V) = \sigma(\mathcal{T}) \subseteq \mathcal{B}(U) \otimes \mathcal{B}(V)$.

Remark. The fact that $\mathcal{B}(\mathbb{R}^d \times \mathbb{R}^d) = \mathcal{B}(\mathbb{R}^d) \otimes \mathcal{B}(\mathbb{R}^d)$ can be seen as follows, in the general context. Let (U, τ) and (V, ν) be two topological spaces. The Borel σ-field $\mathcal{B}(U \times V)$ is the smallest σ-filed making the projections $U \times V \to U$ and $U \times V \to V$ measurable. As such, $\mathcal{B}(U) \otimes \mathcal{B}(V) \subseteq \mathcal{B}(U \times V)$. Suppose that both U and V are second countable. Let $\mathbb{B}(U)$ and $\mathbb{B}(V)$ be countable bases of U, V, respectively. Then, $\Theta = \{A \times B : A \in \mathbb{B}(U), B \in \mathbb{B}(V)\}$ is a countable base for the product topology $\mathcal{T}$ of $U \times V$, and we have $\Theta \subseteq \mathcal{B}(U) \otimes \mathcal{B}(V)$. On the other hand, each open set in $\mathcal{T}$ is a countable union of open sets in Θ, and hence $\mathcal{T} \subseteq \mathcal{B}(U) \otimes \mathcal{B}(V)$ implying that $\mathcal{B}(U \times V) = \sigma(\mathcal{T}) \subseteq \mathcal{B}(U) \otimes \mathcal{B}(V)$.

In view of the fact that a metric space is second countable if and only if it is separable, we arrive at the natural setting for the concept of convergence in probability, namely *separable metric spaces*. See the excellent text of Dudley [24]. Note also that the *convergence in distribution* of random elements is formulated on metric spaces Billingsley [10]. But on separable metric spaces, the convergence in distribution of random elements (i.e., weak convergence of their probability laws) can be metrizable by the *Prohorov metric*.

Specifically, let (U, ρ) be a metric space. Let $\mathcal{M}(U)$ denote the class of all probability measures on the Borel σ-field $\mathcal{B}(U)$ of U. The *Prohorov metric* Δ on $\mathcal{M}(U)$ is defined as

$$\Delta(\mu, \nu) = \inf\{\varepsilon > 0 : \mu(F) \leq \nu(F^\varepsilon) + \varepsilon \text{ for all } F \in \mathcal{F}(U)\}.$$

If (U, ρ) is *separable* then for $\mu_n, \mu \in \mathcal{M}(U)$,

$$\mu_n \to \mu \text{ weakly } \Leftrightarrow \Delta(\mu_n, \mu) \to 0.$$

Remark. The following *Skorohod representation* for random elements with values in separable metric spaces is useful for statistics with coarse data in Chapter 8.

If Q_n, Q are probability measures on $\mathcal{B}(U)$, where (U, ρ) is a separable metric space, such that $Q_n \to Q$ in the Prohorov metric, then there exists a probability space $(\Omega, \mathcal{A}, P)$ on which are defined U-valued random elements X_n, X whose probability laws are Q_n, Q, respectively, and $X_n \to X$, P-a.s., as $n \to \infty$. For a proof, see Ethier and Kurtz [29].

Now in the context of random closed sets on $\mathbb{R}^d$, or more generally, on a Hausdorff, locally compact and second countable space, the space $\mathcal{F}$, with the hit-or-miss topology $\mathcal{T}$, is a *separable metric space*. Thus, the concept of *convergence in probability* of random closed sets is formulated just in the general theory of random elements taking values in a separable metric space.

Let ∇ be any metric on $\mathcal{F}$ compatible with the topology $\mathcal{T}$. Then the sequence of random closed sets S_n converges in probability to the random closed set S, in symbol, $S_n \overset{P}{\to} S$, if

$$\forall \varepsilon > 0, \quad P(\nabla(S_n, S) > \varepsilon) \to 0$$

as $n \to \infty$.

The following *equivalent* definition of the convergence in probability of random closed sets (see Molchanov [79]) seems to be more tractable.

Recall that if X, X_n, $n \geq 1$, are random variables, then $X_n \overset{a.s.}{\longrightarrow} X$ implies that, for any $\varepsilon > 0$,

$$\lim_{n \to \infty} P(|X_n - X| > \varepsilon) = 0.$$

This last property means that X_n tends to X with "high probability" (or with probability tending to one). This weaker concept is termed the convergence in probability of X_n to X.

Now, on $\mathbb{R}$,

$$\{\omega : |X_n(\omega) - X(\omega)| > \varepsilon\} =$$

$$\{\omega : X(\omega) > X_n(\omega) + \varepsilon\} \cup \{\omega : X_n(\omega) > X(\omega) + \varepsilon\} =$$

$$\{\omega : X(\omega) \notin B(X_n(\omega), \varepsilon)\} \cup \{\omega : X_n(\omega) \notin B(X(\omega), \varepsilon)\},$$

where for $x \in \mathbb{R}$, $B(x, r) = \{y \in \mathbb{R} : |x - y| \leq r\}$.

Thus, when replacing X, X_n by random closed sets S, S_n on a *metric space* like $(\mathbb{R}^d, \rho)$, we look at

$$E_{n, \varepsilon} = (S \setminus S_n^\varepsilon) \cup (S_n \setminus S^\varepsilon),$$

where for $A \subseteq \mathbb{R}^d$ and $\varepsilon > 0$,

$$A^\varepsilon = \{x \in \mathbb{R}^d : \rho(x, A) \leq \varepsilon\},$$

$$\rho(x, A) = \inf\{\rho(x, y) : y \in A\}.$$

Now it can be shown that (see Salinetti and Wets [108]) the following are equivalent:

α) $S_n \xrightarrow{a.s.} S$

β) For each $\varepsilon > 0$, $E_{n,\varepsilon} \xrightarrow{a.s.} \emptyset$.

γ) For every compact $K \in \mathcal{K}(\mathbb{R}^d)$, $E_{n,\varepsilon} \cap K = \emptyset$, a.s., for all n sufficiently large.

Moreover, if $E_{n,\varepsilon} \xrightarrow{a.s.} \emptyset$, then for every $K \in \mathcal{K}(\mathbb{R}^d)$, $P[E_{n,\varepsilon} \cap K \neq \emptyset] \to 0$ as $n \to \infty$.

In view of all these motivations, we consider the following concept.

DEFINITION 7.2   *A sequence of random closed sets S_n is said to converge in probability to a random closed set S if for every $\varepsilon > 0$ and compact $K \in \mathcal{K}(\mathbb{R}^d)$,*

$$\lim_{n\to\infty} P\{[(S \setminus S_n^\varepsilon) \cup (S_n \setminus S^\varepsilon)] \cap K \neq \emptyset\} = 0.$$

Recall that, in the case of random variables, the convergence in probability can be expressed in terms of a metric on the space of equivalence classes of random variables, i.e.,

$$X \sim Y \qquad \text{iff} \qquad X = Y \quad a.s.$$

Specifically, let $\widetilde{X}$ denote the equivalence class of X. Then, for $X \in \widetilde{X}$ and $Y \in \widetilde{Y}$,

$$r(\widetilde{X}, \widetilde{Y}) = E\left(\frac{|X - Y|}{1 + |X - Y|}\right)$$

is a metric, and

$$X_n \xrightarrow{P} X \quad \text{if and only if} \quad r(\widetilde{X}_n, \widetilde{Y}) \to 0.$$

We will prove the last assertion, leaving the verification of the properties of r as a metric as an exercise!

LEMMA 7.1   *Let $g : \mathbb{R} \to [0, \infty)$, measurable, even and nondecreasing on $[0, \infty)$ Let X be any random variable. Then, for any $a > 0$,*

$$\frac{E[g(X)] - g(a)}{\text{ess.}\sup g(x)} \leq P(|X| \geq a) \leq \frac{E[g(X)]}{g(a)},$$

where $\text{ess.}\sup |Y|$, the essential supremum or a.s. supremum of a random variable Y, is $\inf\{c \geq 0 : P(|Y| > c) = 0\}$.

Proof. Note that

$$E\left(g(X)\right) = \int_\Omega g(X)dP = \int_{\{|X|\geq a\}} g(X)dP + \int_{\{|X|<a\}} g(X)dP.$$

Since

$$0 \leq \int_{\{|X|<a\}} g(X)dP \leq g(a)$$

and

$$g(a)P(|X| \geq a) \leq \int_{\{|X|\geq a\}} g(X)dP \leq (a.s. \sup g(X))P(|X| \geq a),$$

we have

$$g(a)P(|X| \geq a) \leq E\left(g(X)\right) \leq (\text{ess. sup } g(X))P(|X| \geq a) + g(a),$$

and hence the desired result follows. □

Now, if we take

$$g(x) = \frac{|x|}{1 + |x|}$$

and consider the sequence $\{X_n - X, n \geq 1\}$, then, for any $\varepsilon > 0$,

$$E\left(\frac{|X_n - X|}{1 + |X_n - X|}\right) - \left(\frac{\varepsilon}{1 + \varepsilon}\right) \leq P\left(|X_n - X| \geq \varepsilon\right)$$

$$\leq \frac{1 + \varepsilon}{\varepsilon} E\left(\frac{|X_n - X|}{1 + |X_n - X|}\right).$$

From these inequalities, if follows that, as $n \to \infty$, $X_n \xrightarrow{P} X$ if and only if

$$\rho(\tilde{X}_n, \tilde{X}) = E\left(\frac{|X_n - X|}{1 + |X_n - X|}\right) \to 0.$$

The convergence in probability of random elements with values in a *separable metric space* (U, ρ) is metrizable by the following *Ky Fan metric*.

Let $\mathcal{L}^0(\Omega, U)$ be the class of $\mathcal{A}$-$\mathcal{B}(U)$-measurable mappings from Ω to U. Define $\alpha_\rho : \mathcal{L}^0(\Omega, U) \times \mathcal{L}^0(\Omega, U) \to \mathbb{R}^+$ by

$$\alpha_\rho(X, Y) = \inf\{\varepsilon \geq 0 : P(\rho(X, Y) > \varepsilon) \leq \varepsilon\}.$$

Then it can be checked that α_ρ is a metric on $\mathcal{L}^0(\Omega, U)$, and α_ρ is called the *Ky Fan metric*. Moreover, $X_n \xrightarrow{P} X$ if and only if $X)n$ converges to X in the metric α_ρ, i.e., $\lim_{n \to \infty} \alpha_\rho(X_n, X) = 0$. In other words, the Ky Fan metric α_ρ metrizes the convergence in probability on separable metric spaces.

The convergence in probability of random elements implies their convergence in distribution is manifested by the fact that, for any random elements X, Y (with values in a separable metric spaces) with probability laws P_X, P_Y, respectively, we have

$$\Delta(P_X, P_Y) \leq \alpha(X, Y),$$

where Δ, α are the Prohorov and the Ky Fan metrics, respectively. For a proof, see Dudley [24].

In our context of random closed sets on $\mathbb{R}^d$, the space (U, ρ) is $(\mathcal{F}(\mathbb{R}^d), \nabla)$ where ∇ is some metric compatible with the topology $\mathcal{T}$.

The Ky Fan metric α_∇ on $\mathcal{L}^0(\Omega, \mathcal{F})$ is defined similarly. For example, let (U, ρ) be a complete metric subspace of $\mathbb{R}^d$. Consider nonempty random closed (compact) sets on U, i.e., with values in $\mathcal{K}_o(U) = \mathcal{K}(U) - \{\emptyset\}$. Let H_ρ denote the Hausdorff metric on $\mathcal{K}_0$. Then, for random compact sets S, S',

$$\alpha(S, S') = \inf\{\varepsilon \geq 0 : P(\omega : H_\rho(S(\omega), S'(\omega)) > \varepsilon) \leq \varepsilon\}.$$

7.2 Convergence in Distribution

We turn now to the concept of convergence in distribution of random closed sets. Although the concept can be investigated in the general setting of *Hausdorff, locally compact and second countable topological spaces*, we consider $\mathbb{R}^d$ for concreteness. Some special structures of $\mathbb{R}^d$ allow us to go deeper in the analysis.

Let us recall the notation of Chapter 5. Having the space $\mathbb{R}^d$ in mind, we drop from the notation for simplicity. $\mathcal{F}$, $\mathcal{G}$, $\mathcal{K}$ denote the spaces of closed, open, and compact subsets of $\mathbb{R}^d$, respectively. $\mathcal{T}$ denotes the hit-or-miss topology on $\mathcal{F}$, and $\mathcal{B}(\mathcal{F})$ denotes the Borel σ-field of subsets of $\mathcal{F}$ generated by $\mathcal{T}$. $\overline{A}$ and $(A)^o$ denote the closure and interior of $A \subseteq \mathcal{F}$, in the topology $\mathcal{T}$, respectively. The boundary of A is denoted as $\partial A = \overline{A} \setminus (A)^o$.

Let $(\Omega, \mathcal{A}, P)$ be a probability space. We will consider random closed sets $(S_n, n \geq 1)$, S, defined on Ω, i.e., $\mathcal{A}$-$\mathcal{B}(\mathcal{F})$-measurable maps from ω to $\mathcal{F}$. Their probability laws are probability measures on $\mathcal{B}(\mathcal{F})$, defined by $Q_n = PS_n^{-1}$, $Q = PS^{-1}$. Their associate capacity functionals are denoted as T_n, T, respectively, where, for $K \in \mathcal{K}$,

$$T_n(K) = P(S_n \cap K \neq \emptyset) = Q_n(\mathcal{F}_K),$$

$$T(K) = P(S_n \cap K \neq \emptyset) = Q(\mathcal{F}_K),$$

where $\mathcal{F}_K = \{F \in \mathcal{F} : F \cap K \neq \emptyset\}$.

Note that the hit-or-miss topology $\mathcal{T}$ is generated by the base $\mathbb{B}$, where

$$\mathbb{B} = \left\{\mathcal{F}^K_{G_1, G_2, \dots, G_n} : K \in \mathcal{K}, G_i \in \mathcal{G}\right\}$$

(when $n = 0$, the elements are $\mathcal{F}^K$, $K \in \mathcal{K}$), with

$$\mathcal{F}^K_{G_1, G_2, \dots, G_n} = \mathcal{F}^K \cap \mathcal{F}_{G_1} \cap \dots \cap \mathcal{F}_{G_n},$$

$$\mathcal{F}_G = \{F \in \mathcal{F} : F \cap G = \emptyset\}.$$

Note also that capacity functionals, defined on $\mathcal{K}$, are extended to the power set of $\mathbb{R}^d$ as usual, i.e., for $G \in \mathcal{G}$,

$$T(G) = \sup\{T(K) : K \in \mathcal{K}, K \subseteq G\},$$

and for $A \subseteq \mathbb{R}^d$,

$$T(A) = \inf\{T(G) : G \in \mathcal{G}, A \subseteq G\}.$$

The range space $\mathcal{F}$ of random closed sets is a topological space which is compact, Hausdorff, and second countable. As such, $\mathcal{F}$ is metrizable (see a text like Engelking [28]), and the concept of convergence in distribution of random closed sets can be neatly formulated in the general setting of convergence of probability measures in metric spaces (see Billingsley [10] or Appendix). Specifically, we say that S_n *converges in distribution* to S (in symbol, $S_n \overset{D}{\to} S$) iff $Q_n(A) \to Q(A)$ for any $A \in \mathcal{B}(\mathcal{F})$ such that $Q(\partial A) = 0$ (i.e., for any Q-*continuous* set A). In view of the Portmanteu Theorem (Appendix), this definition is equivalent to either:

(i) $\int_{\mathcal{F}} f(F)dQ_n(F) \to \int_{\mathcal{F}} f(F)dQ(F)$, as $n \to \infty$, for any $f \in C_b(\mathcal{F})$, the set of real-valued, continuous and bounded functions defined on $\mathcal{F}$ (i.e., Q_n converges *weakly* to Q, in symbol, $Q_n \overset{W}{\longrightarrow} Q$), or

(ii) For any open set $G \in \mathcal{G}$,

$$Q(G) \le \liminf_{n \to \infty} Q_n(G).$$

Remark. The concept of convergence in (i) on the space of all probability measures on $\mathcal{B}(\mathcal{F})$, denoted as $\mathcal{M}(\mathcal{F})$, is termed the *weak convergence* of probability measures, and can be topologized (see Exercises 7.1 and 7.2). In the context of functional analysis, it is the relative topology on $\mathcal{M}(\mathcal{F})$ induced by the *weak-star* (weak*) topology on the dual of the Banach space $C_b(\mathcal{F})$, i.e., the space of all finite signed measures on $\mathcal{B}(\mathcal{F})$. See, e.g., Walters [126]. Thus, we also say that Q_n converges *weakly* to Q, in symbol, $Q_n \overset{W}{\longrightarrow} Q$, when S_n converges in distribution to S.

This section is devoted to the study of convergence in distribution of random closed sets in terms of their capacity functionals, which play the role of distribution functions in the setting of random vectors. Note that, like the case of random vectors, such a study reduces the complicated situation of "set functions" to the simpler one of "point functions", where here, point functions mean functions defined on $\mathcal{F}$ rather than on subsets of $\mathcal{F}$. In the next section, we will establish an equivalence of (i) above with the *Choquet integral* with respect to capacity functionals.

The material in this section is drawn mainly from Salinetti and West [110]. The main result is this. A sequence of random closed sets S_n converges in distribution to a random closed set S on $\mathbb{R}^d$ if and only if the sequence of capacity functionals T_n converges pointwise to the capacity functional T of S on some subclass of $\mathcal{K}(\mathbb{R}^d)$.

Having the capacity functional T of the random closed set S playing the role of the distribution function of a random vector, we first establish some sort of "continuity set" for T. A set of "continuity points" $\mathcal{C}(T)$ for T is intended to show that $Q_n \xrightarrow{W} Q \Leftrightarrow T_n \to T$ on $\mathcal{C}(T)$.

Now, in view of the correspondence $T(K) = Q(\mathcal{F}_K)$ on $K \in \mathcal{K}$, it is natural to look for K such that $\mathcal{F}_K$ is a Q-continuity set, i.e., $Q(\partial \mathcal{F}_K) = 0$.

LEMMA 7.2 $\{K \in \mathcal{K} : Q(\partial \mathcal{F}_K) = 0\} = \{K \in \mathcal{K} : T(K) = T(K^o)\}$.

Proof. For $K \in \mathcal{K}$, $\mathcal{F}^K$ is open in the $\mathcal{T}$ topology since it is a member of the subbase of $\mathcal{T}$. Thus, $\mathcal{F}_K = (\mathcal{F}_K)^c = \mathcal{F} \setminus \mathcal{F}^K$ is a closed set in $\mathcal{T}$, and hence $\partial(\mathcal{F}_K) = \mathcal{F}_K \setminus (\mathcal{F}_K)^o = \mathcal{F}_K \setminus \mathcal{F}_{K^o}$. (The fact that $(\mathcal{F}_K)^o = \mathcal{F}_{K^o}$ is left as Exercise 7.4.)

If $Q(\partial \mathcal{F}_K) = 0$ then $Q(\mathcal{F}_K) = Q(\mathcal{F}_{K^o})$, i.e., $T(K) = Q(\mathcal{F}_{K^o})$. Lemma 7.2 is proved if

$$T(K^o) = Q(\mathcal{F}_{K^o}),$$

where $T(K^o) = \sup\{T(K') : K' \in \mathcal{K}, K' \subseteq K^o\}$.

This fact is true since for any open set in $\mathbb{R}^d$, such as K^o, there exists an increasing sequence of compact sets K'_n such that $K'_n \subseteq K^o$ and $K^o = \bigcup_{n \geq 1} K'_n$.

Details are left in Exercise 7.7.

Remark. The capacity functional T is defined on the topological space $\mathcal{F}$ and as such we can also consider the continuity of T in the topology $\mathcal{T}$, i.e., T is continuous at $K \in \mathcal{K}$ if $K_n \to K$ in $\mathcal{T}$, then $T(K) = \lim_{n \to \infty} T(K_n)$. It is not clear whether T is continuous at K will imply that $Q(\partial \mathcal{F}_K) = 0$. On the other hand, if K_n is an increasing sequence of compact sets in $\mathbb{R}^d$, $K_n \to K$ in $\mathcal{T}$ means $K = \overline{\left(\bigcup_{n \geq 1} K_n \right)}$. We say that a sequence is *regularly increasing* to K if, in addition, $K^o = \bigcup_{n \geq 1} K_n$. Let $\mathcal{U}(T)$ denote the class of compact sets K such that for any sequence of compact sets K_n, regularly increasing to K, we have $T(K) = \lim_{n \to \infty} T(K_n)$. It turns out that if $K \in \mathcal{U}(T)$ and $K = \overline{(K^o)}$, e.g., K is a finite union of closed balls in $\mathbb{R}^d$, then $Q(\partial \mathcal{F}_K) = 0$. Indeed, let K'_n be a sequence regularly increasing to K, i.e.,

$$K = \overline{\left(\bigcup_{n \geq 1} K'_n \right)}, \quad K^o = \bigcup_{n \geq 1} K'_n.$$

We have

$$Q(\mathcal{F}_{K^o}) = \lim_{n \to \infty} Q(\mathcal{F}_{K_n}) = \lim_{n \to \infty} T(K_n) = T(K) = Q(\mathcal{F}_k)$$

and hence $Q(\partial \mathcal{F}_K) = 0$.

Conversely, if $Q(\partial \mathcal{F}_K) = 0$, then $K \in \mathcal{U}(T)$. Indeed, when $Q(\partial \mathcal{F}_K) = 0$, we have $T(K) = Q(\mathcal{F}_K) = Q(\mathcal{F}_{K^o})$.

Let K_n be a sequence regularly increasing to K, then

$$T(K) = Q(\mathcal{F}_{K^o}) \leq Q\left(\mathcal{F}_{\bigcup\limits_{n \geq 1} K_n}\right) = \lim_{n \to \infty} Q(\mathcal{F}_{K_n}) = \lim_{n \to \infty} T(K_n).$$

On the other hand, since

$$T(K_n) = Q(\mathcal{F}_{K_n}) \leq Q(\mathcal{F}_k) = T(K),$$

we have $T(K) = \lim\limits_{n \to \infty} T(K_n)$.

In view of this analysis, the class of compact sets K in $\mathcal{U}(T)$ and $K = \overline{(K^o)}$ can be used to generate a *convergence-determining* class of the weak convergence of probability measures on $\mathcal{B}(\mathcal{F})$. For details on this, see Salinetti and Wets [110].

LEMMA 7.3 *Let*

$$\mathcal{D}(T) = \{\mathcal{F}_{K_1,\dots,K_n}^{K_0} : K_i \in \mathcal{K}, T(K_i) = T(K_i^o), i = 0, 1, \dots, n\},$$

then:

(i) $\mathcal{D}(T)$ *is closed under finite intersections,*

(ii) *for every closed set F, and any neighborhood V of F (in $\mathcal{T}$), there is an $A \in \mathcal{D}$ such that $F \in A^o \subseteq A \subseteq V$.*

Proof.

(i) Let $\mathcal{F}_{K_1,\dots,K_n}^{K_0}$ and $\mathcal{F}_{L_1,\dots,L_m}^{L_0}$ in $\mathcal{D}(T)$. Then,

$$\mathcal{F}_{K_1,\dots,K_n}^{K_0} \cap \mathcal{F}_{L_1,\dots,L_m}^{L_0} = \mathcal{F}_{K_1,\dots,K_n,L_1,\dots,L_m}^{K_0 \cup L_0}.$$

It remains to verify that

$$T(K_0 \cup L_0) = T((K_0 \cup L_0)^o).$$

In view of Lemma 7.2, it suffices to show that $Q(\partial \mathcal{F}_{K_0 \cup L_0}) = 0$. But it is obvious since

$$\mathcal{F}_{K_0 \cup L_0} = \mathcal{F}_{K_0} \cup \mathcal{F}_{L_0}$$

and $\partial(\mathcal{F}_{K_0} \cup \mathcal{F}_{L_0}) \subseteq \partial(\mathcal{F}_{K_0}) \cup \partial(\mathcal{F}_{L_0})$.

(ii) The proof of (ii) is cumbersome! We refer the reader to page 416 in Salinetti and Wets [110]. □

Remark. Since the topological space $(\mathcal{F}, \mathcal{T})$ is separable, Lemma 7.3 says that the class $\mathcal{D}(T)$ is a convergence-determining class. Thus in view of Corollary 1 (p. 14) in Billingsley [10], to establish the weak convergence of Q_n to Q, it suffices that $Q_n \to Q$ on $\mathcal{D}$.

Note also that if $\mathcal{F}^{K_0}_{K_1,\ldots,K_n} \in \mathcal{D}(T)$, then $Q(\partial \mathcal{F}^{K_0}_{K_1,\ldots,K_n}) = 0$. Indeed,

$$\mathcal{F}^{K_0}_{K_1,\ldots,K_n} = \mathcal{F}^{K_0} \cap \mathcal{F}_{K_1} \cap \ldots \cap \mathcal{F}_{K_n}$$

so that

$$\partial(\mathcal{F}^{K_0}_{K_1,\ldots,K_n}) \subseteq \bigcup_{i=0}^{n} \partial \mathcal{F}_{K_i}$$

by noting that $\partial(\mathcal{F}^{K_0}) = \partial(\mathcal{F}_{K_0})$. Thus, if $T(K_i) = T(K_i^o)$, $i = 0, 1, \ldots, n$, then, in view of Lemma 7.2, $Q(\partial \mathcal{F}_{K_i}) = 0$ and hence,

$$0 \leq Q(\partial \mathcal{F}^{K_0}_{K_1,\ldots,K_n}) \leq \sum_{i=0}^{n} Q(\partial \mathcal{F}_{K_i}) = 0.$$

THEOREM 7.1 Let Q, Q_n be probability laws on the random closed sets S, S_n on $\mathbb{R}^d$, and T, T_n be their associated capacity functionals. Then $Q_n \xrightarrow{W} Q$ if and only if $T_n \to T$ on

$$\mathcal{C}(T) = \{K \in \mathcal{K} : T(K) = T(K^o)\}.$$

Proof.

(i) *Necessity.* In view of Lemma 7.2, if $K \in \mathcal{C}(T)$, then $\mathcal{F}_K$ is a Q-continuity set, and hence $Q_n(\mathcal{F}_K) \to Q(\mathcal{F}_K)$, i.e., $T_n(K) \to T(K)$.

(ii) *Sufficiency.* In view of the above remark, it suffices to show that the pointwise convergence of T_n to T on $\mathcal{C}(T)$ imply that $Q_n \to Q$ on $\mathcal{D}(T)$. This is achieved by induction, by noting that
For $\mathcal{F}^{K_0} \in \mathcal{D}(T)$,

$$Q_n(\mathcal{F}^{K_0}) = 1 - Q_n(\mathcal{F}_{K_0}) = 1 - T_n(K_0) \to$$

$$1 - T(K_0) = 1 - Q(\mathcal{F}_{K_0}) = Q_n(\mathcal{F}^{K_0}).$$

For $\mathcal{F}^{K_0}_{K_1} \in \mathcal{D}(T)$

$$Q_n(\mathcal{F}^{K_0}_{K_1}) = Q_n(\mathcal{F}_{K_0 \cup K_1}) - Q_n(\mathcal{F}^{K_0}) = T_n(K_0 \cup K_1) - T_n(K_0) \to$$

$$T(K_0 \cup K_1) - T(K_0) = Q(\mathcal{F}_{K_0 \cup K_1}) - Q(\mathcal{F}^{K_0}) = Q(\mathcal{F}^{K_0}_{K_1}).$$

□

7.3 Weak Convergence of Capacity Functionals

The convergence in distribution of random elements with values in *metric spaces* is formulated neatly in the *Portmanteau theorem* (see Appendix). The large deviations principle (LDP) can be viewed as a generalization of weak convergence of probability measures (see Appendix). Specifically, Let U be a *polish* space (i.e., a complete, separable metric space), and $\mathcal{U}$ its Borel σ-field. A *rate function* is a function $I : U \to [0, \infty]$ which is *lower semicontinuous* (l.s.c.). A family $\{P_\varepsilon, \varepsilon > 0\}$ of probability measures on $\mathcal{U}$ is said to satisfy the *large deviations principle* with rate function I if, for all $A \in \mathcal{U}$,

$$- \inf_{A^\circ} I \leq \liminf_{\varepsilon \to 0} \varepsilon \log P_\varepsilon(A) \leq \limsup_{\varepsilon \to 0} \varepsilon \log P_\varepsilon(A) \leq - \inf_{\bar{A}} I. \qquad (7.1)$$

Remark. For a sequence of probability measures, we have $\varepsilon = 1/n$. If, in addition, the rate function I is assumed to have *compact level-sets*, then $\inf_{u \in U} I(u) = 0$ is attained at some u_0 since then I attains its infimum on a *closed* set.

As stated earlier, (7.1) is equivalent to

$$\limsup_{\varepsilon \to 0} \varepsilon \log P_\varepsilon(F) \leq -I(F), \qquad \text{for} \quad F \quad \text{closed} \qquad (7.2)$$

and

$$\liminf_{\varepsilon \to 0} \varepsilon \log P_\varepsilon(G) \geq -I(G), \qquad \text{for} \quad G \quad \text{open}, \qquad (7.3)$$

where $I(A) = \inf_{u \in A} I(u)$.

Now, let us take a closer look at (7.2) and (7.3). The set function τ on $\mathcal{U}$, defined by $\tau(A) = e^{-I(A)}$ is an *idempotent probability*, i.e.,

$$\tau(A) = \sup_{u \in A} \tau(\{u\}), \qquad \tau(\{u\}) = \tau(u) = e^{-I(u)},$$

which is *upper semicontinuous* (u.s.c.). Next, $\{P_\varepsilon^\varepsilon, \varepsilon > 0\}$ is a family of *subprobability measures*. Rewrite (7.2) and (7.3) as:

$$\limsup_{\varepsilon \to 0} [P_\varepsilon(F)]^\varepsilon \leq \tau(F) \qquad (7.4)$$

and

$$\liminf_{\varepsilon \to 0} [P_\varepsilon(G)]^\varepsilon \geq \tau(G) \qquad (7.5)$$

and take this as a *definition* for the convergence of the family of subprobability measures $\{P_\varepsilon^\varepsilon, \varepsilon > 0\}$ to the idempotent probability τ. Then, clearly, (7.4) and (7.5) remind us of the *weak convergence of probability measures*, in which, if $P_\varepsilon^\varepsilon$, τ are replaced by probability measures, then (7.4) and (7.5) are equivalent. Thus, the above concept of convergence of subprobabilities to idempotent probabilities (which is another way of stating the large deviations principle (LDP)) is a generalization of weak convergence of probability measures.

Note that u.s.c. "densities" of idempotent probability measures are *rate functions* for LDP. Such idempotent probability measures are capacity functionals of random closed sets on $\mathbb{R}^d$. By *subprobability measures*, we mean here set functions $\nu : \mathcal{B}(U) \to [0,1]$ such that

a) $\nu(\emptyset) = 0$.

b) $\nu(A) = \inf\{\nu(G) : A \subseteq G \text{ open}\}$, for any $A \in \mathcal{B}(U)$.

c) $\nu\left(\overset{\infty}{\underset{n=1}{\cup}} A_n\right) \leq \overset{\infty}{\underset{n=1}{\sum}} v(A_n)$.

d) For any open G, $\nu(G) = \lim_{\delta \to 0} v(G^{-\delta})$, where $G^{-\delta} = ((G^c)_\delta)^c$ with $A_\delta = \{u \in U : \rho(u, A) < \delta\})$.

Note that ν is monotone increasing in view of b). Also, the space $\mathcal{M}(U)$ of all subprobability measures contains all idempotent probability measures (also called *supmeasures*) and set functions of the form P^ε with $\varepsilon \in (0,1]$ and P a probability measure. The space $\mathcal{M}(U)$ is relevant to *capacity functionals* of *random closed sets* in locally compact spaces.

In the following, we are going to consider another generalization of weak convergence of probability measures.

In Section 7.2, we studied the convergence in distribution of random closed sets by looking at the pointwise convergence of their capacity functionals. In this section we view capacity functionals as generalizations of probability measures on $\mathcal{B}(\mathbb{R}^d)$ and as such, we generalize the concept of weak convergence of probability measures to the case of capacity functional, where the Choquet integral will replace the Lebesgue integral in its formulation. We then establish criteria for the convergence in distribution of random closed sets in terms of this generalized weak convergence concept, called the *Choquet weak convergence*. Students interested in a general framework for weak convergence of bounded, monotonic set functions could read Girotto and Holzer [37].

Random sets considered in this section are random closed sets on Euclidean spaces or more generally, on Hausdorff, second countable, locally compact spaces (HSLC). As random elements, probability laws of random closed sets are probability measures on the Borel σ-field (generated by the hit-or-miss topology) of subsets of the space of closed sets of a HSLC space. Since this space is metrizable (and compact), the convergence in distribution of sequences of random closed sets can be formally studied in the lines of weak convergence of probability measures on metric spaces (Billingsley [10]). However, due to the complexity of the spaces of sets, one would like to be able to study this type of convergence at some simpler level. This is indeed possible, thanks to Choquet's theorem [15] characterizing probability laws of random closed sets in terms of their corresponding capacity functionals. In other words, it is possible to study the convergence in distribution of random closed sets by looking at the convergence of capacity functionals as set functions. This was done by several authors, including Molchanov [77], Norberg [93, 94], Salinetti and Wets [110], and Vervaat [123]. Here, we investigate the convergence in distribution of random closed sets also by looking at their capacity functionals, but from a different point of view. We regard capacity functionals are generalizations of probability measures, and as such, we first define a generalized weak convergence type for them, called Choquet weak convergence, in which ordinary Lebesgue integral (of continuous and bounded functions) with respect to probability measures is replaced by an integral with respect to nonadditive set functions, namely, the Choquet integral. We next investigate conditions on this type of Choquet weak convergence of capacity functionals in order to obtain the convergence in distribution of underlying random closed sets.

In the following, the Borel σ-field on U will be denoted as $\mathcal{B}(U)$; the spaces of closed, compact, and open subsets of U are denoted, respectively, as $\mathcal{F}$, $\mathcal{K}$, and $\mathcal{G}$. The space $\mathcal{F}$ is topologized, as usual, by the hit-or-miss topology generating the borel σ-field $\mathcal{B}(\mathcal{F})$ on $\mathcal{F}$. A random closed set is a measurable map S from a probability space $(\Omega, \mathcal{A}, P)$ to $(\mathcal{F}, \mathcal{B}(\mathcal{F}))$. The probability law of S is the probability measure $Q_S = PS^{-1}$, or simply Q, when no confusion is possible. According to Choquet's theorem, Q_S is characterized by its capacity functional T_S, or simply T, defined on $\mathcal{K}$ by $T_S(K) = P(S \cap K \neq \emptyset) = Q(\mathcal{F}_K)$, where $\mathcal{F}_K = \{F \in \mathcal{F} : F \cap K \neq \emptyset\}$. Since $\mathcal{F}$ is Hausdorff, second countable and compact, it is metrizable. As such, convergence in distribution of sequences of random closed sets is defined as weak convergence of probability measures on $(\mathcal{F}, \mathcal{B}(\mathcal{F}))$. However, as in the case of random vectors, convergence of corresponding capacity functionals seems simpler. When U is metric, convergence of probability measures on $(U, \mathcal{B}(U))$ is defined as: $\mu_n \xrightarrow{W} \mu$ (weakly convergent) iff $\int_U f(u) d\mu_n(u) \to \int_U f(u) d\mu(u)$ for any $f \in C_b(U)$, the space of continuous, bounded real-valued functions on U. Since capacity functionals (when extended to $\mathcal{B}(U)$ via $T(A) = \sup\{T(K) : K \in \mathcal{K}, K \subseteq A\}$) are more general than probability measures on $\mathcal{B}(U)$, we could extend the weak convergence of probability measures to capacity functionals by simply replacing the integral with respect to measures $d\mu$ by an integral with respect to a nonadditive set-function T. Such an integral is known as the Choquet integral. Recall that (see Chapter 6) the Choquet integral of a measurable function f, with respect to a capacity functional T, is defined as

$$\int_U f dT = \int_0^\infty T(f \geq t) dt + \int_{-\infty}^0 [T(f \geq t) - T(U)] dt.$$

A sequence of capacity functionals T_n converges, in the Choquet weak sense, or Choquet weakly converges, to a capacity functional T iff $\int_U f(u)dT_n(u) \to \int_U f(u)dT(u)$, for any $f \in C_b(U)$, where the integrals are taken as the Choquet integrals, in symbol, $T_n \longrightarrow_{cw} T$.

While this seems to be a natural, and a convenient way of studying the convergence in distribution of a sequence of random closed sets, S_n to S, i.e., the weak convergence of their probability laws Q_n to Q on the metrizable space $(\mathcal{F}, \mathcal{B}(\mathcal{F}))$, it is not clear whether $Q_n \overset{W}{\longrightarrow} Q$ is equivalent to $T_n \overset{C-W}{\longrightarrow} T$, where T_n, T are associated capacity functionals of Q_n, Q, respectively. It turns out that, unlike the neat situation for random vectors where weak convergence of distribution functions is equivalent to weak convergence of their associated Stieltjes probability measures, the Choquet weak convergence of capacity functionals is a little stronger than the weak convergence of their associated probability laws. Thus, the rest of this chapter is devoted to the investigation of relationships between these two related types of convergence.

In all of the following, we always denote T, T_n the capacity functionals associated with the probability measures Q, Q_n.

First, here is an example showing that $Q_n \overset{W}{\longrightarrow} Q$ does not necessarily imply $T_n \overset{C-W}{\longrightarrow} T$. Consider $U = \mathbb{R}$ and a sequence of random variables X_n with distributions functions

$$F_n(x) = \left(\frac{x}{x+1} \right)^n \quad \text{for } x \geq 0, \text{ and zero for } x < 0.$$

Since for each $x \in \mathbb{R}$, $F_n(x) \to 0$, as $n \to \infty$, X_n does not converge weakly. Let $S_n = \{X_n\}$ be the sequence of singleton random closed sets with capacity functionals

$$T_n(K) = P(S_n \cap K \neq \emptyset) = P(X_n \in K)$$

For any $K \in \mathcal{K}$, $P(X_n \in K) \leq P(X_n \leq \max K) = F_n(\max K) \to 0$ as $n \to \infty$, we have that $T_n(K) \to 0$ as $n \to \infty$. Now, $T = 0$, identically, is a capacity functional, and as such, according to the Choquet theorem, there is a random closed set S with probability law Q determined by T. Note that Q is the Dirac probability measure at the point $\emptyset$ of $\mathcal{F}(\mathbb{R})$. Let Q_n denote the probability laws of S_n, $n \geq 1$. Then $Q_n \overset{C-W}{\longrightarrow} Q$ since T_n converges to T on $\mathcal{K}$, see e.g., Molchanov ([77], p. 8). However, T_n does not converge to T in the Choquet weak sense. This can be seen as follows.

Let $f(x) = 1$ for any $x \in \mathbb{R}$. Then, for any n, $\int f(x)dT_n = 1$, since for $0 < t < 1$, $(f \geq t) = \mathbb{R}$, and $T_n(\mathbb{R}) = \sup\{T_n(K) : K \in \mathcal{K}\} = P(X_n \in \mathbb{R}) = 1$, since probability measures on $\mathbb{R}$ are tight, whereas $\int f(x)dT = 0$.

It is interesting to note that, for HSLC space U, the weak convergence of $Q_n \overset{W}{\longrightarrow} Q$ does imply that $\int_U f(u)dT_n(u) \to \int_U f(u)dT(u)$ for $f \in C_s^+(U)$, the space of nonnegative continuous functions with compact supports on U, a subset of $C_b(U)$.

THEOREM 7.2 Let U be a HSLC space. Then $Q_n \xrightarrow{W} Q$ implies $\int_U f(u)dT_n(u) \to \int_U f(u)dT(u)$ for $f \in C_s^+(U)$.

Proof. For $f \in C_b(U)$, we write $\|f\|$ for $\sup\{|f(x)| : x \in U\}$. Now let $f \in C_s^+(U)$. First, observe that for each $t \in (0, \|f\|)$, the closed set $\{f \geq t\} \subseteq K$, a compact containing $\{f \neq 0\}$, and hence compact, so that $\mathcal{F}_{\{f \geq t\}} = \mathcal{F} - \mathcal{F}^{\{f \geq t\}}$ is closed in $\mathcal{F}$. Moreover, $\mathcal{F}_{\{f > t\}}$ is open in $\mathcal{F}$. We denote by $\partial(.)$ and $(.)^o$ the boundary and interior operators, respectively. Then

$$\partial(\mathcal{F}_{\{f \geq t\}}) = \mathcal{F}_{\{f \geq t\}} - (\mathcal{F}_{\{f \geq t\}})^o \subseteq \mathcal{F}_{\{f \geq t\}} - \mathcal{F}_{\{f > t\}}$$

Thus,

$$0 \leq \int_0^{\|f\|} Q[\partial(\mathcal{F}_{\{f \geq t\}})]dt \leq \int_0^{\|f\|} Q[\mathcal{F}_{\{f \geq t\}} - \mathcal{F}_{\{f > t\}}]dt$$
$$= \int_0^{\|f\|} [T(f \geq t) - T(f > t)]dt = 0,$$

meaning that $\mathcal{F}_{\{f \geq t\}}$ is a Q-continuity set, for almost all $t \in (0, \|f\|)$. Now, assuming $Q_n \longrightarrow_w Q$, and using Lebesgue's Dominated Convergence theorem (noting that $|Q_n| \leq 1$, $|Q| \leq 1$), together with Billingsley's Portmanteau theorem, we have

$$\lim_{n \to \infty} \left[\int f dT_n - \int f dT \right] = \lim_{n \to \infty} \int_0^{\|f\|} [T_n(f \geq t)dt - T(f \geq t)]dt$$
$$= \lim_{n \to \infty} \int_0^{\|f\|} [Q_n(\mathcal{F}_{\{f \geq t\}}) - Q(\mathcal{F}_{\{f \geq t\}})]dt$$
$$= \int_0^{\|f\|} \lim_{n \to \infty} [Q_n(\mathcal{F}_{\{f \geq t\}}) - Q(\mathcal{F}_{\{f \geq t\}})]dt = 0.$$

$\square$

In the special case of *metric* spaces like $U = \mathbb{R}^d$, it turns out that Theorem 7.2 admits a converse.

THEOREM 7.3 Let $U = \mathbb{R}^d$. Then $\int_U f(u)dT_n(u) \to \int_U f(u)dT(u)$ for any $f \in C_s^+(\mathbb{R}^d)$ implies $Q_n \xrightarrow{W} Q$.

Proof. Let ρ denote the euclidean distance on $\mathbb{R}^d$. For $K \in \mathcal{K}_o = \mathcal{K} \setminus \emptyset$, and $j \geq 1$, define $\Psi_K^j : \mathbb{R}^d \to [0, 1]$ by $\Psi_K^j(x) = \max\{0, 1 - j\rho(x, K)\}$. Clearly, $\Psi_K^j \in C_s^+(\mathbb{R}^d)$, so that, by hypothesis,

$$\lim_{n \to \infty} \int \Psi_K^j(x)dT_n = \int \Psi_K^j(x)dT, \text{ for any } K \in \mathcal{K}_o \text{ and } j \geq 1$$

and, in particular,

$$\lim_{n\to\infty} \int \Psi_K^j(x)dT_n = \int \Psi_K^j(x)dT, \text{ for any } K \in \mathcal{S}(T) \setminus \emptyset, \text{ and } j \geq 1$$

where $\mathcal{S}(T) = \{K \in \mathcal{K} : T(K) = T(K^o)\}$. Since $\mathcal{F}$ is compact metrizable, the space of (Borel) probability measures $\mathcal{M}(\mathcal{F})$ on $\mathcal{F}$ is also compact metrizable in the weak topology. Thus, for each subsequence $Q_{n'}$ of the sequence Q_n, there exists a further subsequence $Q_{n'_i}$ weakly converging to some probability measure W of $\mathcal{M}(\mathcal{F})$. Let R be the corresponding capacity functional of W. Since $Q_{n'_i} \xrightarrow{W} W$, we have, by Theorem 7.2, that $\lim_{i\to\infty} \int \Psi_K^j(x)dT_{n'_i} = \int \Psi_K^j(x)dR$, for any $K \in \mathcal{S}_R \setminus \emptyset$, and $j \geq 1$. Thus, $\int \Psi_K^j(x)dT = \int \Psi_K^j(x)dR$ for any $K \in \mathcal{S}_R \cap \mathcal{S}_T \setminus \emptyset$, and $j \geq 1$. But then, by Lemma 7.4 below, $T = R$ on $\mathcal{K}$ which, in turn, implies that $Q = W$, by the Choquet theorem. □

LEMMA 7.4 *Let T and R be two capacity functionals on $\mathcal{K}(\mathbb{R}^d)$. If $\int \Psi_K^j(x)dT = \int \Psi_K^j(x)dR$ for any $K \in \mathcal{S}_R \cap \mathcal{S}_T \setminus \emptyset$, and $j \geq 1$, then $T = R$ on $\mathcal{K}$.*

Proof. For $K \in \mathcal{S}_R \cap \mathcal{S}_T \setminus \emptyset$, we have

$$K \subseteq \{x : \Psi_K^j(x) \geq t\} = \{x : \rho(x, K) \leq (1-t)/j\} \text{ for any } t \in (0, 1]$$

i.e., the sequence of compacts $\{x : \Psi_K^j(x) \geq t\}$ decreases to K as $j \to \infty$. By Lebesgue's Dominated Convergence theorem and upper semicontinuity of capacity functionals, we have

$$R(K) = \int_0^1 \lim_{j\to\infty} R(\{\Psi_K^j \geq t\})dt = \lim_{j\to\infty} \int_0^1 \Psi_K^j(x)dR$$

$$= \lim_{j\to\infty} \int_0^1 \Psi_K^j(x)dT = \int_0^1 \lim_{j\to\infty} T(\{\Psi_K^j \geq t\})dt = T(K)$$

i.e., $T = R$ on $\mathcal{S}_R \cap \mathcal{S}_T$. Now, for $K \in \mathcal{K}_o$, let

$$f(x) = \max\{0, 1 - \rho(x, K)\}.$$

Since $\int_0^1 T(f \geq t)dt = \int_0^1 T(f > t)dt$, we have $T(f \geq t) = T(f > t)$ almost everywhere in $t \in (0, 1)$.

But $T(f \geq t) - T(f \geq t)^o \leq T(f \geq t) - T(f > t)$ and

$$(f \geq t) = \{x : \rho(x, K) \leq 1 - t\}$$

is compact, so we have $(f \geq t) \in \mathcal{S}_T$, a.e. $t \in (0, 1)$. Similarly for R at the place of T, we have $(f \geq t) \in \mathcal{S}_R \cap \mathcal{S}_T$, a.e. $t \in (0, 1)$. Choose a sequence $t_i \subseteq (0, 1)$ such that $(f \geq t_i) \in \mathcal{S}_R \cap \mathcal{S}_T$, $i \geq 1$ and $t_i \nearrow 1$ as $i \to \infty$. Since

$(f \geq t_i) \searrow K$ as $i \to \infty$, as before, we have $R(f \geq t_i) = T(f \geq t_i)$, for $i \geq 1$. By semi-upper-continuity,

$$R(K) = \lim_{i \to \infty} R(f \geq t_i) = \lim_{i \to \infty} T(f \geq t_i) = T(K).$$

$\square$

In order to prepare for the following results, we need to spell out some concepts as well as some known results. In this section, U will be a HSLC space.

First, if $K \in \mathcal{K}$, then $\mathcal{F}_K$ is closed in the hit-or-miss topology on $\mathcal{F}$, and $\mathcal{F}_{K^o}$ is open, and $\partial(\mathcal{F}_K) \subseteq \mathcal{F}_K - \mathcal{F}_{K^o}$. For a given capacity functional T with its associated probability measure Q, we have

$$Q(\partial(\mathcal{F}_K)) \leq Q(\mathcal{F}_K - \mathcal{F}_{K^o}) = T(K) - T(K^o).$$

Thus if $T(K) = T(K^o)$, then $\mathcal{F}_K$ is a Q-continuity set. We let $\mathcal{C}(T) = \{K \in \mathcal{K} : T(K) = T(K^o)\}$.

Let $\mathcal{W} = \{A \in \mathcal{B}(U) : \overline{A} \in \mathcal{K}\}$, the class of Borel sets with compact closure. Following Norberg [93], a subclass $\mathcal{H}$ of $\mathcal{W}$ is said to be *separating* if for $K \in \mathcal{K}$, $G \in \mathcal{G}$, with $K \subset G$, there is some $A \in \mathcal{H}$ such that $K \subset A \subset G$.

We need three lemmas.

LEMMA 7.5 (Norberg [93]) *Let $\{T_n\}$ be a sequence of capacity functionals, with associated $\{Q_n\}$. If there exists a separating subclass $\mathcal{H}$ of $\mathcal{W}$ and a capacity functional R such that*

$$R(A^o) \leq \liminf_{n \to \infty} T_n(A) \leq \limsup_{n \to \infty} T_n(A) \leq R(\overline{A}) \text{ for any } A \in \mathcal{H},$$

then there exists a probability measure Q with associated capacity functional T satisfying $Q_n \xrightarrow{W} Q$ and $T = R$ on $\mathcal{K}$.

LEMMA 7.6 (Norberg [93]) *If $T_n \xrightarrow{C-W} T$, then $\limsup\limits_{n \to \infty} T_n(F) \leq T(F)$ for any $F \in \mathcal{F}$.*

Proof. Let $\emptyset \neq F \in \mathcal{F}$. Let $G_k \in \mathcal{G}$ such that $G_k \searrow F$ and $T(G_k) \searrow T(F)$. For each k, let f_k be a continuous function from $U \to [0,1]$, zero on $U \setminus G_k$ and equal 1 on F (which exists by Urysohn lemma, valid on the HSLC space U since a LC space is regular, and, together with second countability, it is Lindelof, and hence normal. See a text on general topology like Engelking [28]). Note that, for $0 \leq t < 1$, $F \subseteq \{f_k > t\} \subseteq \{f_k > 0\} \subseteq G_k$, it follows that, for each k, in view of $T_n \xrightarrow{C-W} T$,

$$\limsup_{n\to\infty} T_n(F) = \limsup_{n\to\infty} \int_0^1 T_n(F) dt$$

$$\leq \limsup_{n\to\infty} \int_0^1 T_n(\{f_k > t\}) dt$$

$$= \limsup_{n\to\infty} \int_0^1 f_k dT_n = \int_0^1 f_k dT = \int_0^1 T(\{f_k > t\}) dt$$

Thus,

$$\limsup_{n\to\infty} T_n(F) \leq \liminf_{k\to\infty} \int_0^1 T(\{f_k > t\}) dt$$

$$\leq \lim_{k\to\infty} \int_0^1 T(G_k) dt = T(F).$$

□

LEMMA 7.7 If $T_n \xrightarrow{C-W} T$, then $\liminf_{n\to\infty} T_n(G) \geq T(G)$ for any $G \in \mathcal{G}$.

Proof. Let $\emptyset \neq G \in \mathcal{G}$. Let $K_k \in \mathcal{K}$ such that $K_k \nearrow G$ and $T(K_k) \nearrow T(G)$. By Urysohn's lemma, let $g_k : U \to [0,1]$ which is continuous, zero on $U \backslash G$ and equal 1 on K_k, for each k. Taking into account the fact that, for $0 \leq t < 1$, we have

$$K_k \subseteq \{g_k \geq t\} \subseteq \{g_k > 0\} \subseteq G$$

and, by hypothesis, $T_n \xrightarrow{C-W} T$, we get

$$\liminf_{n\to\infty} T_n(G) = \liminf_{n\to\infty} \int_0^1 T_n(G) dt$$

$$\geq \liminf_{n\to\infty} \int_0^1 T_n(\{g_k \geq t\}) dt$$

$$= \liminf_{n\to\infty} \int_0^1 g_k dT_n = \int_0^1 g_k dT = \int_0^1 T(\{g_k \geq t\}) dt.$$

Thus,

$$\liminf_{n\to\infty} T_n(G) \geq \limsup_{k\to\infty} \int_0^1 T(\{g_k \geq t\}) dt$$

$$\geq \limsup_{k\to\infty} \int_0^1 T(K_k) dt = T(G).$$

□

We set out now to show that the Choquet weak convergence of capacity functionals is stronger than the weak convergence of their associated probability laws.

THEOREM 7.4 Let U be an HSLC space. Then $T_n \xrightarrow{C-W} T$ implies $Q_n \xrightarrow{W} Q$.

Proof. The class $\mathcal{C}(T)$ is separating. Indeed, for $K \in \mathcal{K}$, $G \in \mathcal{G}$, with $K \subset G$, since U is HSLC, by the compactness of K, there exists $A \in \mathcal{G}$, such that $\overline{A} \in \mathcal{K}$, and $K \subset A \subset G$. Since K and $U \setminus A$ are closed disjoint sets, there is, by Urysohn's lemma, a continuous function $f : U \to [0,1]$ zero on $U - A$, and equal 1 on K, so that

$$K \subseteq \{f \geq t\} \subseteq A \subseteq G \text{ for any } t \in (0,1).$$

Note also that $T\{f \geq t\} = T\{f > t\}$ almost everywhere on $t \in (0,1)$. Choose $B = \{f \geq t_o\}$ for some $t_o \in (0,1)$ with $T\{f \geq t_o\} = T\{f > t_o\}$. Keeping in mind that $\{f > t_o\} \subseteq B^o \subseteq B$, we have $T(B) = T(B^o) = T(\{f > t_o\})$. Moreover, $B = B^- \subseteq A^- \in \mathcal{K}$ implies that $B \in \mathcal{K}$ and thus, $B \in \mathcal{C}(T)$. On the other hand, in view of Lemmas 7.6 and 7.7, we have, when $T_n \xrightarrow{C-W} T$, for any K in $\mathcal{K}$,

$$\limsup_{n\to\infty} T_n(K) \leq T(K)$$
$$\liminf_{n\to\infty} T_n(K^o) \geq T(K^o),$$

hence, for any $K \in \mathcal{C}(T)$,

$$T(K^o) \leq \liminf_{n\to\infty} T_n(K^o) \leq \liminf_{n\to\infty} T_n(K) \leq \limsup_{n\to\infty} T_n(K) \leq T(K).$$

By Lemma 7.5, there exists a random set S' with probability law Q' and corresponding capacity functional R satisfying $Q_n \xrightarrow{W} Q'$ and $T = R$ on $\mathcal{K}$. It then suffices to use the Choquet theorem to conclude that $Q = Q'$, i.e., $Q_n \longrightarrow_w Q$. □

Remark. We can use this result to show that the Choquet weak limit of capacity functionals is unique. Indeed, since $\mathcal{F}$ with the hit-or-miss topology is compact and metrizable, the space of all probability measures on its borel σ-field, equipped with the weak topology (generated by the weak convergence of probability measures), is compact, metric (and hence Hausdorff). It $T_n \xrightarrow{C-W} T$, and $T_n \xrightarrow{C-W} T'$, then $Q_n \Longrightarrow Q$ and $Q_n \xrightarrow{W} Q'$. But then $Q = Q'$, and hence, $T = T'$, by uniqueness in the Choquet theorem.

To find out how strong $T_n \xrightarrow{C-W} T$ is with respect to $Q_n \xrightarrow{W} Q$, we introduce the concept of *tightness* in the context of capacity functionals as follows. Let

T be a capacity functional. A family Γ of capacity functionals is said to be *T-tight* if for each set of the form $A = \{x \in U : f(x) \geq a\}$, for some $f \in C_b(U)$ and $a \in \mathbb{R}$, such that $T(f \geq a) = T(f > a)$ (such a set A is called a *T-continuous functional closed set*) and each $\varepsilon > 0$, there exists a compact set $K(A, \varepsilon)$ such that

$$\sup\{R(A) - R(A \cap K) : R \in \Gamma\} < \varepsilon.$$

Note that U is obviously a T-continuous functional closed set for any T, there exists a compact $K(U, \varepsilon)$ such that

$$\sup\{R(U) - R(K) : R \in \Gamma\} < \varepsilon.$$

Also, finite families of capacity functionals are T-tight, for any T.

THEOREM 7.5 If $T_n \xrightarrow{C-W} T$ then the sequence $\{T_n, n \geq 1\}$ is T-tight.

Proof. Since U is HSLC, there exists a sequence $O_k \in \mathcal{G}$ such that $O_k^- \in \mathcal{K}$ and $O_k \nearrow U$. Let $A = \{f \geq a\}$ be a T-continuous functional closed set. Since $O_k \cap \{f > a\} \nearrow \{f > a\}$, we have, by upper semicontinuity of T,

$$T(O_k \cap \{f > a\}) \nearrow T(\{f > a\}).$$

For each $\varepsilon > 0$, there is a k_o such that

$$0 \leq T(\{f > a\}) - T(O_k \cap \{f > a\}) < \varepsilon/2 \text{ for } k \geq k_o.$$

By Lemmas 7.6 and 7.7, we get

$$\limsup_{n \to \infty}[T_n(A) - T_n(O_k^- \cap A)] \leq \limsup_{n \to \infty}[T_n(A) - T_n(O_k \cap \{f > a\})] \leq$$

$$\limsup_{n \to \infty} T_n(A) - \liminf_{n \to \infty} T_n(O_k \cap \{f > a\}) \leq$$

$$T(A) - T(O_k \cap \{f > a\}) = T(\{f > a\}) - T(O_k \cap \{f > a\}) < \varepsilon/2.$$

Since any finite family of capacity functionals is T-tight, we obtain that $\{T_n, n \geq 1\}$ is T-tight. $\square$

THEOREM 7.6 $T_n \xrightarrow{C-W} T$ if and only if $Q_n \xrightarrow{W} Q$ and $\{T_n, n \geq 1\}$ is T-tight.

Proof. The necessity comes from Theorems 7.4 and 7.5. The sufficiency follows from Lemmas 7.8 and 7.9 below. $\square$

LEMMA 7.8 If $Q_n \xrightarrow{W} Q$ and $\{T_n, n \geq 1\}$ is T-tight, then $\liminf_{n \to \infty} \int f dT_n \geq \int f dT$, for any $f \in C_b(U)$.

Proof. For $f \in C_b(U)$, we have, for $n \geq 1$,

$$\int f dT_n = \int_{-\|f\|}^{\|f\|} T_n(f > t)dt - T_n(U)\|f\|$$

$$\int f dT = \int_{-\|f\|}^{\|f\|} T(f > t)dt - T(U)\|f\|.$$

Since $Q_n \xrightarrow{W} Q$, we have, by Billingsley's Portmanteau theorem, noting that $\mathcal{F}_{(f>t)}$ is open in $\mathcal{F}$,

$$\liminf_{n\to\infty} Q_n(\mathcal{F}_{(f>t)}) \geq Q(\mathcal{F}_{(f>t)}) \text{ for } t \in (-\|f\|, \|f\|).$$

On the other hand, for each $\varepsilon > 0$, since $\{T_n, n \geq 1\}$ is T-*tight*, there exists a compact K such that $\sup_n[T_n(U) - T_n(K)] < \varepsilon$. Since $\mathcal{F}_K$ is closed in $\mathcal{F}$, we have $\limsup_{n\to\infty} Q_n(\mathcal{F}_K) \leq Q(\mathcal{F}_K)$. By the Choquet theorem and the Fatou lemma,

$$\liminf_{n\to\infty} \int f dT_n \geq \liminf_{n\to\infty} \int_{-\|f\|}^{\|f\|} T_n(f > t)dt - \limsup_{n\to\infty} T_n(U)\|f\| \geq$$

$$\liminf_{n\to\infty} \int_{-\|f\|}^{\|f\|} Q_n(\mathcal{F}_{(f>t)})dt - \limsup_{n\to\infty}[T_n(K) + \varepsilon]\|f\| \geq$$

$$\int_{-\|f\|}^{\|f\|} \liminf_{n\to\infty} Q_n(\mathcal{F}_{(f>t)})dt - \limsup_{n\to\infty}[Q_n(\mathcal{F}_K)\|f\| - \varepsilon]\|f\| \geq$$

$$\int_{-\|f\|}^{\|f\|} Q(\mathcal{F}_{(f>t)})dt - Q(\mathcal{F}_K)\|f\| - \varepsilon]\|f\| \geq$$

$$\int_{-\|f\|}^{\|f\|} Q(\mathcal{F}_{(f>t)})dt - Q(\mathcal{F}_U)\|f\| - \varepsilon]\|f\| =$$

$$\int_{-\|f\|}^{\|f\|} T(f > t)dt - T(U)\|f\| - \varepsilon\|f\| = \int f dT - \varepsilon\|f\|.$$

Since ε is arbitrary, we have that $\liminf_{n\to\infty} \int f dT_n \geq \int f dT$. □

LEMMA 7.9 If $Q_n \xrightarrow{W} Q$ and $\{T_n, n \geq 1\}$ is T-*tight*, then $\limsup_{n\to\infty} \int f dT_n \leq \int f dT$, for any $f \in C_b(U)$.

Proof. For any fixed $f \in C_b(U)$, the decreasing function $t \to T(f \geq t)$ is Riemann integrable on $[-\|f\|, \|f\|]$. Since $(f \geq t)$ is a T-continuous functional closed set, a.e. on $(-\|f\| - 1, \|f\| + 1)$, there exists, for each $\varepsilon > 0$, a subdivision

$$-\|f\| = t_0 < t_1 < \ldots < t_m = \|f\|$$

of $[-\|f\|, \|f\|]$ such that

(i) $(f \geq t_j)$ is a T-continuous functional closed set for $j = 0, 1, \ldots, m-1$,

(ii) $\sum_{j=0}^{m-1} T(f \geq t_j)(t_{j+1} - t_j) - T(U) \|f\| < \int f dT + \varepsilon.$

Since, by hypothesis, $\{T_n, n \geq 1\}$ is T-tight, there is a compact K such that

$$\sup_n [T_n(f \geq t_j) - T_n((f \geq t_j) \cap K)] < \varepsilon, \text{ for } j = 0, 1, \ldots, m-1.$$

Since $\mathcal{F}_U$ is open and $\mathcal{F}_{(f \geq t_j) \cap K}$ is closed in $\mathcal{F}$, we have, by the Billingsley's Portmanteau theorem,

$$\liminf_{n \to \infty} Q_n(\mathcal{F}_U) \geq Q(\mathcal{F}_U)$$

and

$$\limsup_{n \to \infty} Q_n(\mathcal{F}_{(f \geq t_j) \cap K}) \leq Q(\mathcal{F}_{(f \geq t_j) \cap K}).$$

Hence, by the increasing monotonicity of T_n and the Choquet theorem,

$$\limsup_{n \to \infty} \int f dT_n$$

$$= \limsup_{n \to \infty} \left[\int_{-\|f\|}^{\|f\|} T_n(f > t) dt - T_n(U) \|f\| \right]$$

$$\leq \limsup_{n \to \infty} \left[\sum_{j=0}^{m-1} \int_{t_j}^{t_{j+1}} T_n(f > t) dt \right] - \liminf_{n \to \infty} Q_n(\mathcal{F}_U) \|f\|]$$

$$\leq \limsup_{n \to \infty} \sum_{j=0}^{m-1} T_n(f > t_j)(t_{j+1} - t_j) - Q(\mathcal{F}_U) \|f\|$$

$$\leq \limsup_{n \to \infty} \sum_{j=0}^{m-1} [T_n((f > t_j) \cap K) + \varepsilon](t_{j+1} - t_j) - T(U) \|f\|$$

$$\leq \limsup_{n \to \infty} \sum_{j=0}^{m-1} [Q_n(\mathcal{F}_{(f \geq t_j) \cap K})(t_{j+1} - t_j) + 2\varepsilon \|f\| - T(U) \|f\|$$

$$\leq \sum_{j=0}^{m-1} Q(\mathcal{F}_{(f \geq t_j) \cap K})(t_{j+1} - t_j) + 2\varepsilon \|f\| - T(U) \|f\|$$

$$= \sum_{j=0}^{m-1} T((f \geq t_j) \cap K)(t_{j+1} - t_j) + 2\varepsilon \|f\| - T(U) \|f\|$$

$$\leq \sum_{j=0}^{m-1} T((f \geq t_j))(t_{j+1} - t_j) + 2\varepsilon \|f\| - T(U) \|f\|$$

$$< \int f dT + \varepsilon + 2\varepsilon \|f\|.$$

Since $\varepsilon > 0$ is arbitrary, we have $\limsup_{n \to \infty} \int f dT_n \leq \int f dT$. $\qquad$ □

Remark. On *compact metric* spaces, the Choquet weak convergence of capacity functionals is *equivalent* to the convergence in distribution of their associated random closed sets, since then, every sequence of capacity functionals is T-tight, for any capacity functional T.

7.4   Exercises

7.1 Let (U, ρ) be a metric space and $\mathcal{M}(U)$ be the space of all probability measures on the Borel σ-filed of U. Consider the following collection of subsets $\mathcal{V}$ of $\mathcal{M}(U)$:

$$A \in \mathcal{V} \text{ iff } A \text{ is of the form}$$

$$\left\{Q \in \mathcal{M}(U) : \left| \int f_i dQ - \int f_i dP \right| < \varepsilon, i = 1, 2, \ldots, k \right\},$$

where $P \in \mathcal{M}(U)$, $\varepsilon > 0$, $f_i \in C_b(U)$, $k \geq 1$.

(i) Verify that $\mathcal{V}$ forms a base for a topology.

(ii) Show that for $P_n, P \in \mathcal{M}(U)$, $P_n \to P$ in the above topology is equivalent to $P_n \to P$ weakly.

7.2 Let (U, ρ) be a metric and separable space. The *Prohorov metric* on $\mathcal{M}(U)$ is defined as follows:

$$\Delta(P, Q) = \inf\{\varepsilon > 0 : P(A) \leq Q(A^\varepsilon) + \varepsilon \text{ for all closed sets } A \text{ of } U\},$$

where

$$A^\varepsilon = \{x \in U : \rho(x, A) < \varepsilon\},$$
$$\rho(x, A) = \inf\{\rho(x, y) : y \in A\}.$$

(i) Verify that Δ is indeed a metric.

(ii) Show that for P_n, P in $\mathcal{M}(U)$,

$$P_n \to P \text{ weakly } \Longleftrightarrow \Delta(P_n, P) \to 0, \text{ as } n \to \infty.$$

7.3 Let $F_n \in \mathcal{F}(\mathbb{R}^d)$. Define

$$\liminf F_n = \{x \in \mathbb{R}^d : x = \lim x_n, x_n \in F_n\},$$

$$\limsup F_n = \{x \in \mathbb{R}^d : x = \lim x_{n(k)},$$

$$x_{n(k)} \in F_{n(k)} \text{ for subsequence } n(k), \quad k \geq 1\}.$$

Define $F_n \to F$ when $F = \liminf F_n = \limsup F_n$.

Show that $F_n \to \emptyset$ if and only if for each compact $K \in \mathcal{K}(\mathbb{R}^d)$, there exists $n(K)$ such that $n \geq n(K)$ implies $F_n \cap K = \emptyset$.

7.4 For $K \in \mathcal{K}(\mathbb{R}^d)$, show that

$$(\mathcal{F}_K)^o = \mathcal{F}_{(K)^o}.$$

7.5 Let Q, Q_n be probability measures on $\mathcal{B}(\mathcal{F})$ with corresponding capacity functionals T, T_n on $\mathcal{K}(\mathbb{R}^d)$. Show that $T_n \to T$ on $\mathcal{K}$ implies $Q_n \xrightarrow{W} Q$.

7.6 Consider the relation $Q(\mathcal{F}_k) = T(K)$ for $K \in \mathcal{K}$. For $K_0, K_1 \in \mathcal{K}$ such that $T(K_0) = T(K_0^o)$, $T(K_1) = T(K_1^o)$, show that $Q\left(\partial \mathcal{F}_{K_1}^{K_0}\right) = 0$, where

$$\mathcal{F}_{K_1}^{K_0} = \mathcal{F}^{K_0} \cap \mathcal{F}_{K_1}.$$

7.7 With the notation of Exercise 7.6, define

$$T(G) = \sup\{T(K) : K \in \mathcal{K}, K \subseteq G\}$$

for any open set G of $\mathbb{R}^d$. Let $K \in \mathcal{K}$, show that

$$T(K^o) = Q(\mathcal{F}_{K^o}).$$

7.8 Let $K_i \in \mathcal{K}(\mathbb{R}^d)$, $i = 0, 1, 2, \ldots$ Show that if $Q(\partial \mathcal{F}_{K_i}) = 0$, then

$$Q\left(\partial \left(\mathcal{F}_{K_1, K_2}^{K_0}\right)\right) = 0,$$

where

$$\mathcal{F}_{K_1, K_2}^{K_0} = \mathcal{F}^{K_0} \cap \mathcal{F}_{K_1} \cap \mathcal{F}_{K_2}.$$

7.9 Show that if $S_n \xrightarrow{a.s.} S$, then $S_n \xrightarrow{P} S$.

7.10 Show that if $S_n \xrightarrow{P} S$, then $S_n \to S$ in distribution.

Chapter 8

Some Aspects of Statistical Inference with Coarse Data

The theory of random sets is developed, within the general theory of probability, for a variety of applications. In this introductory text, we will only touch upon its applications to statistics. Specifically, having coarse data analysis as a potential application of random set theory, we will present in the following sections some preliminary frameworks for random set observations.

8.1 Expectations and Limit Theorems

In previous chapters, basic concepts of random closed sets, as a specific type of random elements taking values in a metric space, have been developed without major difficulties. With capacity functionals playing the role of probability measures, we define the integral of real-valued functions with respect to capacity functionals by the Choquet integral. Now, let Q be the probability law of a random closed set S on $\mathbb{R}^d$, say. While it is intuitive and in fact desirable to consider the "expected set" of S, it is not clear how this generalization of the concept of an expected value of a random variable can be formulated. Locally speaking, the technical difficulty comes from the fact that we are dealing with *set-valued functions*, rather than with point-valued functions. Integration of set-valued functions is a difficult mathematical topic, due mainly to the structure of sets. There are attempts to formulate such a theory of integration such as in mathematical economics (e.g., Debreu [18]), as well as embedding sets into Banach spaces and considering abstract integration in such spaces. From an application viewpoint, the above approaches do not seem to be practical.

A simpler way to address this problem is to directly define a concept of *expectation of a random set* generalizing expected value of a random variable. There are different ways of doing so as in any generalization problem. For a survey of different concepts of expectation of random sets, see Molchanov [79].

The most popular definition of the expectation of a random closed set is based on *Aumann integral* (Aumann [6]). The development of the Aumann integral has the flavor of coarse data analysis in which the random set under consideration plays the role of a coarsening scheme.

Let S be a random closed set, defined on $(\Omega, \mathcal{A}, P)$, and taking values in $\mathcal{F}(\mathbb{R}^d)$, say. Let $L^1(\Omega, \mathcal{A}, P)$, or, more generally, $L^p(\Omega, \mathcal{A}, P)$, for $1 \leq p \leq \infty$, denote the space of (equivalent classes of) *integrable* random vectors $X : \Omega \to \mathbb{R}^d$. Specifically, let $X : \Omega \to \mathbb{R}^d$ be a d-dimensional random vector, i.e.,

$$X(\omega) = \begin{pmatrix} X_1(\omega) \\ \dots \\ X_d(\omega) \end{pmatrix},$$

where $X_i : \Omega \to \mathbb{R}$, $i = 1, 2, \dots, d$.

For $x = (x_1, \dots, x_d) \in \mathbb{R}^d$, the euclidean norm of x is

$$\|x\| = \left(\sum_{i=1}^{d} x_i^2 \right)^{1/2}.$$

The random set X is said to be *integrable* if $E(\|X\|) < +\infty$, where, as usual,

$$E(\|X\|) = \int_{\Omega} \|X\|(\omega) dP(\omega),$$

$$X = (X_1, X_2, \dots, X_d), \quad X_i : \Omega \to \mathbb{R}, \quad i = 1, 2, , \dots, d.$$

The class of a.s. selectors of S (i.e., random vectors $X : \Omega \to \mathbb{R}^d$ such that $P(\omega : X(\omega) \in S(\omega)) = 1$) is denoted as $\mathcal{S}(S)$. The class

$$\mathcal{T}(S) = \mathcal{S}(S) \cap L^1(\Omega, \mathcal{A}, P)$$

is the class of integrable selectors of S.

Remark. In view of Norberg's theorem (Chapter 5), the study of a.s. selectors of S (as models for population with coarsening scheme S) can be reduced to the study of the core(T) of the capacity functional T of S.

When $\mathcal{T}(S) \neq \emptyset$, the *Aumann integral* of S is defined to be

$$E(S) = \{EX : X \in \mathcal{T}(S)\}.$$

Remarks.

(i) Since S takes values in $\mathcal{F}(\mathbb{R}^d)$, $E(S)$ should be in $\mathcal{F}(\mathbb{R}^d)$. In a general setting (such as separable Banach space), $E(S)$ might not be a closed set, the *expectation* of S should be taken as the closure $\overline{\{EX : X \in \mathcal{T}(S)\}}$. However, on finite-dimensional spaces like $\mathbb{R}^d$, $E(S)$ is a closed subset of $\mathbb{R}^d$ (see Molchanov [79]).

(ii) The expectation of a random closed set S on $\mathbb{R}^d$ is taken to be the Aumann integral ES above when $\mathcal{T}(S) \neq \emptyset$. By analogy, we say that the random closed set S is *integrable* if $\mathcal{T}(S) \neq \emptyset$. A necessary and sufficient condition for $\mathcal{T}(S) \neq \emptyset$ is that the random variable $\omega \to \{\inf\{\|x\| : x \in S(\omega)\}$ is integrable (see Molchanov [79]).

(iii) Note that S, being a (nonempty) random closed set on $\mathbb{R}^d$, $\mathcal{S}(S) \neq \emptyset$, and moreover S admits a *Catsaing representation*, i.e., S is the closure of a sequence of its selectors.

The estimation of $E(S)$ from a random sample $S_1, S_2, \ldots, S_n$ drawn from S is typically justified by the strong law of large numbers. If we look at the sample mean of a random sample $X_1, X_2, \ldots, X_n$ of a random variable X, i.e.,

$$\frac{X_1 + X_2 + \ldots + X_n}{n},$$

then we realize that we need to define addition and multiplication (with scalars) for random sets. On the vector space $\mathbb{R}^d$, these operations can be taken as Minkowski operations for sets.

For $A, B \subseteq \mathbb{R}^d$ and $\alpha \in \mathbb{R}$, we put

$$A \oplus B = \{x + y : x \in A, y \in B\},$$

$$\alpha A = \{\alpha x : x \in A\}.$$

Note that if A and B are compact then $A \oplus B$ is also compact, but a Minkowski addition of arbitrary closed sets might not be a closed set. Thus, in a simple case, we shall consider (nonempty) *compact* random sets in the setting of laws of large numbers.

On the other hand, the convergence of sequences of random vectors can be generalized to the convergence of nonempty compact subsets of $\mathbb{R}^d$ in the sense of the *Hausdorff distance* on $\mathcal{K}(\mathbb{R}^d) \setminus \{\emptyset\}$. Let ρ denote the euclidean distance on $\mathbb{R}^d$, i.e., for $x = (x_1, x_2, \ldots, x_d)$, $y = (y_1, y_2, \ldots, y_d)$ in $\mathbb{R}^d$,

$$\rho(x, y) = \left(\sum_{i=1}^{d} (x_i - y_i)^2 \right)^{1/2}.$$

The *Hausdorff metric* H_ρ on $\mathcal{K}(\mathbb{R}^d) \setminus \{\emptyset\}$ is defined as follows. For $A, B \in \mathcal{K}(\mathbb{R}^d) \setminus \{\emptyset\}$,

$$H_\rho(A, B) = \max \left\{ \sup_{x \in A} \rho(x, B), \sup_{x \in B} \rho(x, A) \right\},$$

where $\rho(x, A) = \inf\{\rho(x, y) : y \in A\}$.

A typical law of large numbers for random sets is this (see, e.g., Artstein and Vitale [5]).

THEOREM (SLLN) Let $S_1, S_2, \ldots, S_n$ be an i.i.d. sample from a random compact set S on $\mathbb{R}^d$. Suppose that $E\|S\| < \infty$, where $\|S\|(\omega) = \sup\{\|x\| : x \in S(\omega)\}$. Then

$$\frac{1}{n}(S_1 \oplus S_2 \oplus \ldots \oplus S_n) \overset{\text{a.s.}}{\to} E(S)$$

as $n \to \infty$, in the H_ρ-sense.

As expected, *central limit theorems* for random sets should also be investigated for the purpose of large sample statistics with random set observations, in particular for coarse data analysis. We simply refer the interested readers to, e.g., Molchanov [79].

8.2 A Statistical Inference Framework for Coarse Data

Recall that the statistical problem for set-valued observations is as follows. Let X be a random vector, defined on some probability space $(\Omega, \mathcal{A}, P)$ with values in $\mathbb{R}^d$. The probability law μ_0 of X is the probability measure PX^{-1} on $\mathcal{B}(\mathbb{R}^d)$, the collection of all Borel subsets of $\mathbb{R}^d$.

Let $X_1, X_2, \ldots, X_n$ be a random sample drawn from X. The Glivenko-Cantelli theorem asserts that μ_0 can be estimated by the empirical measures dF_n, where

$$F_n(x) = \frac{1}{n}\#\{1 \leq i \leq n : X_i \leq x\}$$

is the empirical distribution based on the sample $X_1, X_2, \ldots, X_n$, where $\#$ denotes the cardinality. In other words, for n sufficiently large, μ_0 is in some neighborhood of dF_n.

Suppose that we cannot observe the X_i's directly, but instead, we observe random sets $S_1, S_2, \ldots, S_n$ with $X_i \in S_i$, $i = 1, 2, \ldots, n$. In this situation, in order to construct an estimator of μ_0 based on $S_1, S_2, \ldots, S_n$, we need to develop an appropriate model. Schreiber [113] assumed that the observed sets $S_1, S_2, \ldots, S_n$ are an i.i.d. (independently and identically distributed) sample from a random set S and the population X is an almost sure selector of S. We are going to extend Schreiber's work from finite sample spaces to *compact metric spaces*.

Thus, we consider now a random element X taking values in a compact metric space (Y, ρ), such as $[0, 1]$ with the usual euclidean metric, i.e., X is $\mathcal{A}$-$\mathcal{B}(Y)$-measurable, where $\mathcal{B}(Y)$ is the Borel-σ-field on Y generated by the metric ρ.

Recall that the core of the capacity functional T of S is defined by

$$\text{core}(T) = \{\mu \in \mathcal{M}(Y) : \mu \preceq T\},$$

where $\mathcal{M}(Y)$ is the collection of all probability measures on $\mathcal{B}(Y)$, and we write $\mu \preceq T$ if $\mu(K) \leq T(K)$ for all $K \in \mathcal{K}_0$. Here $\mathcal{K}_0$ denotes the collection of all nonempty compact subsets of Y.

It is well known that $\mathrm{core}(T) \neq \emptyset$ if Y is compact metric space. It follows that the set of almost sure selectors X of S, i.e., random elements with values in Y such that $P(X \in S) = 1$, is nonempty since $\mathrm{core}(T)$ is the set of probability laws of selectors of S (see Norberg [96]). Thus, the population X, viewed as an almost sure selector of S, is adequate. Let us elaborate a little more on this. Essentially, probabilistic models are proposed to model observed data when uncertainty is present. Depending on the type of observed data, statistical procedures are derived in order to make inference about the random phenomenon under study.

Set-valued observations are an example of a new type of data, generalizing point-valued observations in standard statistical applications. They arise in several different contexts. Traditionally, statistics of random sets was investigated to study random patterns of objects such as the Boolean model (Molchanov [73, 78]). Here, random compact sets in euclidean spaces are used to model observed sets (as a generalization of point processes), and the associated statistics is concerned with the inference about various parameters of the random patterns under study such as the expected area, the expected perimeter, and the distribution of the random set model. Note that random set data can arise even in standard framework of multivariate statistics. This is exemplified by the problem of probability density estimation using Hartigan's excess mass approach (Hartigan [50], Polonik [100]). In biostatistics, set-valued observations arise as coarse data: Heijtian and Rubin [51], Gill et al. [36]. Here, the random variable of interest X is not observable, but instead, one observes the values of some random set S containing X almost surely. From a modeling point of view, X is an almost sure selector of S, i.e., $P(X \in S) = 1$. In the above cited works in biostatistics, the emphasis is on models of S that make inference about X feasible. This is the essence of the CAR model. In a related direction, Schreiber [113] set out to investigate a general framework for inference with set-valued observations. His result were reported for the case where the random variable of interest X takes a finite number of values. As in Schreiber's work, our emphasis is on models based upon the capacity functional T of the observed random set S. As an example, consider the the simplest model in Van der Vaart [121]. Let the observed random variable be V which takes values in the compact subspace $[0, M] \subset \mathbb{R}$. The random variable of interest X is not directly observable, but $P(X \in S) = 1$ where $S(\omega) = [0, V(\omega)]$, i.e., $P(X \leq V) = 1$. The stochastic ordering $X \leq V$ (a.s.), in the context of coupling, is equivalent to $F_V(.) \leq F_X(.)$, where F_X and F_V denote the distribution functions of V and X, respectively. If X is assumed to be independent of V, the independence assumption restricts further the class of distribution functions dominating F_V, namely,

$$P(X \leq V) = \int_0^M F_X(v)dF_V(v) = 1 \tag{8.1}$$

Let Θ be the class of distribution functions F such that $F(x) = 0, \forall x < 0$, $F_V \leq F$ and satisfying (8.1). Then Θ is the parameter space for F_X. Now, in our setting, the capacity functional T of the random set $S = [0, V]$ is defined on Borel sets of $[0, M]$ by

$$T(A) = \sup\{T(K), K \subset A, K \in \mathcal{K}\} \tag{8.2}$$

where $\mathcal{K}$ denotes the class of all compact subsets of $[0, M]$, and here

$$T(K) = P(S \cap K \neq \emptyset) = P(V \geq \min K) = 1 - F_V(\min K)$$

where $\min K$ denote the greatest lower bound of the compact subset K. Thus,

$$\text{core}(T) = \{dF : dF \preceq T\} = \{dF : F \in \Theta\},$$

where, again, dF denotes the probability measure associated with the distribution function F.

Now, back to our general framework, the empirical capacity function $T^{(n)}$ based on the i.i.d. random set sample $S_1, S_2, \ldots, S_n$ is defined on $\mathcal{K}$ as

$$T^{(n)}(K) = \frac{1}{n}\#\{1 \leq i \leq n : S_i \cap K \neq \emptyset\}.$$

Clearly by the strong law of large numbers, $T^{(n)}(K) \to T(K)$ almost surely as $n \to \infty$ for any $K \in \mathcal{K}$. Note that $dF_n \in \text{core}(T^{(n)})$ a.s. for any n.

The counterpart of the empirical measure dF_n is the $\text{core}(T^{(n)})$, which is a subset of $\mathcal{M}(Y)$. Since $\mu_0 \in \text{core}(T)$, the estimation of μ_0 based on $S_1, S_2, \ldots, S_n$ is based on the approximation of $\text{core}(T)$ by $\text{core}(T^{(n)})$. We will also show that the rate of convergence of $\text{core}(T^{(n)})$ to $\text{core}(T)$ is exponential.

Remark. In applications, we assume, as in Schreiber [113], that the unknown μ_0 belongs to *a priori* known class Ξ of probability measures on $\mathcal{B}(Y)$. In the special case where $\Xi \cap \text{core}(T) = \{\mu_0\}$, the approximation of $\text{core}(T)$ by $\text{core}(T_n)$ leads to a consistent estimator of μ_0.

The following material is drawn mainly from Feng and Feng [33]. Let (Y, ρ) be a compact metric space. Denote by $\mathcal{K}_0 = \mathcal{K}(Y)$ the collection of all non-empty compact subsets of Y. Endow $\mathcal{K}_0$ with the Hausdorff metric H_ρ. It is well known that $(\mathcal{K}_0, H_\rho)$ is also a compact metric space (see, e.g., Mathéron [73]). Let $\mathcal{B}(Y)$ and $\mathcal{B}(\mathcal{K}_0)$ be the collections of Borel sets in (Y, ρ) and $(\mathcal{K}_0, H_\rho)$, respectively.

Denote by $\mathcal{M}(Y)$ the collection of all probability measures on Y, and $C(Y)$ the space of all continuous real functions on Y endowed with the uniform topology. Since Y is compact, there exists a sequence $\{f_i\}$ of continuous real functions dense in $C(Y)$. Define a metric Δ on $\mathcal{M}(Y)$ by

$$\Delta(\mu, \nu) = \sum_{i=1}^{\infty} \frac{\left| \int f_i d\mu - \int f_i d\nu \right|}{2^i \|f_i\|}, \tag{8.3}$$

where $\|f\| := \max_{y \in Y} |f(y)|$. Then the metric Δ on $\mathcal{M}(Y)$ gives the weak-star topology, and $(\mathcal{M}(Y), \Delta)$ is a compact space (see, e.g., Walters [126], Theorems 6.4–6.5).

Let $\mathcal{K}_0(\mathcal{M}(Y))$ denote the collection of all nonempty compact subsets of $\mathcal{M}(Y)$ endowed with the Hausdorff metric H_Δ, and $\mathcal{B}(\mathcal{K}_0(\mathcal{M}(Y)))$ the collection of all Borel sets in $\mathcal{K}_0(\mathcal{M}(Y))$.

Denote by $\mathcal{A}(Y)$ the class of all nonempty sets $E \subset Y$ such that

$$\{K \in \mathcal{K}_0 : K \cap E \neq \emptyset\} \in \mathcal{B}(\mathcal{K}_0).$$

We will see that the class $\mathcal{A}(Y)$ contains all the nonempty compact sets and open sets in (Y, ρ).

Recall that a nonempty compact *random set* S is a map defined on a probability space $(\Omega, \mathcal{A}, P)$ taking values in $\mathcal{K}_0$, and measurable with respect to $\mathcal{A}$–$\mathcal{B}(\mathcal{K}_0)$. The *capacity functional* of S is denoted as T_s or simply T.

First, we define a pseudometric on the space of all capacity functionals on $\mathcal{A}(Y)$. For any $\varepsilon > 0$ and $E \subset Y$, denote by $B_\varepsilon(E)$ the ε-neighborhood of E in Y. That is,

$$B_\varepsilon(E) := \{y \in Y : \exists x \in E \text{ with } \rho(x, y) < \varepsilon\}.$$

For simplicity, we denote $B_\varepsilon(y) = B_\varepsilon(\{y\})$ for $y \in Y$. A set $E \subset Y$ is called an ε-*spanning set* of Y if $B_\varepsilon(E) = Y$. By the compactness of Y, for each $\varepsilon > 0$ there exists an ε-spanning set consisting of finitely many points. For $n \geq 1$, we choose a $1/n$-spanning set H_n of Y such that H_n is a finite set. Define

$$\mathcal{O}_n = \left\{ B_{1/n}(E) : E \subset H_n \right\}, \qquad n \geq 1. \tag{8.4}$$

For two capacity functionals T and T', define

$$\Lambda(T, T') = \sum_{n \geq 1} 2^{-n - \#H_n} \sum_{W \in \mathcal{O}_n} |T(W) - T'(W)|. \tag{8.5}$$

We establish some lemmas that will be needed in subsequent analysis.

LEMMA 8.1

(i) If $\{E_n\}$ is an increasing sequence of sets in $\mathcal{A}(Y)$, then $\lim_n E_n \in \mathcal{A}(Y)$.

(ii) $E \in \mathcal{A}(Y)$ for any $E \in \mathcal{K}$.

(iii) $E \in \mathcal{A}(Y)$ for each open subset E of Y.

Proof. Note that if $\{E_n\}$ is an increasing sequence of sets in $\mathcal{A}(Y)$, then

$$\{K \in \mathcal{K} : K \cap \lim_n E_n \neq \emptyset\} = \lim_n \{K \in \mathcal{K} : K \cap E_n \neq \emptyset\}, \qquad (8.6)$$

from which (i) follows. For (ii) and (iii), see Sections 1.1, 1.2, and 2.1 of Mathéron [73] $\square$

LEMMA 8.2

(i) If $\{E_n\}$ is an increasing sequence of sets in $\mathcal{A}(Y)$, then $T\left(\lim_n E_n\right) = \lim_n T(E_n)$.

(ii) If $\{E_n\}$ is a decreasing sequence of sets in $\mathcal{K}$, then $T\left(\lim_n E_n\right) = \lim_n T(E_n)$.

Proof. To see (i), note that if $\{E_n\}$ is an increasing sequence of sets in $\mathcal{A}(Y)$,

$$T\left(\lim_n E_n\right) = P\left(S^{-1}(\{K \in \mathcal{K} : K \cap \lim_n E_n \neq \emptyset\})\right)$$

$$= P\left(\bigcup_n S^{-1}(\{K \in \mathcal{K} : K \cap E_n \neq \emptyset\})\right)$$

$$= \lim_n P\left(S^{-1}(\{K \in \mathcal{K} : K \cap E_n \neq \emptyset\})\right)$$

$$= \lim_n T(E_n).$$

For (ii), see Sections 1.1 and 1.2 of Molchanov [77]. $\square$

As a direct corollary, we get

COROLLARY 8.1 For any $E \in \mathcal{K}$, we have

$$\lim_{\varepsilon \to 0} T\left(\overline{B_\varepsilon(E)}\right) = \lim_{\varepsilon \to 0} T\left(B_\varepsilon(E)\right) = T(E).$$

Proof. It suffices to show $\lim_{n \to \infty} T\left(\overline{B_{1/n}(E)}\right) = T(E)$. To see this, note that $\left\{\overline{B_{1/n}(E)}\right\}$ is a decreasing sequence of compact sets in Y with $\lim_n \overline{B_{1/n}(E)} = E$. It follows from Lemma 8.2 that $\lim_n T\left(\overline{B_{\frac{1}{n}}(E)}\right) = T(E)$. $\square$

PROPOSITION 8.1 If $\mu \in \text{core}(T)$, then for any $E \in \mathcal{B}(Y)$, there exists $F \subset E$ with $F \in \mathcal{A}(Y)$ such that $\mu(E) \leq T(F)$.

Proof. Since Y is compact, μ is a Radon measure. Thus for any $E \in \mathcal{B}(Y)$, there exists an increasing sequence of compact sets $K_n \subset E$ with $\mu(E) = \lim_n \mu(K_n) = \mu\left(\lim_n K_n\right)$. Denote by $K = \lim_n K_n$. By Lemma 8.1, we have $K \in \mathcal{A}(Y)$. By Lemma 8.2, we have

$$T(K) = \lim_n T(K_n) \geq \lim_n \mu(K_n) = \mu(E).$$

$\square$

PROPOSITION 8.2 $\Lambda(T, T') = 0 \iff T(K) = T'(K), \forall K \in \mathcal{K}.$

Proof. First we prove "$\Longleftarrow$." Assume $T(K) = T'(K)$ for all $K \in \mathcal{K}$. For any $W \in \mathcal{O}_n$, there exists an increasing sequence of sets $K_i \in \mathcal{K}$ such that $W = \lim_i K_i$. By Lemma 8.2,

$$T(W) = \lim_i T(K_i) = \lim_i T'(K_i) = T'(W),$$

from which $\Lambda(T, T') = 0$ follows.

Now we prove "$\Longrightarrow$." Assume $\Lambda(T, T') = 0$, that is, $T(W) = T'(W)$ for any n and $W \in \mathcal{O}_n$. For any $K \in \mathcal{K}$ define $E_n = B_{1/n}(K) \cap H_n$ and $W_n = B_{1/n}(E_n)$. It can be checked that $W_n \in \mathcal{O}_n$ and $W_n \supset K$. Moreover the sequences $\{\overline{W_n}\}$ and $\{W_n\}$ are decreasing and they converge to K. Therefore

$$T(K) = \lim_n T(\overline{W_n}) = \lim_n T(W) = \lim_n T'(W) = \lim_n T'(\overline{W_n}) = T'(K).$$

$\square$

For any $n \geq 1$, let H_n be the finite $1/n$-spanning set of Y defined previously. Define $\pi_n : 2^Y \to 2^{H_n}$ by

$$\pi_n(E) = \left\{ x \in H_n : \exists y \in E \text{ with } d(x, y) < \frac{1}{n} \right\}, \qquad \forall E \subset Y. \qquad (8.7)$$

The set $\pi_n(E)$ may be considered as the projection of E onto H_n. Since H_n is a finite set, there exists a map θ_n from 2^{H_n} (the class of all subsets of H_n) to H_n such that $\theta_n(E) \in E$ for any $\emptyset \neq E \subset H_n$.

LEMMA 8.3 For each $n \geq 1$,

(i) $\pi_n|_{\mathcal{K}}$ (the restriction of π_n on $\mathcal{K}$) is measurable from $\mathcal{K}$ to H_n.

(ii) θ_n is continuous from 2^{H_n} to H_n.

(iii) For any $E \subset H_n$ and $\varepsilon > 0$, the set $\left\{ x \in Y : \pi_n\left(\overline{B_\varepsilon(x)}\right) = E \right\}$ is a measurable subset of Y.

Proof. To see (i), note that for each nonempty subset $E \subset H_n$,

$$\pi_n|_{\mathcal{K}}(K) = E \iff d(x, K) < \frac{1}{n} \text{ for any } x \in E \text{ and } d(y, K) \geq \frac{1}{n} \text{ for any } y \in H_n \backslash E,$$

Therefore

$$(\pi_n|_{\mathcal{K}})^{-1}(E) =$$

$$\left(\bigcap_{x \in E} \left\{ K \in \mathcal{K} : d(x, K) < \frac{1}{n} \right\} \right) \cap \left(\bigcap_{y \in H_n \backslash E} \left\{ K \in \mathcal{K} : d(y, K) \geq \frac{1}{n} \right\} \right).$$

Note that for each $x \in H_n$, the set $\{ K \in \mathcal{K} : d(x, K) < \frac{1}{n} \}$ is an open set in $\mathcal{K}$, while $\{ K \in \mathcal{K} : d(x, K) \geq \frac{1}{n} \}$ is closed in $\mathcal{K}$. Thus $(\pi_n|_{\mathcal{K}})^{-1}(E)$ is $\mathcal{B}(\mathcal{K})$ measurable.

The proof (ii) is trivial since the topology on finite set 2^{H_n} is the discrete topology. The proof of (iii) is similar to that of (i). To make proof more precise, note that

$$\left\{ x \in Y : \pi_n \left(\overline{B_\varepsilon(x)} \right) = E \right\} =$$

$$\left(\bigcap_{y \in E} \left\{ x \in Y : d \left(y, \overline{B_\varepsilon(x)} \right) < \frac{1}{n} \right\} \right) \cap \left(\bigcap_{z \in H_n \backslash E} \left\{ x \in Y : d \left(z, \overline{B_\varepsilon(x)} \right) \geq \frac{1}{n} \right\} \right).$$

Note that

$$\left\{ x \in Y : \rho \left(y, \overline{B_\varepsilon(x)} \right) < 1/n \right\}$$

is open in Y and $\left\{ x \in Y : \rho \left(z, \overline{B_\varepsilon(x)} \right) \geq 1/n \right\}$ is closed in Y, we obtain the desired result. □

The following result is known (see Molchanov [77], p. 102), we include a proof for convenience of the readers.

PROPOSITION 8.3 *There exists a map $q : \mathcal{K} \rightarrow Y$ such that q is $\mathcal{B}(\mathcal{K})$–$\mathcal{B}(Y)$ measurable and $q(K) \in K, \forall K \in \mathcal{K}$.*

Proof. Let H_n, π_n and θ_n $(n \geq 1)$ be defined as above. Now define a sequence $\{q_n\}_{n \geq 1}$ of maps from $\mathcal{K}$ to Y by induction as follows.

$$q_1(K) = \theta_{2^1} \left(\pi_{2^1}(K) \right), \qquad \forall K \in \mathcal{K}$$

and for any $k \geq 2$,

$$q_k(K) = \theta_{2^k} \left(\pi_{2^k}(K) \cap \pi_{2^k} \left(\overline{B_{2^{-(k-1)}} \left(q_{k-1}(K) \right)} \right) \right), \qquad \forall K \in \mathcal{K}.$$

One may show by induction that

$$\rho \left(q_k(K), K \right) \leq 2^{-k}, \quad \rho \left(q_{k+1}(K), q_k(K) \right) \leq 2^{-k} + 2^{-k+1}$$

for each $k \geq 1$ and $K \in \mathcal{K}$. Therefore we can define a map $q : \mathcal{K} \to Y$ by $q(K) = \lim_n q_n(K)$. It is clear $q(K) \in K$ for each $K \in \mathcal{K}$. Now we prove that q is $\mathcal{B}(\mathcal{K})$–$\mathcal{B}(Y)$ measurable. To show this it suffices to prove q_n is $\mathcal{B}(\mathcal{K})$–$\mathcal{B}(Y)$ measurable for each $n \geq 1$. By Lemma 8.3, q_1 is $\mathcal{B}(\mathcal{K})$–$\mathcal{B}(Y)$ measurable. Suppose q_{k-1} is $\mathcal{B}(\mathcal{K})$–$\mathcal{B}(Y)$ measurable for some $k \geq 2$. Then by Lemma 8.3, for any $F, E \subset H_n$, the set

$$\left\{ K \in \mathcal{K} : \pi_{2^k}(K) = F, \pi_{2^k} \left(\overline{B_{2^{-(k-1)}}(q_{k-1}(K))} \right) = E \right\}$$

is a Borel set in $\mathcal{B}(\mathcal{K})$. This implies that q_k is measurable. By induction we conclude that q_n is measurable for each $n \geq 1$. ☐

PROPOSITION 8.4 *Suppose* $(\Omega, \mathcal{A}, P)$ *is a probability space, and* $S : \Omega \to \mathcal{K}$, $y : \Omega \to Y$ *are* $\mathcal{A}$–$\mathcal{B}(\mathcal{K})$ *and* $\mathcal{A}$–$\mathcal{B}(Y)$ *measurable, respectively. Assume there exists* $\varepsilon > 0$ *such that* $\rho(y(\omega), S(\omega)) < \varepsilon$, $\forall \omega \in \Omega$. *Then there exists a* $\mathcal{A}$–$\mathcal{B}(Y)$ *measurable map* x *from* Ω *to* Y *such that*

$$x(\omega) \in S(\omega) \text{ and } \rho(x(\omega), y(\omega)) < 6\varepsilon, \qquad \forall \omega \in \Omega.$$

Proof. Let k_0 be the integer so that $\varepsilon \leq 2^{-k_0} < 2\varepsilon$. We construct a sequence of maps $\{q_k\}_{k \geq k_0}$ from Ω to Y by

$$q_{k_0}(\omega) = y(\omega),$$
$$q_{k_0+1}(\omega) = \theta_{2^{k_0+1}} \left(\pi_{2^{k_0+1}}(S(\omega) \cap \pi_{2^{k_0+1}} \left(\overline{B_{2^{-k_0}}(q_{k_0}(\omega))} \right) \right),$$

and

$$q_k(\omega) = \theta_{2^k} \left(\pi_{2^k}(S(\omega) \cap \pi_{2^k} \left(\overline{B_{2^{-(k-1)}}(q_{k-1}(\omega))} \right) \right)$$

for any $k \geq k_0 + 1$. It can be checked that

$$\rho(q_{k+1}(\omega), q_k(\omega)) \leq 3 \cdot 2^{-(k+1)}, \qquad \rho(q_k(\omega), S(\omega)) \leq 2^{-k} \qquad (8.8)$$

for any $k \geq k_0$.

Take $q(\omega) = \lim_k q_k(\omega)$. By (8.8), $q(\omega) \in S(\omega)$ and

$$\rho(q(\omega), y(\omega)) \leq \sum_{k \geq k_0} \rho(q_{k+1}(\omega), q_k(\omega)) \leq 3 \cdot 2^{-k_0} < 6\varepsilon.$$

By Lemma 8.3, q_k is $\mathcal{A}$–$\mathcal{B}(Y)$ measurable for any $k \geq 1$, which implies that q is $\mathcal{A}$–$\mathcal{B}(Y)$ measurable. ☐

THEOREM 8.1 *Let* Y *be a compact metric space. For each capacity functional* T *on* $\mathcal{A}(Y)$, core(T) *is a nonempty compact convex subset of* $\mathcal{M}(Y)$.

Proof.

(i) We prove that $\text{core}(T)$ is compact and convex. The convexity is trivial. To see the compactness, suppose $\mu_n \in \text{core}(T)$ and μ_n converges to μ. It suffices to show $\mu \in \text{core}(T)$. Note that for each $K \in \mathcal{K}$ and $\varepsilon > 0$

$$\mu(K) \leq \mu\left(B_\varepsilon(K)\right) \leq \limsup_n \mu_n\left(B_\varepsilon(K)\right) \leq$$

$$\limsup_n \mu_n\left(\overline{B_\varepsilon(K)}\right) \leq T\left(\overline{B_\varepsilon(K)}\right).$$

Letting $\varepsilon \downarrow 0$, by Corollary 8.1 we have $\mu(K) \leq T(K)$. Thus $\mu \in \text{core}(T)$.

(ii) We prove that $\text{core}(T)$ is nonempty. Let q be defined as in Proposition 8.3. Then $q(S(\omega))$ is a random variable taking values in Y. Denote by μ the distribution of $q(S(\omega))$ on Y, that is,

$$\mu(E) = P(\omega : q(S(\omega)) \in E), \qquad \forall E \in \mathcal{B}(Y).$$

By the definition of T one can check directly that $\mu \in \text{core}(T)$. □

Now we need to study the perturbation of $\text{core}(T)$, i.e., the way in which the $\text{core}(T)$ depends on T, in the finite case.

LEMMA 8.4 *Suppose Y is a finite set. Let T_1 and T_2 be two capacity functionals on 2^Y. Assume*

$$|T_1(E) - T_2(E)| < \delta, \qquad \forall E \subset Y, \tag{8.9}$$

Then

$$\rho_\Delta\left(\text{core}(T_1), \text{core}(T_2)\right) < \#Y \cdot 4^{\#Y} \cdot \delta.$$

Proof. Define

$$\phi_i(E) = \sum_{F \subset E, F \neq \emptyset} (-1)^{\#(E \backslash F)} T(E), \qquad \forall E \in 2^Y \backslash \{\emptyset\}, i = 1, 2.$$

By (8.9),

$$|\phi_1(E) - \phi_2(E)| < \#(2^E) \cdot \delta \leq \#(2^Y) \cdot \delta = 2^{\#Y} \cdot \delta.$$

Let $\mu_1 \in \text{core}(T_1)$. By the characterization of $\text{core}(T_1)$ (see Chapter 5), there is a map $p : 2^Y \backslash \{\emptyset\} \times Y \to \mathbb{R}^+$ satisfying

(i) $p(E, x) = 0$ if $x \notin E$;

(ii) $\sum\limits_{x \in E} p(E, x) = 1$ for any $E \in 2^Y \setminus \{\emptyset\}$.

such that

$$\mu_1(\{x\}) = \sum_{E \ni x} p(E, x)\phi_1(E), \qquad \forall x \in Y.$$

Now define a probability measure μ_2 on Y by

$$\mu_2(\{x\}) = \sum_{E \ni x} p(E, x)\phi_2(E), \qquad \forall x \in Y.$$

Using the characterization of $\text{core}(T)$ again, we know that $\mu_2 \in \text{core}(T_2)$. Since for any $x \in Y$,

$$|\mu_1(\{x\}) - \mu_2(\{x\})| \le \sum_{E \ni x} |\phi_1(E) - \phi_2(E)| \le 2^\# \cdot 2^\# \cdot \delta = 4^{\#Y}\delta,$$

by (8.3),

$$\Delta(\mu_1, \mu_2) \le \sum_{x \in Y} |\mu_1(\{x\}) - \mu_2(\{x\})| \le \#Y \cdot 4^{\#Y} \cdot \delta.$$

This implies that

$$\text{core}(T_1) \subset B_\varepsilon(\text{core}(T_2)),$$

where $\varepsilon = \#Y \cdot 4^{\#Y} \cdot \delta$. In a similar way, we can prove

$$\text{core}(T_2) \subset B_\varepsilon(\text{core}(T_1)).$$

Therefore we have $\rho_\Delta(\text{core}(T_1), \text{core}(T_2)) < \#Y \cdot 4^{\#Y} \cdot \delta.$ $\quad\square$

To study the perturbation of $\text{core}(T)$ in the case where Y is an arbitrary compact space, we use a technique to approach $\text{core}(T)$ in the following way.

Suppose Y is a compact space. T is the capacity functional of a random set $S : \Omega \to \mathcal{K}$. Let H_n, π_n and θ_n be defined before. Define a capacity $\Theta_n(T)$ on 2^{H_n} as follows:

$$\Theta_n(T)(E) = P\{\omega \in \Omega : \pi_n(S(\omega)) \cap E \ne \emptyset\}, \qquad \forall E \in H_n, \tag{8.10}$$

It is clear that

$$\Theta_n(T)(E) = T\left(B_{1/n}(E)\right), \qquad \forall E \in H_n, \tag{8.11}$$

As we know, $\Theta_n(T)$ is the capacity functional of the random set $\pi_n S$, taking values in the finite collection of all subsets of H_n. And $\text{core}(\Theta_n(T))$ is a set of probability measures on H_n. Since every probability measure on H_n can be viewed as a Borel probability measure on Y, $\text{core}(\Theta_n(T))$ can be treated as a compact subset of $\mathcal{M}(Y)$. In the following we consider the distance between $\text{core}(\Theta_n(T))$ and $\text{core}(T)$ in the Hausdorff metric H_ρ.

LEMMA 8.5 *Suppose Y is a metric compact space. Let $\{f_i\}$ be defined as in (8.3). T and $\Theta_n(T)$ are given as above. Then*

$$H_\rho\left(\text{core}(\Theta_n(T)), \text{core}(T)\right) \leq \sum_{i=1}^{\infty} \frac{C_{f_i}(6/n)}{2^i \|f_i\|}, \tag{8.12}$$

where $C_f(\varepsilon) = \sup\{|f(x) - f(y)| : d(x,y) \leq \varepsilon\}$.

Proof. The proof will be divided into two steps.

(i) $\text{core}(\Theta_n(T)) \subset B_\delta(\text{core}(T))$ with

$$\delta = \sum_{i=1}^{\infty} \frac{C_{f_i}(1/n)}{2^i \|f_i\|}.$$

To show this for each $\mu \in \text{core}(T)$, by Norberg's theorem (Chapter 5) there exists a probability space $(\Omega_1, \mathbf{S_1}, P_1)$, a random set $S_1 : \Omega_1 \to \mathcal{K}$ and a random variable $x_1 : \Omega_1 \to Y$ such that S_1 has the same distribution as S, x_1 has the distribution μ and moreover $x_1 \in S_1$ almost surely. Now define $S_2 : \Omega_1 \to 2^{H_n}$ and $x_2 : \Omega_1 \to H_n$ by

$$S_2(\omega_1) = \pi_n(S_1(\omega_1)) \text{ and } x_2(\omega_1) = \theta_n(\pi_n(x_1(\omega_1))).$$

Denote by μ_2 the distribution of x_2. Since $\Theta_n(T)$ is the capacity functional of S_2 and $x_2 \in S_2$ almost surely, $\mu_2 \in \text{core}(\Theta_n(T))$. And by (8.3),

$$\begin{aligned}
\Delta(\mu_2, \mu) &= \sum_{i=1}^{\infty} \frac{\left| \int f_i(x_2(\omega_1)) dP_1(\omega_1) - \int f_i(x_1(\omega_1)) dP_1(\omega_1) \right|}{2^i \|f_i\|} \\
&\leq \sum_{i=1}^{\infty} \frac{\int |f_i(x_2(\omega_1)) - f_i(x_1(\omega_1))| \, dP_1(\omega_1)}{2^i \|f_i\|} \\
&\leq \sum_{i=1}^{\infty} \frac{C_{f_i}(1/n)}{2^i \|f_i\|},
\end{aligned}$$

from which we get the desired result.

(ii) $\text{core}(T) \subset B_\delta(\text{core}(\Theta_n(T)))$ with

$$\delta = \sum_{i=1}^{\infty} \frac{C_{f_i}(6/n)}{2^i \|f_i\|}.$$

To show this, assume $\mu \in \text{core}(\Theta_n(T))$. Define ϕ by

$$\phi(E) = P\{\omega \in \Omega : \pi_n(S(\omega)) = E\}, \qquad \forall \, E \subset H_n.$$

By the characterization of $\text{core}(T)$, there exists a map $p : 2^{H_n} \backslash \{\emptyset\} \times H_n \to \mathbb{R}^+$ satisfying

(i) $p(E, x) = 0$ if $x \notin E$;

(ii) $\sum\limits_{x \in E} p(E, x) = 1$ for any $E \in 2^{H_n} \setminus \{\emptyset\}$.

such that

$$\mu(\{x\}) = \sum_{E \ni x} p(E, x)\phi(E), \qquad \forall x \in H_n.$$

For any $\emptyset \neq E \subset H_n$, define

$$\Omega_E = \{\omega \in \Omega : \pi_n(S(\omega)) = E\}.$$

By Lemma 8.3, $\Omega_E \in \mathcal{A}$. Construct

$$\Omega_1 = \bigcup_{\emptyset \neq E \subset H_n} \Omega_E \times \{(E, x) : x \in E\}.$$

Define a σ-algebra $\mathbf{S_1}$ such that each element of $\mathbf{S_1}$ is the finite union of elements of following form:

$$A_E \times \{(E, x)\}, \qquad E \subset H_n, x \in E, A_E \subset \Omega_E, A_E \in \mathcal{A}.$$

By Kolmogorov's consistency theorem, there is a unique probability measure on the measurable space $(\Omega_1, \mathbf{S_1})$ such that

$$P_1(A_E \times \{(E, x)\}) = P(A_E)p(E, x), \qquad E \subset H_n, x \in E, A_E \subset \Omega_E, A_E \in \mathcal{A}.$$

Define $S_1 : \Omega_1 \to \mathcal{K}$, $S_2 : \Omega_1 \to 2^{H_n}$ and $x_2 : \Omega_1 \to H_n$, respectively, by

$$S_1(\omega, E, x) = S(\omega), \quad S_2(\omega, E, x) = E, \quad x_2(\omega, E, x) = x.$$

One can check that S_1 has the same distribution as S, S_2 has the same distribution as $\pi_n S$ and that induces the capacity $\Theta_n(T)$. Further, x_2 has the distribution μ. Moreover for each $(\omega, E, x) \in \Omega_1$,

$$x \in E = \pi_n(S(\omega)),$$

hence $d(x_2(\omega, E, x), S_1(\omega, E, x)) \le 1/n$. By Proposition 8.4, there exists a $\mathbf{S_1}$–$\mathcal{B}(Y)$ measurable map $y : \Omega_1 \to Y$ such that

$$y(\omega, E, x) \in S_1(\omega, E, x), \qquad d(y(\omega, E, x), x_2(\omega, E, x)) \le \frac{6}{n}$$

for any $(\omega, E, x) \in \Omega_1$. Let μ_1 denote the distribution of y, then $\mu_1 \in \mathrm{core}(T)$. Furthermore

$$\Delta(\mu_1, \mu) = \sum_{i=1}^{\infty} \frac{\left| \int f_i(x_2(\omega, E, x))dP_1 - \int f_i(y(\omega, E, x))dP_1 \right|}{2^i \|f_i\|}$$

$$\le \sum_{i=1}^{\infty} \frac{\int |f_i(x_2(\omega, E, x)) - f_i(y(\omega, E, x))| \, dP_1}{2^i \|f_i\|}$$

$$\le \sum_{i=1}^{\infty} \frac{C_{f_i}(6/n)}{2^i \|f_i\|},$$

from which we get the desired result. □

THEOREM 8.2 *Suppose Y is a metric compact space. Let $\{f_i\}$ be a sequence of continuous real-valued functions dense in $\mathcal{C}(Y)$ and T_1 and T_2 are two capacity functionals on $\mathcal{A}(Y)$. Then for any $n \geq 1$,*

$$\rho_\Delta \left(\text{core}(T_1), \text{core}(T_2) \right) \leq$$

$$\#H_n \cdot 4^{\#H_n} \cdot \max_{W \in \mathcal{O}_n} |T_1(W) - T_2(W)| + 2 \sum_{i=1}^\infty \frac{C_{f_i}(6/n)}{2^i \|f_i\|}, \qquad (8.13)$$

where $\mathcal{O}_n$ is defined as in (8.4), and $C_f(\varepsilon) = \sup\{|f(x) - f(y)| : d(x,y) \leq \varepsilon\}$.

Proof. Fix an integer n. Let $\Theta_n(T_1)$ and $\Theta_n(T_2)$ be defined as in (8.11). By Lemma 8.5, we have

$$\rho_\Delta \left(\text{core}(\Theta_n(T_j)), \text{core}(T_j) \right) \leq \sum_{i=1}^\infty \frac{C_{f_i}(6/n)}{2^i \|f_i\|}, \qquad j = 1, 2. \qquad (8.14)$$

In Lemma 8.4, replacing Y, T_1, T_2, respectively, by $H_n, \Theta_n(T_1)$ and $\Theta_n(T_2)$ we obtain that

$$\rho_\Delta \left(\text{core} \left(\Theta_n(T_1) \right), \text{core} \left(\Theta_n(T_2) \right) \right) \leq$$

$$\#H_n \cdot 4^{\#H_n} \cdot \max_{E \subset H_n} |\Theta_n(T_1)(E) - \Theta_n(T_2)(E)| =$$

$$\#H_n \cdot 4^{\#H_n} \cdot \max_{W \in \mathcal{O}_n} |T_1(W) - T_2(W)|.$$

Combining this with (8.14), we have

$$\rho_\Delta \left(\text{core}(T_1), \text{core}(T_2) \right) \leq$$

$$\sum_{j=1}^2 \rho_\Delta \left(\text{core}(\Theta_n(T_j)), \text{core}(T_j) \right) + \rho_\Delta \left(\text{core} \left(\Theta_n(T_1) \right), \text{core} \left(\Theta_n(T_2) \right) \right) \leq$$

$$\#H_n \cdot 4^{\#H_n} \cdot \max_{W \in \mathcal{O}_n} |T_1(W) - T_2(W)| + 2 \sum_{i=1}^\infty \frac{C_{f_i}(6/n)}{2^i \|f_i\|}.$$

□

COROLLARY 8.2 *Suppose Y is a compact space. Let T_k $(k \geq 1)$ and T be capacity functionals on $\mathcal{A}(Y)$. If $\lim_k \Lambda(T_k, T) = 0$, then*

$$\lim_k \rho_\Delta (\text{core}(T_k), \text{core}(T)) = 0.$$

By the definition of the Hausdorff metric, under the condition of the above result, for any $\mu \in \text{core}(T)$, there exists a sequence $\mu_k \in \text{core}(T_k)$, $k \geq 1$, such that $\Delta(\mu_k, \mu) \to 0$ and thus μ_k converges to μ weakly.

The above results have an important application in the analysis of the convergence property of empirical capacity functionals.

Now suppose T is the capacity functional on $\mathcal{A}(Y)$ of a random set $S : \Omega \to \mathcal{K}_0$. And $\{S_n\}$ is a sequence of i.i.d. random sets with the same distribution as S. For each $\omega \in \Omega$, define a sequence of set functions $\chi_i(\omega, \cdot)$ on $\mathcal{A}(Y)$ by

$$\chi_i(\omega, E) = \begin{cases} 1, & \text{if } S_i(\omega) \cap E \neq \emptyset, \\ 0, & \text{otherwise.} \end{cases} \qquad \forall E \in \mathcal{A}(Y), \qquad (8.15)$$

and define $T_\omega^{(n)}$ ($n \in \mathbb{N}$) on $\mathcal{A}(Y)$ by

$$T_\omega^{(n)}(E) = \frac{1}{n} \sum_{i=1}^n \chi_i(\omega, E) = \frac{1}{n} \#\{1 \leq i \leq n, S_i(\omega) \cap E \neq \emptyset\}. \qquad (8.16)$$

$T_\omega^{(n)}$ is called the nth *empirical capacity functional* based on $\{S_n\}_{n \geq 1}$. In fact, for each ω, $T_\omega^{(n)}$ is really the capacity functional of a random set. To see this, define a probability space $(\Omega_n, 2^{\Omega_n}, P_n)$ by $\Omega_n = \{1, 2, \ldots, n\}$ and

$$P_n(E) = \frac{\#E}{n}, \qquad \forall E \subset \Omega_n.$$

The random set $V_\omega : \Omega_n \to \mathcal{K}$ is defined by

$$V_\omega(i) = S_i(\omega).$$

One may check that

$$T_\omega^{(n)}(E) = P_n\{i : V_\omega(i) \cap E \neq \emptyset\}, \qquad \forall E \in \mathcal{A}(Y).$$

Moreover if X is a selector of S, i.e., X is a random variable on Ω taking values in Y and $P(X \in S) = 1$. And X_i is a sequence of i.i.d. random variables with the same distribution as X and $P(X_i \in S_i) = 1$. The empirical measure based on $X_1, \ldots, X_n$ is

$$dF_n(\omega) = \frac{1}{n} \sum_{i=1}^n \delta_{x_i(\omega)},$$

where δ_y denotes the Dirac measure at y. Set

$$v_\omega(i) = X_i(\omega).$$

v_ω is a random variable on Ω_n which has the distribution $dF_n(\omega)$. Since $P(\omega \in \Omega : v_\omega \in S_i(\omega)) = 1$, we have

$$dF_n(\omega) \in \text{core}(T_\omega^{(n)}), \qquad a.s.\omega.$$

LEMMA 8.6   Fix an integer $m \geq 1$. For any $\delta > 0$, there exists N_δ and $R_\delta > 0$ such that

$$P\left\{\omega \in \Omega : \sum_{W \in \mathcal{O}_m} \left|T_\omega^{(n)}(W) - T(W)\right|^2 > \delta^2\right\} \leq e^{-nR_\delta}$$

for all $n \geq N_\delta$.

Proof. For any $i \geq 1$, let $\chi_i(\omega, W)$ be defined as in (8.15). Denote by $Y_i(\omega)$ the $\#\mathcal{O}_m$ dimensional vector $(\chi_i(\omega, W))_{W \in \mathcal{O}_m}$ indexed by $W \in \mathcal{O}_m$. It is clear that $\{Y_i\}$ is a sequence of i.i.d random vector taking values in $\mathbb{R}^{\#\mathcal{O}_m}$, with a distribution supported on finitely many points. Note that

$$\frac{1}{n}\sum_{i=1}^{n} Y_i(\omega) = \left(T_\omega^{(n)}(W)\right)_{W \in \mathcal{O}_m}, \qquad E(Y_1) = (T(W))_{W \in \mathcal{O}_m}.$$

Applying the classical Cramer Principle to the i.i.d random vector $\{Y_i\}$ (see, e.g., Theorem 3.5.1 of Dupuis and Ellis [25], p. 87), we get the desired result. See also the Appendix for *large deviations principle*. □

THEOREM 8.3   Suppose Y is a compact metric space, T is the capacity functional on $\mathcal{A}(Y)$ of a random set $S : \Omega \to \mathcal{K}_0$, and $\{T_\omega^{(n)}\}_{n \geq 1}$ is a sequence of empirical capacity functionals generated by a sequence of i.i.d. random sets $\{S_n\}_{n \geq 1}$ (with the same distribution as S). Then $\lim_{n} \rho_\Delta(\mathrm{core}(T_\omega^{(n)}), \mathrm{core}(T)) = 0$ almost surely. Moreover, for any $\varepsilon > 0$, there exist n_ε and $L_\varepsilon > 0$ such that

$$P\left\{\omega \in \Omega : \rho_\Delta\left(\mathrm{core}(T_\omega^{(n)}), \mathrm{core}(T)\right) > \varepsilon\right\} \leq e^{-nL_\varepsilon}$$

for all $n \geq n_\varepsilon$.

Proof. Fix $\varepsilon > 0$. Choose a large integer m such that

$$\sum_{i=1}^{\infty} \frac{C_{f_i}\left(\dfrac{6}{m}\right)}{2^i \|f_i\|} < \frac{\varepsilon}{4}.$$

Set

$$\delta = \frac{\varepsilon}{2\#H_m 4^{\#H_m}}.$$

By Lemma 8.6, there exists n_δ and $R_\delta > 0$ such that

$$P\left\{\omega \in \Omega : \sum_{W \in \mathcal{O}_m} \left|T_\omega^{(n)}(W) - T(W)\right|^2 > \delta^2\right\} \leq e^{-nR_\delta}$$

for all $n \geq n_\delta$.

By Theorem 8.2,

$$\rho_\Delta\left(\mathrm{core}\left(T_\omega^{(n)}\right), \mathrm{core}(T)\right) > \varepsilon \implies \max_{W \in \mathcal{O}_m} |T_\omega^{(n)}(W) - T(W)| \geq \delta.$$

Therefore

$$P\left\{\omega \in \Omega : \rho_\Delta\left(\mathrm{core}(T_\omega^{(n)}), \mathrm{core}(T)\right) > \varepsilon\right\} \leq e^{-nR_\delta}$$

for all $n \geq n_\delta$. Define $n_\varepsilon = N_\delta$ and $L_\varepsilon = R_\delta$. Then the above fact implies the desired result.   □

We remark that under the conditions of the above theorem, for any $\mu_0 \in \mathrm{core}(T)$ and for almost all $\omega \in \Omega$, there exists a sequence $\mu_n(\omega) \in \mathrm{core}(T_\omega^{(n)})$, $n \geq 1$, such that $\mu_n(\omega)$ converges to μ in the weak-star topology. Moreover, the above $\mu_n(\omega)$ can clearly be constructed using the steps in the proofs of Lemma 8.4 and Lemma 8.5.

Let Ξ be a nonempty subset of $\mathcal{M}(Y)$ considered as a statistical model in Schreiber [113]. For any capacity functional T on $\mathcal{A}(Y)$, define

$$\widehat{\Delta}(T|\Xi) = \inf\{\Delta(\mu, \nu) : \mu \in \mathrm{core}(T), \nu \in \Xi\}.$$

For two capacity functionals T_1 and T_2, it is easy to check that

$$\left|\widehat{\Delta}(T_1|\Xi) - \widehat{\Delta}(T_2|\Xi)\right| \leq \rho_\Delta(\mathrm{core}(T_1), \mathrm{core}(T_2)).$$

COROLLARY 8.3   *Under the conditions of above, we have for any* $\emptyset \neq \Xi \subset \mathcal{M}(Y)$, $\lim_n \widehat{\Delta}(T_\omega^{(n)}|\Xi) = \widehat{\Delta}(T|\Xi)$ *almost surely. Moreover, for any* $\varepsilon > 0$, *there exist* n_ε *and* $L_\varepsilon > 0$ *such that*

$$P\left(\omega \in \Omega : \left|\widehat{\Delta}(T_\omega^{(n)}|\Xi) - \widehat{\Delta}(T|\Xi)\right| > \varepsilon\right) \leq e^{-nL_\varepsilon}$$

for all $n \geq n_\varepsilon$.

8.3   A Related Statistical Setting

The analysis in the previous section follows closely the work of Schreiber [113] in the finite case, leading to the obtainment of the rate of convergence via the large deviations principle. Now, since the space $\mathcal{F}(\mathbb{R}^d)$ is a *separable metric space*, the statistical inference with random closes sets can be carried out in the setting of such topological spaces. We elaborate this setting in this section when we evoke the concept of *Choquet weak convergence* of capacity functionals.

Let (U, ρ) be a *compact subspace of* $\mathbb{R}^d$. Let S be a nonempty random closed (compact) set with values in $\mathcal{K}_0(U) = \mathcal{K}(U) \setminus \{\emptyset\}$. The relative hit-or-miss topology $\mathcal{T}$ on $\mathcal{K}_0$ is equivalent to the topology generated by the Hausdorff metric H_ρ. The metric space $(\mathcal{K}_0, H_\rho)$ is separable (compact). Let T denote the capacity functional of S. In the context of coarse data analysis, S is a coarsening scheme of our random vector of interest X whose probability law lies in $\mathrm{core}(T)$ which is a nonempty compact (convex) subset of $\mathcal{M}(U)$, the space of all probability measures on the Borel σ-filed $\mathcal{B}(U)$ of U. As in Section 7.1, the space $\mathcal{M}(U)$ is topologized by the Prohorov metric Δ.

Let $S_1, S_2, \ldots, S_n$ be an i.i.d. random sample from S. Let $T^{(n)}$ denote the empirical capacity functional based on $S_1, S_2, \ldots, S_n$, i.e.,

$$T^{(n)}(K) = \frac{1}{n} \# \{1 \leq i \leq n : S_i \cap K \neq \emptyset\}.$$

By the strong law of large numbers, $T^{(n)} \to T$ almost surely on $\mathcal{K}$ and hence (see Section 7.2), $Q^{(n)} \to Q$ weakly, where $Q^{(n)}$, Q denote the probability measures associated with $T^{(n)}$, T, respectively. In view of Section 7.3, we have that $Q^{(n)} \xrightarrow{W} Q \Leftrightarrow T^{(n)} \to T$ in the Choquet weak sense.

The Hausdorff metric on $\mathcal{K}(\mathcal{M}(U))$ generated by the Prohorov metric Δ on $\mathcal{M}(U)$ is denoted as H_Δ. We are going to prove that

$$\lim_{n \to \infty} H_\Delta(\mathrm{core}(T^{(n)}), \mathrm{core}(T)) = 0, \quad a.s.,$$

implying that there exist $\mu_n \in \mathrm{core}(T^{(n)})$, $n \geq 1$, such that $\Delta(\mu_n, \mu) \to 0$, a.s.

More generally, with (U, ρ) being a compact metric subspace of $\mathbb{R}^d$, let T_n, T be capacity functionals on U, such that $T(U) = T_n(U) = 1$, $n \geq 1$, then

$$H_\Delta(\mathrm{core}(T_n), \mathrm{core}(T)) = 0 \text{ if and only if } T_n \to T \text{ in Choquet weak sense.}$$

The following results will allow us to carry on the above program.

First, in view of *Skorohod representation* (of random closed sets) and related topological properties, we have:

LEMMA 8.7 *Let T, T_n, $n \geq 1$, be capacity functionals on U such that $T_n \xrightarrow{C-W} T$, then there exists a probability space $(\Omega, \mathcal{A}, P)$ on which are defined random closed sets S, S_n, $n \geq 1$, having T, T_n as capacity functionals, respectively, such that $S_n \to S$, a.s., in the Hausdorff metric H_ρ of $\mathcal{K}(U)$.*

LEMMA 8.8 *Let $S : \Omega \to \mathcal{F}_0(\mathbb{R}^d)$ and $X : \Omega \to \mathbb{R}^d$ be random elements. Define the projection of X on S as*

$$\pi_S(X) : \Omega \to \text{power set of } \mathbb{R}^d$$

by

$$\pi_S(X)(\omega) = \{x \in \mathbb{R}^d : \rho(x, X(\omega)) = \rho(X(\omega), S(\omega))\}.$$

Then $\pi_S(X)$ is a *nonempty random closed set.*

Proof. For each ω, $S(\omega)$ is a nonempty closed set of $\mathbb{R}^d$. Thus, $\rho(X(\omega), S(\omega))$ is attained, and hence $\pi_S(X)(\omega)$ is a nonempty closed set. We need to verify that $\pi_S(X)$ is $\mathcal{A}$-$\mathcal{B}(\mathcal{F})$-measurable.

Let $h(\omega, y) = \rho(y, X(\omega)) - \rho(X(\omega), S(\omega))$. h is $\mathcal{A} \otimes \mathcal{B}(\mathcal{F})$-measurable since it is measurable in ω and continuous in y. Note that

$$\pi_S(X)(\omega) = S(\omega) \cap \{y : h(\omega, y) = 0\}.$$

Now $W(\omega) = \{y \in \mathbb{R}^d : h(\omega, y) = 0\}$ is a random closed set on $\mathbb{R}^d$ because it can be written as

$$W(\omega) = (J \circ L)(\omega),$$

where $L(\omega) = (X(\omega), S(\omega))$ and

$$J : \mathbb{R}^d \times \mathcal{F}_0(\mathbb{R}^d) \to \mathcal{K}_0(\mathbb{R}^d) :$$

$$J(x, F) = \{y \in \mathbb{R}^d : \rho(x, y) = \rho(x, F)\},$$

and by noting that

$$H_\rho(J(x_1, F_1), J(x_2, F_2)) \leq \rho(x_1, x_2) + |\rho(x_1, F_1) - \rho(x_2, F_2)|,$$

implying that J is continuous.

The result follows from the following lemma:

LEMMA 8.9 *The intersection of two random closed sets is a random closed set.*

Proof. Let $S, W : (\Omega, \mathcal{A}, P) \to (\mathcal{F}(\mathbb{R}^d), \mathcal{B}(\mathcal{F}))$ be two random closed sets on $\mathbb{R}^d$. Clearly the mapping $\omega \to (S(\omega), W(\omega))$ from $(\Omega, \mathcal{A})$ into

$$(\mathcal{F}(\mathbb{R}^d) \times \mathcal{F}(\mathbb{R}^d), \mathcal{B}(\mathcal{F}) \otimes \mathcal{B}(\mathcal{F}))$$

is measurable. Moreover, the mapping $\varphi : (F_1, F_2) \to F_1 \cap F_2$ from $\mathcal{F}(\mathbb{R}^d) \times \mathcal{F}(\mathbb{R}^d)$ into $\mathcal{F}(\mathbb{R}^d)$ is upper-semicontinuous (i.e., $\varphi^{-1}(\mathcal{F}^K)$ is open for any $K \in \mathcal{K}(\mathbb{R}^d)$), and hence measurable. □

LEMMA 8.10 *Let $S : \Omega \to \mathcal{F}_0(\mathbb{R}^d)$ and $X : \Omega \to \mathcal{F}_0(\mathbb{R}^d)$ be random elements. Then there exists a random vector $Y : \Omega \to \mathbb{R}^d$ such that*

(i) $Y \in S$, a.s.;

(ii) $\rho(X(\omega), Y(\omega)) = \rho(X(\omega), S(\omega))$, a.s.

Proof. It suffices to take Y as a selector of the nonempty random closed set $\pi_S(X)$, which is also a selector of S.   □

PROPOSITION 8.5 Let $S : \Omega \to \mathcal{F}_0(\mathbb{R}^d)$ with capacity functional T, and $X : \Omega \to \mathbb{R}^d$ be a random vector with the probability law μ on $\mathcal{B}(\mathbb{R}^d)$. Then

$$\Delta(\mu, core(T)) \leq \alpha_\rho(X, S).$$

Proof. Let Y be a selector of S as in Lemma 8.10. Then the probability law P_Y of Y is in the $core(T)$. In view of the dominance of the Ky Fan metric over the Prohorov metric (Section 7.1), we have

$$\Delta(\mu, core(T)) \leq \Delta(\mu, P_Y) \leq \alpha_\rho(X, Y) \leq \alpha_\rho(\{X\}, S),$$

where

$$\alpha_\rho(X, Y) = \inf\{\varepsilon > 0 : P(\rho(X, Y) > \varepsilon) \leq \varepsilon\} =$$
$$\inf\{\varepsilon > 0 : P(\rho(X, S) > \varepsilon) \leq \varepsilon\} \leq \alpha_\rho(\{X\}, S)$$

with α_ρ standing for the Ky Fan metric on the space $\mathcal{L}(\Omega, \mathcal{F}_0(\mathbb{R}^d))$.   □

PROPOSITION 8.6 Let (U, ρ) be a compact subspace of $\mathbb{R}^d$. Let S and S' be two nonempty compact random sets on U, with capacity functionals T, T', respectively. Then,

$$H_\Delta(core(T), core(T')) \leq \alpha(S, S'),$$

where α is the Ky Fan metric on $\mathcal{L}(\Omega, \mathcal{K}_0(\mathbb{R}^d))$.

Proof. For each $\mu \in core(T)$, there is a selection Y of S' with law μ. By Lemma 8.10, there exists a selector X of S such that $\rho(X(\omega), Y(\omega)) = \rho(X(\omega), S(\omega))$, a.s., so that

$$\rho(X(\omega), Y(\omega)) = \rho(X(\omega), S(\omega)) \leq H_\rho(S(\omega), S'(\omega)).$$

Let ν denote the law of X. Then $\nu \in core(T)$ and hence, by Proposition 8.5,

$$\Delta(\mu, core(T)) \leq \Delta(\mu, \nu) \leq \alpha_\rho(X, Y) \leq \alpha(S, S').$$

Similarly, for each $\upsilon \in core(T)$, we have

$$\Delta(\upsilon, core(T')) \leq \alpha(S, S'),$$

and hence

$$H_\Delta(core(T), core(T')) \leq \alpha(S, S').$$

          □

THEOREM 8.4 *Let* (U, ρ) *be a compact subspace of* $\mathbb{R}^d$ *and* T, T_n, $n \geq 1$, *be capacity functionals on* U *such that* $T_n(U) = T(U) = 1$, $\forall n \geq 1$. *Then* $\lim\limits_{n \to \infty} H_\Delta(\mathrm{core}(T_n), \mathrm{core}(T)) = 0$ *if and only if* $T_n \xrightarrow{C-W} T_\infty$.

Proof.

a) *Sufficiency.* Note that $\mathcal{F}$, with the hit-or-miss topology, is compact and metrizable. Let δ be a metric on $\mathcal{F}$. By using the fact that $T_n \xrightarrow{C-W} T$ if and only if $Q_n \xrightarrow{W} Q$ and Ethier and Kurtz ([29], Theorem 3.1, p. 108), we obtain $T_n \xrightarrow{C-W} T$ iff $\Delta(Q_n, Q) \to 0$. By virtue of Skorohod's Representation theorem (see Theorem 1.8 in Ethier and Kurtz [29]), there exists a probability space $(\Omega, \mathcal{A}, \nu)$ on which are defined $\mathcal{F}$-valued random variables S_n, $n = 1, 2, \ldots$, and S with distributions Q_n, $n = 1, 2, \ldots$, and Q, respectively, such that $\lim\limits_{m \to \infty} S_m = S$ in $(\mathcal{F}, \delta)$ a.s. Now we need to prove that S_n, $n = 1, 2, \ldots$, and S have capacity functionals T_n, $n = 1, 2, \ldots$, and T, respectively.

To see this, simply note that, for each $K \in \mathcal{K}$, the Choquet's theorem implies

$$\nu\{\omega : S_n(\omega) \cap K \neq \varnothing\} = \nu\{\omega : S_n(\omega) \in \mathcal{F}_K\} = Q_n(\mathcal{F}_K) = T_n(K),$$

for $n \geq 1$. Next step is to show that all S_n, $n = 1, 2, \ldots$, and S are a.s. nonempty *random compact sets* and $\lim\limits_{n \to \infty} S_n = S$ a.s. in $(\mathcal{K}_0, H_\rho)$. We note $\mathcal{F} = \mathcal{K}$ by compactness of U. Keeping in mind that $T_n(U) = T(U) = 1$, we can easily obtain $\nu\{S_n = \emptyset\} = \nu\{S = \emptyset\} = 0$. By virtue of Proposition 1-4-4 in Mathéron [73], the topology on $\mathcal{K}_0$ defined by the Hausdorff metric H_ρ is equivalent to the relative hit-or-miss topology on $\mathcal{K}_0$. That is to say, $\lim\limits_{n \to \infty} S_n = S$ in $(\mathcal{K}_0, \delta|_{\mathcal{K}_0})$ if and only if $\lim\limits_{n \to \infty} S_n = S$ in $(\mathcal{K}_0, H_\rho)$. Hence $\lim\limits_{n \to \infty} \alpha_{H_\rho}(S_n, S) = 0$.

Finally, since $T_n \xrightarrow{C-W} T$ if and only if $Q_n \xrightarrow{W} Q$, by using Proposition 8.6, we have

$$0 \leq \lim_{n \to \infty} H_\Delta(\mathrm{core}(T_n), \mathrm{core}(T)) \leq \lim_{n \to \infty} \alpha_{H_\rho}(S_n, S) = 0.$$

b) *Necessity.* By using the fact that $T_n \xrightarrow{C-W} T$ if and only if $Q_n \xrightarrow{W} Q$, we only need to prove $Q_n \xrightarrow{W} Q$. Since the space $\mathcal{F}$, with the hit-or-miss topology, is a compact metric space, the space of all Borel probability measures is weakly compact. For each subsequence $\{Q_{n'}\}_{n'=1}^\infty$ of the sequence $\{Q_n\}_{n=1}^\infty$, there exists a further subsequence $\{Q_{n'_j}\}_{j=1}^\infty$ weakly converging to some Borel probability measure P on $\mathcal{F}$. Let R be the the corresponding capacity functional on $\mathcal{K}$ determined by P. Since $Q_{n'_j} \xrightarrow{W} P$ is equivalent to $T_{n'_j} \xrightarrow{C-W} R$, we have $\lim\limits_{j \to \infty} H_\Delta(\mathrm{core}(T_{n'_j}), \mathrm{core}(R)) = 0$. So we obtain

$H_\Delta(\mathrm{core}(T), \mathrm{core}(R)) = 0$, i.e., $\mathrm{core}(T) = \mathrm{core}(R)$. By the following Lemma 8.11, we obtain $T = R$ on $\mathcal{K}$. So, it implies $Q = P$ by the Choquet theorem, i.e., $Q_{n'_j} \xrightarrow{W} Q$. Thus $Q_n \xrightarrow{W} Q$ by Theorem 2.3 in Billingsley [10]. $\qquad\square$

LEMMA 8.11 *Let U be a compact subspace of $\mathbb{R}^d$, and T be a capacity functional with $T(U) = 1$. Then $T(K) = \max\{\mu(K) : \mu \in \mathrm{core}(T)\}$, $K \in \mathcal{K}$.*

Proof. Let $\emptyset \neq K \in \mathcal{K}$. Let S_T be a random closed set in $\mathbb{R}^m$ with *the capacity functional T*. Then we have

$$T(U) = P\{\omega : S_T(\omega) \cap U \neq \emptyset\} = 1,$$

i.e., $P\{S_T = \emptyset\} = 1 - P\{\omega : S_T(\omega) \cap U \neq \emptyset\} = 0$. Note that $\mathrm{core}(T)$ is nonempty. Choose $\mu \in \mathrm{core}(T)$. Then there exist a probability space $(\Omega, \mathcal{A}, P)$, a random compact set $S : \Omega \to \mathcal{K}$ with the same probability law as S_T and a random variable $X : \Omega \to U$ with law μ such that

$$P\{\omega : X(\omega) \in S(\omega)\} = 1.$$

Let $S_1 : \Omega \to \mathcal{K}$ defined by

$$S_1(\omega) = \begin{cases} S(\omega) \cap K, & \text{if } S(\omega) \cap K \neq \emptyset \\ \{X(\omega)\}, & \text{if } S(\omega) \cap K = \emptyset. \end{cases}$$

Then S_1 is a nonempty random compact set. Then there exists a variable $Y : \Omega \to U$ such that $Y \in S_1$ a.s. Hence $Y \in S_1 \subset S$ a.s. Let λ denote the probability distribution of Y. Then $\lambda \in \mathrm{core}(T)$. Now,

$$\lambda(K) = P\{\omega : Y(\omega) \in K\} = P\{\omega : S(\omega) \cap K \neq \emptyset\} = T(K).$$

$\qquad\square$

COROLLARY 8.4 *Let $U \subset \mathbb{R}^d$ be a compact subspace, $S_1, S_2, \ldots, S_n, \ldots$ be a random sample from a random closed set S in U. If T is the capacity functional of S, and T_n is the empirical capacity functionals based on $S_1, S_2, \ldots, S_n$, then*

$$\lim_{n \to \infty} H_\Delta(\mathrm{core}(T_n), \mathrm{core}(T)) = 0 \text{ a.s.}$$

8.4 A Variational Calculus of Set Functions

As exemplified by the excess mass approach to density estimation (Chapter 2), random sets appear as set estimators and an analog of the maximum likelihood principle in standard statistics needs algorithms to produce set estimators. We present in this section a proposed variational calculus of set functions for this purpose.

Optimizing (maximizing or minimizing) set functions can occur in many situations of scientific investigations.

Example 1: *Neyman-Pearson Lemma.* The Neyman-Pearson lemma in testing statistical hypotheses is a problem of optimization of set function S.

Let the observable X take values in $\mathbb{R}^n$. The null hypothesis H_0 specifies the distribution P_0 for X, whereas the alternative hypothesis specifies the distribution P_a for X. For simplicity, assume that

$$P_0(dx) = f_0(x)dx, \quad P_a(x) = f_a(x)dx.$$

If B is a critical region of a test, then the two types of errors are

$$\alpha = P(X \in B|H_0) = \int_B f_0(x)dx,$$

$$\beta = P(X \notin B|H_a) = \int_{B^c} f_a(x)dx.$$

The most powerful test is the one whose critical region is $B_* \in \mathcal{B}(\mathbb{R}^n)$ where B_* is the solution of the following problem:

$$\text{Find } B_* \text{ in } \{B \in \mathcal{B}(\mathbb{R}^n : \int_B f_0(x)dx \le \alpha\} = \mathcal{C}$$

$$\text{so that } \int_{B_*} f_0(x)dx = \alpha$$

$$\text{and } \int_{B_*^c} f_a(x)dx \le \int_{B^c} f_a(x)dx, \quad \forall B \in \mathcal{C},$$

which requires maximizing the set function $B \in \mathcal{B}(\mathbb{R}^n) \to \int_B f_a(x)dx$ subject to $B \in \mathcal{C}$.

Example 2: *Bayes tests between capacities* (Huber and Strassen [54]). Consider the following situation in robust statistics. Let $\mathcal{M}$ denote the class of all probability measures on a complete, separable and metrizable space. Consider testing the null hypothesis H_0 against the alternative H_a, where

$$H_0 = \{P \in M : P \le T_0\}$$
$$H_a = \{P \in M : P \le T_1\}$$

where T_0, T_1 are 2-alternating capacities.

Let A be a critical region for the above test, i.e., rejecting H_0 if $x \in A$ is observed. Then the upper probability of falsely rejecting H_0 is $T_0(A)$, and of falsely accepting H_1 is $T_1(A')$. If we assume that H_0 is true with prior probability $\frac{t}{1+t}$, $0 \le t \le +\infty$, then the (upper) Bayes risk of A is

$$\frac{t}{1+t}T_0(A) + \frac{1}{1+t}T_1(A') = \frac{t}{1+t}T_0(A) + \frac{1}{1+t}(1 - V_1(A)),$$

where V_1 is the conjugate of T_1, i.e., $V_1(A) = 1 - T_1(A')$. Thus, one is led to minimizing the set function

$$A \to tT_0(A) - V_1(A), \text{ for } t \text{ fixed.}$$

Example 3: *Dividing a territory.* Let U be an open bounded set of $\mathbb{R}^d$, and $v_i : U \to \mathbb{R}^+$, $i = 1, 2$, continuous. Find a closed set $F \subseteq U$ such that

$$\int_F v_1(x)dx \cdot \int_{U\setminus F} v_2(x)dx$$

is maximal among all closed subsets of U.

Example 4: *Statistics of random sets.* The excess mass approach to density estimation (see Chapter 2) leads to the maximization of the set function

$$A \in \mathcal{C} \to \mathcal{E}_{\alpha,n}(A) = (F_n - \alpha\tau)(A).$$

As far as we know, there seems to be no appropriate available variational calculus for the above type of problems. As such, below is our attempt to provide a tool for solving optimization problems of set functions. We are going to define a concept of derivative DF for a set function F, defined on some class of subsets of a set U so that when $DF(A) = 0$, $F(A)$ will be a stationary value for F. Although this can be formulated in a fairly general context, we restrict ourself to the concrete case of $U = \mathbb{R}^d$, and the domain of F is $\mathcal{B}(\mathbb{R}^d)$, i.e., $F : \mathcal{B} \to \mathbb{R}$.

In order to extend the classical notion of derivatives of functions to set functions, we choose to use the symmetric form of derivatives (also called Schwartz derivatives).

For $f : \mathbb{R} \to \mathbb{R}$, defined in some neighborhood of x, and provided that the limit exists, the symmetric (Schwartz) derivative of f at x is by definition

$$f'(x) = \lim_{h \to 0} \frac{f(x+h) - f(x-h)}{2h}.$$

Remark. This type of symmetric derivatives is useful in numerical estimation since its truncation error is of order of $O(h^2)$. Recall that if

$\lim\limits_{h\to 0} \dfrac{f(x+h)-f(x)}{h}$ exists, then $f'(x)$ leads to the same value. However, the converse is not true: it is possible that f does not have a derivative in the usual sense, but yet it has a derivative $f'(x)$ above.

Our second idea about extending $f'(x)$ to the setting of set functions is based upon an analogy with partial derivative of a function $f(x_1, \ldots, x_n)$ of several variables, we need to specify two things: the point $(x_1, \ldots, x_n)$ at which the derivative is computed, and which variable x_i (among $x_1, \ldots, x_n$) over which we differentiate. Thus, for $F : \mathcal{B} \to \mathbb{R}$, the *analogy* is this. Each $A \in \mathcal{B}$ is characterized by its indicator function $1_A : \mathbb{R}^d \to \{0, 1\}$. To determine A from 1_A we need to evaluate $1_A(t)$ for all $t \in \mathbb{R}^d$. Viewing the $1_A(t)$'s as "variables," we need also to specify a particular variable $1_A(t)$, or equivalently a $t \in \mathbb{R}^d$. Therefore we are led to consider the notion of derivative of F at (A, t) for $A \in \mathcal{B}$, $t \in \mathbb{R}^d$. See Nguyen and Kreinovich [85].

DEFINITION 8.1   *Let $F : \mathcal{B}(\mathbb{R}^d) \to \mathbb{R}$, and τ be the Lebesgue measure on $\mathcal{B}$. The derivative of F at (A, t), for $A \in \mathcal{B}$, $t \in \mathbb{R}^d$ is defined to be*

$$DF(A, t) = \lim_{H \to \{t\}} \frac{F(A \cup H) - F(A \cap H')}{\tau(H)}$$

provided the limit exists, where the limit is taken over H such that $\tau(H) \neq 0$, and $H \to \{t\}$ in the Hausdorff metric sense.

Remark. If $\|\cdot\|$ denotes the usual norm on $\mathbb{R}^d$, then the (extended, pseudo) Hausdorff metric on nonempty subsets of $\mathbb{R}^d$ is

$$d(A, B) = \max\{\sup_{a \in A} \inf_{b \in B} \|a - b\|, \ \sup_{b \in B} \inf_{a \in A} \|a - b\|\}.$$

Clearly, the definition of $DF(A, t)$ is inspired from $f'(x)$ in which its numerator $f(x + h) - f(x - h)$ is replaced by $F(A \cup H) - f(A \setminus H)$, whereas its denominator $2h$ (the length of $[-h, h]$) is replaced by $\tau(H)$ with $\tau(H) \neq 0$. ($[-h, h] \to \{0\}$, a singleton, corresponds to $H \to \{t\}$).

Note also that if F is *additive*, then $F(A \cup H) - F(A \cap H') = F(H)$, and hence the definition of $DF(A, t)$ turns into the so-called "general derivative" $DF(t)$ (see e.g., Shilov and Gurevich [116]):

$$DF(t) = \lim_{H \to \{t\}, \ \tau(H) \neq 0} \frac{F(H)}{\tau(H)}.$$

If F is a measure that is absolutely continuous with respect to τ, then its Radon-Nikodym derivative $\dfrac{dF}{d\tau}(t)$ is equal almost everywhere to $DF(t)$.

DEFINITION 8.2   *The set function $F : \mathcal{B} \to \mathbb{R}$ is said to be differentiable if it has a derivative at every point $(A, t) \in \mathcal{B} \times \mathbb{R}^d$. F is said to be continuously* *differentiable if it is differentiable and the map $t \to DF(A,t)$ is continuous for every $A \in \mathcal{B}(\mathbb{R}^d)$.*

Examples.

(i) In the territorial division problem, the set functions F_1, F_2, defined on subset $A \subseteq U$ (an open bounded subset of $\mathbb{R}^d$), are continuously differentiable, where

$$F_1(A) = \int_A v_1(x) d\tau(x), \quad F_2(A) = \int_U v_2(x) d\tau(x) - \int_A v_2(x) d\tau(x)$$

with

$$DF_1(A,t) = v_1(t), \quad DF_2(A,t) = -v_2(t).$$

And hence $F(A) = F_1(A)F_2(A)$ is also continuously differentiable with

$$DF(A,t) = v_1(t)F_2(A) - v_2(t)F_1(A).$$

(ii) The excess mass set function (see Chapter 2)

$$\mathcal{E}_\alpha(A) = (F - \alpha\tau)(A)$$

is continuously differentiable if the density $f(x) = \dfrac{dF}{d\tau}(x)$ is continuous. As such,

$$D\mathcal{E}_\alpha(A,t) = f(t) - \alpha.$$

In the following, the interior, closure and boundary of a set A is denoted as A°, $\bar{A}$, $\delta(A)$, respectively. Of course, we assume $A^\circ \neq \emptyset$.

Observe that if F attains its maximum at A, then $DF(A,t) \geq 0$ or ≤ 0 for all $t \in A^\circ$ or all $t \in (A^c)^\circ$, respectively. Indeed, assume that A is a set at which F is maximum. For $t \in A^\circ$, the open ball $B(t,r) = \{x \in \mathbb{R}^d : ||x-t|| < r\} \subseteq A$ for r sufficiently small. Thus, $F(A \cup B(t,r)) = F(A)$ and hence $DF(A,t) \geq 0$ since $F(A)$ is a maximum value. Similarly, for $t \in (A^c)^\circ$, $B(t,r) \subseteq A^c$ for small r and hence $A \backslash B(t,r) = A$, implying that $DF(A,t) \leq 0$. The situation for minimum is dual: $DF(A,t) \geq 0$ or ≤ 0 according to $t \in A^\circ$ or $t \in (A^c)^\circ$.

The following result is useful for optimization of set functions.

THEOREM 8.5 *Let F be continuously differentiable. If F attains its maximum (or minimum) at some set A (with A° and $(A^c)^\circ$ nonempty), then necessarily $DF(A,t) = 0$ for all $t \in \delta(A^\circ) \cap \delta((A^c)^\circ)$.*

Proof. Suppose F attains its maximum at A. For $t \in \delta(A^\circ)$, t is limit of a sequence $t_n \in A^\circ$ with $DF(A, t_n) \geq 0$, for each n, as observed above. Thus, $DF(A, t) \geq 0$, by continuity. Similarly, for $t \in \delta((A^c)^\circ)$, t is limit of a sequence $s_n \in (A^c)^\circ$ with $DF(A, s_n) \leq 0$ for each n, and hence $DF(A, t) \leq 0$. $\square$

For applications of the above theorem to examples mentioned at the beginning of this section, see Nguyen and Kreinovich [85].

8.5   Exercises

8.1 Let S be a random closed set on $\mathbb{R}^d$. Show that the intersection of the set of all a.s. selectors $\mathcal{S}(S)$ of S with $L^1(\Omega, \mathcal{A}, P)$ is a closed subset of $L^1(\Omega, \mathcal{A}, P)$.

8.2 Let S be a random closed set on $\mathbb{R}^d$. Let $Y(\omega) = \sup\{\|x\| : x \in S(\omega)\}$, where $\|x\|$ is the norm of $x \in \mathbb{R}^d$.

(i) Show that Y is a random variable.

(ii) Show that if Y is integrable, then $\mathcal{S}(S) \neq \emptyset$ and $\mathcal{S}(S) \subseteq L^1(\Omega, \mathcal{A}, P)$.

8.3 The Hausdorff metric H_ρ on the set $\mathcal{K}_0$ of nonempty compact subsets of $(\mathbb{R}^d, \rho)$ is defined as

$$H_\rho(A, B) = \max\{\sup_{x \in A} \rho(x, B), \sup_{x \in B} \rho(x, A)\},$$

where $\rho(x, A) = \inf\{\rho(x, y) : y \in A\}$. For $\varepsilon \geq 0$ and $A \in \mathcal{K}_0$, let

$$A^\varepsilon = \{x \in \mathbb{R}^d : \rho(x, A) \leq \varepsilon\}.$$

Show that:

(i) $H_\rho(A, B) = \max\{\inf\{\varepsilon \geq 0 : A \subseteq B^\varepsilon\}, \inf\{\varepsilon \geq 0 : B \subseteq A^\varepsilon\}\}$,

(ii) $H_\rho(A, B) = \inf\{\varepsilon \geq 0 : A \subseteq B^\varepsilon, B \subseteq A^\varepsilon\}$.

8.4 Let S be a random set taking values as subsets of $\{0, 1\}$, with density $f(A) = P(S = A)$, $A \subseteq \{0, 1\}$ given by

$$f(\{0\}) = f(\{1\}) = f(\{0, 1\}) = \frac{1}{3}.$$

Specify the core(T) of its capacity functional T.

8.5 Let $S_1, \ldots, S_n$ be a random sample from a random closed set S on $[0,1]$. Let $S_1, \ldots, X_n$ be a random sample from an a.s. selector X of S. The empirical capacity functional T_n is defined as

$$T_n(K) = \frac{1}{n}\#\{i : S_i \cap K \neq \emptyset\},$$

and the empirical probability measure is

$$dF_n = \frac{1}{n}\sum_{i=1}^{n} \delta_{X_i},$$

where δ_{X_i} is the random Dirac measure at X_i.
    Show that, $\forall n \geq 1$, $dF_n \in \text{core}(T_n)$, a.s.

8.6 Let $\mathcal{C}$ denote the class of (nonempty) compact convex subsets of $\mathbb{R}^d$. Let $L(dx)$ denote the Lebesgue measure on $\mathcal{B}(\mathbb{R}^d)$. Let $\Delta_L(A, B) = L(A\Delta B)$ for $A, B \in \mathcal{C}$, and $A\Delta B = (A \cap B^c) \cup (B \cap A^c)$.
    Show that if $A_n, A \in \mathcal{C}$ and $A_n \to A$ in the Hausdorff metric H_ρ (where ρ is the euclidean metric on $\mathbb{R}^d$), then $A_n \to A$ in the metric Δ_L.

8.7 Let (U, ρ) be a metric space. Let $\mathcal{M}(U)$ denote the space of all probability measures on the Borel σ-field $\mathcal{B}(U)$ of U. For $A \subseteq U$ and $\varepsilon > 0$, let

$$A^\varepsilon = \{x \in U : \rho(x, A) < \varepsilon\}.$$

Define $\Delta_\rho : \mathcal{M}(U) \times \mathcal{M}(U) \to \mathbb{R}^+$ by

$$\Delta_\rho(\mu, \nu) = \inf\{\varepsilon > 0 : \mu(A) \leq \nu(A^\varepsilon) + \varepsilon, \text{ for all } A \in \mathcal{F}(U)\},$$

where $\mathcal{F}(U)$ is the class of closed sets of U.
    Verify that Δ_ρ is a metric on $\mathcal{M}(U)$.

8.8 (continuation of 8.7) Suppose in addition that the metric space (U, ρ) is separable. Show that $\mu_n \to \mu$ in the metric Δ_ρ if and only if $\mu_n \to \mu$ weakly (i.e., $\mu_n(A) \to \mu(A)$, $\forall A \in \mathcal{B}(U)$ such that $\mu(\partial A) = 0$).

8.9 Let U be an open bounded set of $\mathbb{R}^d$ and $f_i : U \to \mathbb{R}^+$, $i = 1, 2$, be continuous.

(i) Show that the set functions

$$A \subseteq U \to F_i(A) = \int_A f_i(x)dx, \quad i = 1, 2,$$

are continuously differentiable in the sense of Definition 8.2 of Section 8.2.

(ii) Compute the derivative of the set function

$$A \rightarrow F(A) = F_1(A)F_2(A^c).$$

(iii) Verify that the set function in (ii) attains its maximum at some subset A of the form

$$\left\{ x \in U : \frac{v_1(x)}{v_2(x)} \geq \alpha \right\},$$

for some $\alpha > 0$.

8.10 Let $f : \mathbb{R}^d \rightarrow \mathbb{R}^d$ be a continuous probability density function. Let $L(dx)$ denote the Lebesgue measure on $\mathcal{B}(\mathbb{R}^d)$. For $\alpha > 0$, the excess mass at $A \in \mathcal{B}(\mathbb{R}^d)$ is defined as

$$\mathcal{E}_\alpha(A) = \int_A f(x)dL(x) - \alpha L(A).$$

(i) Verify that the set function $\mathcal{E}_\alpha(\cdot)$ is differentiable and compute its derivative.

(ii) Verify that $\mathcal{E}_\alpha(\cdot)$ attains its maximum at some set in $\mathcal{B}(\mathbb{R}^d)$ of the form $\{x \in \mathbb{R}^d : f(x) \geq \alpha\}$.

Appendix

Basic Concepts and Results of Probability Theory

A.1 Probability Spaces

A probability space $(\Omega, \mathcal{A}, P)$ is a mathematical model for a random experiment or phenomenon, in which Ω is a set (representing outcomes of an experiment, called the *sample space*), $\mathcal{A}$ is a σ-*field* of subsets of Ω (representing events), and $P : \mathcal{A} \to [0, 1]$ is a *probability measure*. Specifically, $\mathcal{A}$ satisfies the following.

(i) $\Omega \in \mathcal{A}$

(ii) If $A \in \mathcal{A}$, then $A^c \in \mathcal{A}$,

(iii) If $A_n \in \mathcal{A}$, $n \geq 1$, then $\bigcup_{n \geq 1} A_n \in \mathcal{A}$.

When (iii) is replaced by the weaker condition: if A, B in $\mathcal{A}$ then $A \cup B \in \mathcal{A}$, the $\mathcal{A}$ is called a *field.*

The map P satisfies the following.

(a) $P(\Omega) = 1$

(b) If $\{A_n, n \geq 1\}$ is a sequence (finite or countably infinite) of pairwise disjoint elements of $\mathcal{A}$ (i.e., $A_n \cap A_m = \emptyset$, $n \neq m$), then

$$P\left(\bigcup_{n \geq 1} A_n \right) = \sum_{n \geq 1} P(A_n).$$

The property (b) is refereed to as σ-*additivity* of P. P is said to be *finitely additive* if (b) is only satisfied for finite sequences.

Let $\mathcal{C}$ be a class of subsets of a set Ω, then the smallest σ-field containing $\mathcal{C}$, denoted by $\sigma(\mathcal{C})$, is called the σ-*field generated by* $\mathcal{C}$. Note that the class of all σ-fields containing any $\mathcal{C}$ is not empty since the power set $\mathcal{P}(\Omega)$ of Ω is such a σ-field. Also, arbitrary intersection of σ-fields is a σ-field. The partial order relation on σ-fields is inclusion. Thus $\sigma(\mathcal{C})$ is the intersection of all σ-fields containing $\mathcal{C}$.

CARATHEODORY THEOREM (extension of probability measures) *Let $\mathcal{C}$ be a field of subsets of an abstract space Ω, and $P : \mathcal{C} \to [0, 1]$ such that*

(i) $P(\Omega) = 1$,

(ii) For $A_n \in \mathcal{C}$, $n \geq 1$, pairwise disjoint, and $\bigcup_{n \geq 1} A_n \in \mathcal{C}$,

$$P\left(\bigcup_{n \geq 1} A_n\right) = \sum_{n \geq 1} P(A_n)$$

(i.e., P is σ-additive on $\mathcal{C}$),

then P is uniquely extended to a probability measure on $\sigma(\mathcal{C})$.

The following result is useful for checking whether a finitely additive set function on a σ-field is a probability measure.

THEOREM *Let $(\Omega, \mathcal{A})$ be a measurable space, and $P : \mathcal{A} \to [0, 1]$ be a finitely additive set function with $P(\Omega) = 1$. Then the following statements are equivalent.*

(i) P is a probability measure,

(ii) If $\{A_n, n \geq 1\}$ is a nondecreasing sequence of elements of $\mathcal{A}$, then

$$\lim_{n \to \infty} P(A_n) = P\left(\bigcup_{n \geq 1} A_n\right)$$

(iii) If $\{A_n, n \geq 1\}$ is a nonincreasing sequence of elements of $\mathcal{A}$, then

$$\lim_{n \to \infty} P(A_n) = P\left(\bigcap_{n \geq 1} A_n\right)$$

(iv) P is continuous at $\emptyset$, i.e., if $A_n \searrow \emptyset$, then $\lim_{n \to \infty} P(A_n) = 0$.

The continuity at $\emptyset$ of an additive probability measure P on a σ-field $\mathcal{A}$ follows from the *finite intersection property*. This is made explicit by Neveu [81], pp. 26–28. Inspired from the finite intersection property of compact sets in a topological space (see A.2 next), we say that a class $\mathcal{K}$ of subsets of a set

Ω is a *compact class* if every sequence $\{K_n, n \geq 1\} \subseteq \mathcal{K}$, such that $\bigcap\limits_{n \geq 1} K_n = \emptyset$

implies the existence of an integer N such that $\bigcap\limits_{n=1}^{N} K_n = \emptyset$.

THEOREM Let $\mathcal{K}$ be a compact subclass of a field $\mathcal{A}$ (of subsets of Ω). Let $P : \mathcal{A} \to [0,1]$ be additive and $P(\Omega) = 1$. If

$$\forall A \in \mathcal{A}, \quad P(A) = \sup\{P(K) : K \in \mathcal{K}, K \subseteq A\},$$

then P is continuous at $\emptyset$, and hence σ-additive on $\mathcal{A}$.

A.2 Topological Spaces

A *topological space* is a set A and a class $\mathcal{T}$ of subsets of A such that

(i) $\emptyset$ and A are in $\mathcal{T}$,

(ii) $\mathcal{T}$ is closed under arbitrary unions,

(iii) $\mathcal{T}$ is closed under finite intersections.

The topological space is denoted as $(A, \mathcal{T})$, the elements of $\mathcal{T}$ are called *open sets*, and $\mathcal{T}$ is called a *topology* on A. For any topological space, the set complement of an open set is called a *closed set*. By DeMorgan's laws, we see that the following dual properties hold for the set $\mathcal{F}$ of closed sets.

(a) $\emptyset$ and A are in $\mathcal{F}$.

(b) $\mathcal{F}$ is closed under arbitrary intersections.

(c) $\mathcal{F}$ is closed under finite unions.

Topologies can be generated from bases. If $\mathcal{T}$ is a topology on A, then a subclass $\mathcal{C}$ of $\mathcal{T}$ is called a *base* for $\mathcal{T}$ if each element of $\mathcal{T}$ is the union of elements of $\mathcal{C}$. For a class $\mathcal{C}$ of subsets of $\mathcal{T}$ to be a basis for some topology, $\mathcal{C}$ has to have the property that for any $B, C \in \mathcal{C}$. and each $x \in B \cap C$, there exists a $D \in \mathcal{C}$ such that $x \in D \subseteq B \cap C$. A class $\mathcal{C}$ of subsets is a *subbase* for a topology $\mathcal{T}$ if the class of finite intersections of elements of $\mathcal{C}$ is a base for $\mathcal{T}$. Thus, a nonempty class $\mathcal{C}$ is the subbase for some topology uniquely determined by $\mathcal{C}$, namely, the smallest topology containing $\mathcal{C}$.

Other basic concepts are these. For $x \in A$, a set U containing x is called a *neighborhood* of x if there is an open set T containing x and contained in U. For the topological space $(\mathbb{R}, \mathcal{T})$, if $x, y \in \mathbb{R}$, and $x \neq y$, then there are disjoint open sets, in fact, disjoint open intervals, one containing x, the other containing y. These open sets are, of course, neighborhoods of x and y, respectively. A topological space having this property, namely that distinct points are contained in distinct disjoint neighborhoods is called a *Hausdorff space*.

The smallest closed set containing a subset S of A is called the *closure* of S, and denoted $\overline{S}$. The closure of S is the intersection of all closed subsets of A containing S, of which A itself is one. Similarly, there is a largest open set contained in S. It is called the *interior* of S, and denoted S°. The *boundary* of S is the closed set $\overline{S} \setminus S^{\circ}$, and denoted $\partial(S)$.

If $\mathbb{Q}$ denotes the set of rational numbers, then $\overline{\mathbb{Q}} = \mathbb{R}$. A set with this property is called a *dense subset*. That is, if D is a subset of a topological space A, and $\overline{D} = A$, then D is dense in A. If D is countable, and $\overline{D} = A$, then the topological space $(A, \mathcal{T})$ is called a *separable space*. For example, the space $(\mathbb{R}, \mathcal{T})$ is separable since $\mathbb{Q}$ is countable.

Moreover, $\mathbb{R}$ is *second countable*, i.e., its topology has a *countable base*, namely the class of intervals with rational endpoints. In general, a topological space $(A, \mathcal{T})$ is said to be second countable if $\mathcal{T}$ has a countable base. Note that second countable topological spaces are separable.

In $\mathbb{R}$, a closed interval $[a, b]$ has the property that any set of open sets whose union contains $[a, b]$ has a finite subset whose union contains $[a, b]$. This is phrased by saying that any open cover of $[a, b]$ has a finite subcover. The set $[a, b]$ is an example of a *compact set*. That is, if C is a subset of topological space A and every open cover of C has a finite subcover, then C is called a compact set. A space is called *locally compact* if every point has a neighborhood that is compact. For example, in $\mathbb{R}$, $[x - \varepsilon, x + \varepsilon]$ is a compact neighborhood of the point x.

The function $\rho : \mathbb{R} \times \mathbb{R} \to \mathbb{R}^{+}$ defined by $\rho(x, y) = |x - y|$ is called a *metric*, or a *distance*, and satisfies the following properties.

(i) $\rho(x, y) = 0$ if and only if $x = y$.

(ii) $\rho(x, y) = \rho(y, x)$.

(iii) $\rho(x, z) \le \rho(x, y) + \rho(y, z)$.

The real line $\mathbb{R}$, together with this metric, denoted by $(\mathbb{R}, \rho)$, is a *metric space*. More generally, any set A with a function $\rho : A \times A \to [0, \infty)$ satisfying (i)–(iii) above is called a metric space. Further, any metric ρ on a set A gives rise to a topology on A by taking as open sets unions of sets of the form $\{y : \rho(x, y) < r\}$, where r ranges over the positive real numbers, and x over the elements of A. In the case of $\mathbb{R}$, the metric above gives rise to the topology discussed on $\mathbb{R}$, since the set $\{y : \rho(x, y) < r\}$ are open intervals. If there is a metric on a topological space that gives rise to its topology, then that topological space is called metrizable. As just noted, the topological space $\mathbb{R}$ is *metrizable*.

A sequence $\{x_n, n \ge 1\}$ in a metric space is said to *converge* to the element x if for each $\varepsilon > 0$, there is a positive integer N such that for $n > N$,

$\rho(x_n, x) < \varepsilon$. The sequence $\{x_n, n \geq 1\}$ is a *Cauchy sequence* if for each $\varepsilon > 0$, there is a positive integer N such that if $m, n > N$, then $\rho(x_n, x_m) < \varepsilon$. If in a metric space, every Cauchy sequence converges, then that metric space is said to be *complete*. The metric space $\mathbb{R}$ is complete. The topological space $\mathbb{R}$ is metrizable, has a countable dense subset, and is complete with respect to a metric that gives rise to its topology. Such a topological space is called a *Polish space*. Note that Polish spaces are natural spaces in probability theory. On a topological space, the smallest σ-field containing the open sets is called the *Borel σ-field* of that space. The Borel σ-field of $\mathbb{R}$ is denoted from now on as $\mathcal{B}(\mathbb{R})$. Elements of $\mathcal{B}(\mathbb{R})$ are called *Borel sets* of $\mathbb{R}$. Examples of Borel sets are singletons $\{a\}$ and all types of intervals such as

$$(a, b), \quad [a, b], \quad [a, b), \quad (-\infty, b), \quad (-\infty, b], \quad (a, \infty), \quad \dots.$$

It is left as exercises to show that $\mathcal{B}(\mathbb{R})$ is also generated by any of the following:

a) all closed sets of $\mathbb{R}$.

b) all open intervals (a, b) with end points a, b rationals.

c) all intervals of the form $(-\infty, a], a \in \mathbb{R}$.

d) all intervals of the form $(a, b], -\infty \leq a < b < \infty$, with $(a, \infty]$ to be taken as (a, ∞).

In summary, on a topological space such as $\mathbb{R}$, the collection of events is its Borel σ-field $\mathcal{B}(\mathbb{R})$, and hence models for random experiments with outcomes in $\mathbb{R}$ are probability spaces of the form $(\mathbb{R}, \mathcal{B}(\mathbb{R}), P)$, where each P is a probability measure on $\mathcal{B}(\mathbb{R})$. Note that, when needed, the extended real line $[-\infty, \infty]$ is topologized appropriately with Borel σ-field generated by $\{(a, b], -\infty \leq a < b \leq \infty\}$.

Now, in practice, how to construct probability measures on $\mathcal{B}(\mathbb{R})$? That is, how to suggest models from empirical observations? It turns out that the key is in the concept of distribution functions.

Let P be a probability measure on $\mathcal{B}(\mathbb{R})$, define $F : \mathbb{R} \to [0, 1]$ by

$$F(x) = P((-\infty, x]), \qquad x \in \mathbb{R}.$$

This function satisfies the following basic properties:

(i) F is monotone non-decreasing, i.e $x \leq y$ implies $F(x) \leq F(y)$.

(ii) $\lim_{x \searrow -\infty} F(x) = 0$ and $\lim_{x \nearrow \infty} F(x) = 1$.

(iii) F is right continuous on $\mathbb{R}$, i.e.,

$$\lim_{y \searrow x} F(y) = F(x) \text{ for any } x \in \mathbb{R}.$$

These simply follow from the basic properties of probability measures. First, note that, as a monotone function, F has left and right limits at all points, i.e., both

$$F(x^-) = \lim_{y \nearrow x} F(y) \quad \text{and} \quad \lim_{y \searrow x} F(y) = F(x^+)$$

exist for each x, and moreover

$$F(x^-) \leq F(x) \leq F(x^+).$$

We usually write $F(-\infty)$ and $F(\infty)$ for the right and left limits in the above (ii).

The upshot is this. All functions on $\mathbb{R}$ which satisfy the above (i)–(iii) properties are distribution functions of probability measures on $\mathcal{B}(\mathbb{R})$, and hence it suffices to look at such functions (which are simpler than probability measures!) when suggesting models for laws of random evolution on $\mathbb{R}$.

Specifically, if we define *a distribution function* on $\mathbb{R}$ as any function F satisfying the above properties (i)–(iii), then we have the following result.

THEOREM (characterization theorem/Lebesgue-Stieltjes) *There exists a bijection between distribution functions F on $\mathbb{R}$ and probability measures P on $\mathcal{B}(\mathbb{R})$ via $F(x) = P((-\infty, x])$.*

Remark. The probability measure P on $\mathcal{B}(\mathbb{R})$ associated with a given distribution function F is also denoted as dF (called the *Stieltjes measure* associated with F). The Stieltjes measure associated with the distribution function

$$F(y) = \begin{cases} 0 & \text{if} \quad y < 0 \\ y & \text{if} \quad 0 \leq y < 1 \\ 1 & \text{if} \quad y \geq 1 \end{cases}$$

is the *Lebesgue measure* on $[0, 1]$. Stieltjes measures can be defined in a more general context, namely, for functions of *bounded variation,* in particular for monotone functions, such as $F(x) = x$ on $\mathbb{R}$ whose associated Stieltjes measure is the *Lebesgue measure* on $\mathbb{R}$. Note that a (nonnegative) *measure* on a measurable space like $(\mathbb{R}, \mathcal{B}(\mathbb{R}))$ is a set function $\nu : \mathcal{B}(\mathbb{R}) \to [0, \infty]$ such that $\nu(\emptyset) = 0$ and σ-additive. The Lebesgue measure ν on $\mathbb{R}$ is the unique measure ν such that $\nu(a, b) = b - a$, for $a, b \in \mathbb{R}$. The Lebesgue measure $\nu(A)$ on a Borel set A of $\mathbb{R}$ is the "length" of A. A property that holds on $\mathbb{R}$ except a Borel set with zero Lebesgue measure is said to hold *almost everywhere* (*a.e.*). The Lebesgue measure μ on $\mathbb{R}$ is σ-*finite,* i.e., there exists a sequence $A_n \in \mathcal{B}\mathbb{R}$, $n \geq 1$, such that $\bigcup_{n \geq 1} A_n = \mathbb{R}$ and $\mu(A_n) < +\infty$ for each n (e.g., $A_n = (-n, n)$).

In multivariate statistical analysis, the sample space of observations is the Cartesian product space $\mathbb{R}^d$. The space $\mathbb{R}^d$ denotes the set of points $\mathbf{x} = (x_1, x_2, \ldots, x_d)$, where each $x_i \in \mathbb{R}$, $i = 1, 2, \ldots, d$. Note that $d \geq 1$ is a fixed integer and the integer n is reserved for "sample size" or index in sequences of events (or variables) where we could let n tend to infinity!

The Euclidean space $\mathbb{R}^d$ is a metric space with metric

$$\rho(\mathbf{x}, \mathbf{y}) = \left[\sum_{i=1}^{d}(x_i - y_i)^2\right]^{1/2},$$

where $\mathbf{x} = (x_1, x_2, \ldots, x_d)$ and $\mathbf{y} = (y_1, y_2, \ldots, y_d)$. The partial order on $\mathbb{R}^d$ is

$$\mathbf{x} \leq \mathbf{y} \text{ if and only if } x_i \leq y_i, \quad i = 1, 2, \ldots, d.$$

We write $\mathbf{y} \searrow \mathbf{x}$ to mean that $y_i \searrow x_i$, $i = 1, 2, \ldots, d$.

The program for $\mathbb{R}^d$ is similar to $\mathbb{R}$, namely, we seek to define a canonical σ-field $\mathcal{A}$ on $\mathbb{R}^d$, and then probability measures on $\mathcal{A}$ via multivariate (or joint) distribution functions. Now $\mathcal{A}$ can arrive from two points of view. We can take $\mathcal{A}$ directly as the Borel σ-field $\mathcal{B}(\mathbb{R}^d)$ of $\mathbb{R}^d$, which is topologized by the above Euclidean metric, or as a *product σ-field* of d identical measurable spaces $(\mathbb{R}, \mathcal{B}(\mathbb{R}))$. It turns out that two approaches coincide since $\mathbb{R}^d$ is a separable metric space. We will elaborate this in a more general setting that is useful in applications.

Let $(\Omega_i, \mathcal{A}_i)$, $i = 1, 2, \ldots, d$ be d measurable spaces and

$$\Omega = \Omega_1 \times \Omega_2 \times \ldots \times \Omega_d,$$

the *product space* of the Ω_i's, i.e.,

$$\omega = (\omega_1, \omega_2, \ldots, \omega_d) \in \Omega \quad \text{iff} \quad \omega_i \in \Omega_i, \quad i = 1, 2, \ldots, d.$$

Let $\mathcal{C}$ denote the class of subsets of Ω of the form $A_1 \times A_2 \times \ldots \times A_d$, where $A_i \in \mathcal{A}_i$, $i = 1, 2, \ldots, d$. Each element of $\mathcal{C}$ is called a d-dimensional "rectangle."

DEFINITION *Given d measurable spaces $(\Omega_i, \mathcal{A}_i)$, $i = 1, 2, \ldots, d$, the product σ-field of the $\mathcal{A}_i$'s, denoted as $\mathcal{A}_1 \otimes \mathcal{A}_2 \otimes \ldots \otimes \mathcal{A}_d$ is the σ-field on $\Omega_1 \times \Omega_2 \times \ldots \times \Omega_d$ generated by $\mathcal{C}$.*

In applications, it happens that the spaces Ω_i's are topological, and the $\mathcal{A}_i$'s are their Borel σ-fields. Then Ω is also topological with the product topology and hence has its own Borel σ-field $\mathcal{B}(\Omega)$. We will leave as exercises for interested students to work out some of the technical results, and simply say here that, in general, $\mathcal{A}_1 \otimes \mathcal{A}_2 \otimes \ldots \otimes \mathcal{A}_d \subseteq \mathcal{B}(\Omega)$. However, if each Ω_i has a countable base for its topology, then $\mathcal{A}_1 \otimes \mathcal{A}_2 \otimes \ldots \otimes \mathcal{A}_d = \mathcal{B}(\Omega)$. This is precisely the case of $\mathbb{R}^d$. Thus $\mathcal{B}(\mathbb{R}^d)$ is also the product σ-field $\mathcal{B}(\mathbb{R}) \otimes \mathcal{B}(\mathbb{R}) \otimes \ldots \otimes \mathcal{B}(\mathbb{R})$ (d times). In fact, $\mathcal{B}(\mathbb{R}^d)$ is generated by d-dimensional rectangles of the form

$$(x_1, y_1] \times (x_2, y_2] \times \ldots \times (x_d, y_d],$$

where $x_i, y_i \in \mathbb{R}$, $i = 1, 2, \ldots, d$.

Thus for $\mathbb{R}^d$, we consider $(\mathbb{R}^d, \mathcal{B}(\mathbb{R}^d))$ and probability models for "random vectors" are probability measures on $\mathcal{B}(\mathbb{R}^d))$.

First, there is a special kind of probability measures on $\mathcal{B}(\mathbb{R}^d))$, or more generally, on $\Omega = \Omega_1 \times \Omega_2 \times \ldots \times \Omega_d$, corresponding to the concept of *independence* of random variables. To be simple, consider $d = 2$, the case $d \geq 3$ is similar. Let $(\Omega_i, \mathcal{B}_i, P_i)$, $i = 1, 2$, be two probability spaces. Let $\Omega = \Omega_1 \times \Omega_2$, and consider $\mathcal{A}_1 \otimes \mathcal{A}_2$ as a σ-field on Ω. Define P on the class $\mathcal{C} = \{A_1 \times A_2 : A_i \in \mathcal{A}_i, i = 1, 2\}$ as

$$P(A_1 \times A_2) = P_1(A_1)P_2(A_2), \tag{A.1}$$

then using arguments that should be familiar to students now, P can be extended to a unique probability measure on $\sigma(\mathcal{C}) = \mathcal{A}_1 \otimes \mathcal{A}_2$.

DEFINITION *The product probability measure $P = P_1 \otimes P_2$ is the probability measure on $\mathcal{A}_1 \otimes \mathcal{A}_2$ satisfying (A.1) above.*

Of course, probability measures on $\mathcal{A}_1 \otimes \mathcal{A}_2$ can be of various different forms! To characterize all of them, we are guided by the situation in $\mathbb{R}$. Let us consider $(\mathbb{R}^d, \mathcal{B}(\mathbb{R}^d))$. Let P be a probability measure on $\mathcal{B}(\mathbb{R}^d))$. The joint (multivariate) distribution function F is defined as

$$F : \mathbb{R}^d \to [0, 1], \qquad F(x_1, x_2, \ldots, x_d) = P\left(\prod_{i=1}^{d}(-\infty, x_i]\right),$$

where $\prod$ stands for product. Then F has the following basic properties

(i) $\lim_{x_j \to -\infty} F(x_1, \ldots, x_d) = 0$ for at least one j and

$$\lim_{x_j \to \infty} F(x_1, \ldots, x_d) = 1$$

for all $j = 1, 2, \ldots d$.

(ii) F is right continuous on $\mathbb{R}^d$,

$$\lim_{\mathbf{y} \searrow \mathbf{x}} F(\mathbf{y}) = F(\mathbf{x}), \qquad \text{for any} \quad \mathbf{x} \in \mathbb{R}^d.$$

(iii) For any $\mathbf{a} = (a_1, \ldots, a_d)$, $\mathbf{b} = (b_1, \ldots, b_d) \in \mathbb{R}^d$ with $\mathbf{a} \leq \mathbf{b}$, define $\Delta_{a_i, b_i} : \mathbb{R}^d \to \mathbb{R}$ as

$$\Delta_{a_i, b_i} F(\mathbf{x}) = F(x_1, \ldots, x_{i-1}, b_i, x_{i+1}, \ldots x_d)$$
$$- F(x_1, \ldots, x_{i-1}, a_i, x_{i+1}, \ldots x_d).$$

Then

$$\Delta_{\mathbf{a},\mathbf{b}} F(\mathbf{x}) = \Delta_{a_1,b_1} \circ \ldots \circ \Delta_{a_d,b_d} F(\mathbf{x})$$

$$= P\left(\prod_{i=1}^{d}(a_i, b_i]\right) = P((\mathbf{a},\mathbf{b}]) \geq 0,$$

where $\circ$ denotes composition of operations. Thus the property (iii) is

$$\Delta_{\mathbf{a},\mathbf{b}} F(\mathbf{x}) \geq 0 \qquad \text{for } \mathbf{a} \leq \mathbf{b} \in \mathbb{R}^d.$$

As in the case of $\mathbb{R}$, the above properties follow from the properties of the probability measure P. As expected, these basic properties will characterize P, so that we consider the following.

DEFINITION *A d-dimensional distribution function is a function* $F : \mathbb{R}^d \to [0, 1]$ *satisfying conditions (i)–(iii) above.*

The following is the characterization theorem (Lebesgue/Stieltjes) in $\mathbb{R}^d$, its proof is similar to the case of $\mathbb{R}$.

THEOREM *Each distribution function F on $\mathbb{R}^d$ is the distribution function of a unique probability measure P on $\mathcal{B}(\mathbb{R}^d)$, i.e.,*

$$F(\mathbf{x}) = P\left(\prod_{i=1}^{d}(-\infty, x_i]\right), \qquad \text{for} \quad \mathbf{x} = (x_1, \ldots, x_d) \in \mathbb{R}^d.$$

A.3   Expectation of a Random Variable

Let $(\Omega, \mathcal{A}, P)$ be a probability space and X is a map from Ω to $\mathbb{R}$ that is $\mathcal{A}\text{-}\mathcal{B}(\mathbb{R})$-measurable. The probability law of X is the probability measure P_X on $\mathcal{B}(\mathbb{R})$ where

$$P_X(A) = P(X^{-1}(A)) = P(\{\omega : X(\omega) \in A\}).$$

Since a probability measure P_X on $\mathcal{B}(\mathbb{R})$ is characterized by a distribution function F via $F(x) = P((-\infty, x])$, we also denote P_X by the notation $dF(x)$, emphasizing the Lebesgue-Stieltjes probability measure associated with F. To make the analysis rigorous, we allow random variables to take values in the extended real line $\overline{\mathbb{R}} = [-\infty, \infty]$. This is necessary to talk about the existence of quantities such as $\sup A$, or limits of monotone sequences of numbers that exist in $\overline{\mathbb{R}}$. For this, $\mathcal{B}(\overline{\mathbb{R}})$ will be taken to be the σ-field generated by $\mathcal{B}(\mathbb{R})$ and $\{-\infty\}, \{\infty\}$. Of course, $\mathcal{B}(\mathbb{R}^+)$ is the trace σ-field of $\mathcal{B}(\mathbb{R})$ on $\mathbb{R}^+$. We also extend multiplication and addition to $[-\infty, \infty]$ as follows.

$$0(\infty) = (\infty)0 = (-\infty)0 = 0(-\infty) = 0,$$
$$x(\infty) = (\infty)x = \infty \quad\quad \text{if} \quad x > 0,$$
$$x(-\infty) = (-\infty)x = \infty \quad\quad \text{if} \quad x < 0,$$
$$x + (\infty) = (\infty) + x = \infty \quad\quad \text{for all} \quad x \in \mathbb{R},$$
$$\infty + (-\infty) \quad\quad \text{is undefined.}$$

We start out with simple random variables. A *simple random variable* is a random variable X, which has a finite set of values. If $x_1, x_2, \ldots, x_k$ are distinct values of X in $\mathbb{R}$, the X can be written in the canonical form

$$X = \sum_{i=1}^{k} x_i I_{A_i},$$

where $A_i = \{\omega : X(\omega) = x_i\}$, noting again that $\{A_1, A_2, \ldots, A_k\}$ form a $\mathcal{A}$-measurable partition of Ω.

DEFINITION *The expected value of a simple random variable X, denoted as $E(X)$, is defined as*

$$E(X) = \sum_{i=1}^{k} x_i P(A_i).$$

This definition does not depend on this particular choice of canonical representation of X.

Next, for *nonnegative random variables*, we make two basic observations:

(a) Let X be a nonnegative random variable. Then there exists a sequence of nonnegative, simple random variables $\{X_n, n \geq 1\}$, such that

$$X_n(\omega) \leq X_{n+1}(\omega), \quad\quad \text{for} \quad \omega \in \Omega$$

(i.e., the sequence $\{X_n, n \geq 1\}$ is nondecreasing), and

$$X(\omega) = \lim_{n \to \infty} X_n(\omega), \quad\quad \text{for} \quad \omega \in \Omega.$$

(b) For *any sequence* of nondecreasing, simple and nonnegative random variables $\{X_n, n \geq 1\}$ converging to a nonnegative random variable X, we have

$$\lim_{n \to \infty} E(X_n) = \sup\{E(Y) : Y \text{ simple and } 0 \leq Y \leq X\}.$$

The implication of the above two observations is this. We can define the *expected value of a nonnegative* random variable X either by a or b given above. If a is chosen, the definition is well-defined since a does not depend on any particular choice of the approximate sequence, i.e., for any two sequences of nonnegative simple random variables such that $X_n \nearrow X$, and $Y_n \nearrow X$, we have

$$\lim_{n \to \infty} E(X_n) = \lim_{n \to \infty} E(Y_n)$$

since they are both equal to b. We choose a for convenience of analysis, namely, for using "monotone convergence property."

DEFINITION *The expected value of a nonnegative random variable X is defined to be*

$$E(X) = \lim_{n \to \infty} E(X_n),$$

where $\{X_n, n \geq 1\}$ is any sequence of nondecreasing, nonnegative, and simple random variables such that $X_n \nearrow X$.

Finally, for *arbitrary random variable X*, we write $X = X^+ - X^-$, where $X^+(\omega) = \max\{X(\omega), 0\}$ (positive part of X) and $X^-(\omega) = -\min\{X(\omega), 0\}$ (negative part of X). Note that both $X^+ \geq 0$ and $X^- \geq 0$. We then define

$$E(X) = E(X^+) - E(X^-)$$

and we write

$$E(X) = \int_\Omega X(\omega) \, dP(\omega),$$

provided, of course, that not both $E(X^+)$ and $E(X^-)$ are ∞.

Remark. Clearly $E(X) < \infty$ when $E(X^+) < \infty$ and $E(X^-) < \infty$. Now $|X(\omega)| = X^+(\omega) + X^-(\omega)$, we see that $E(X) < \infty$ if and only if $E|X| < \infty$, in which case, we say that X has a finite expected value, or X is *integrable* (with respect to P). If only one of $E(X^+)$, $E(X^-)$ is ∞, then X has an infinite expected value, and if both $E(X^+)$ and $E(X^-)$ are ∞, then we say that X does not have an expected value or $E(X)$ does not exist.

MONOTONE CONVERGENCE THEOREM Let $\{X_n, n \geq 1\}$ be *a nondecreasing sequence of nonnegative random variables with measurable limit X a.s., i.e.,*

$$P\left(\omega : \lim_{n \to \infty} X_n(\omega) \neq X(\omega)\right) = 0,$$

then

$$\lim_{n \to \infty} \int_A X_n(\omega) \, dP(\omega) = \int_A X(\omega) \, dP(\omega).$$

FATOU'S LEMMA If $\{X_n, \, n \geq 1\}$ is a sequence of nonnegative random variables such that $\liminf\limits_{n \to \infty} X_n = X$ a.s., then

$$\int_A X(\omega) \, dP(\omega) \leq \liminf_{n \to \infty} \int_A X_n(\omega) \, dP(\omega).$$

DOMINATED CONVERGENCE THEOREM Let $\{X_n, \, n \geq 1\}$ be a sequence of random variables with $\lim\limits_{n \to \infty} X_n = X$ a.s., and $|X_n| \leq Y$ for any $n \geq 1$, with $\int_A Y(\omega) \, dP(\omega) < \infty$. Then

$$\int_A |X(\omega)| \, dP(\omega) < \infty.$$

and

$$\int_A X(\omega) \, dP(\omega) = \lim_{n \to \infty} \int_A X_n(\omega) \, dP(\omega).$$

FUBINI'S THEOREM Let μ_i be σ-finite measures on $(\Omega_i, \mathcal{A}_i)$, $i = 1, 2$ and $h : \Omega_1 \times \Omega_2 \to \mathbb{R}$ measurable (i.e., $\mathcal{A}_1 \otimes \mathcal{A}_2$-$\mathcal{B}(\mathbb{R})$-measurable).

(i) If h is nonnegative, then

$$\int_{\Omega_1 \times \Omega_2} h(\omega_1, \omega_2) \, d(\mu_1 \otimes \mu_2)(\omega_1, \omega_2)$$

$$= \int_{\Omega_1} \left(\int_{\Omega_2} h(\omega_1, \omega_2) \, d\mu_2(\omega_2) \right) d\mu_1(\omega_1)$$

$$= \int_{\Omega_2} \left(\int_{\Omega_1} h(\omega_1, \omega_2) \, d\mu_1(\omega_1) \right) d\mu_2(\omega_2).$$

(ii) If h is arbitrary (not necessarily nonnegative), but integrable with respect to $\mu_1 \otimes \mu_2$, i.e.,

$$\int_{\Omega_1 \times \Omega_2} |h(\omega_1, \omega_2)| \, d(\mu_1 \otimes \mu_2)(\omega_1, \omega_2) < \infty,$$

then the equalities in (i) hold.

Remarks. (a) The hypothesis of "σ-finite measures" is used to establish the existence and measurability of quantities involved in the above formula.

(b) The practical example is $(\Omega_i, \mathcal{A}_i) = (\mathbb{R}, \mathcal{B}(\mathbb{R}))$, $i = 1, 2$ and $\mu_1 = \mu_2 =$the Lebesgue measure dx on $\mathbb{R}$. The Fubini's theorem says that, for appropriate h, the double integral of h is computed by an iterated integral in *any order* of preference. But it should be noted that if h is *not* integrable (with respect to $\mu_1 \otimes \mu_2$), it can happen that the above two iterated integrals (in different orders) are not equal. Also, these two integrals can be equal to but h is not integrable. See exercises.

(c) To apply Fubini's theorem in computations, besides the case where $h \geq 0$, we should first check the *integrability* of h with respect to the product measure $\mu_1 \otimes \mu_2$. This is done as follows. Since $|h| \geq 0$, the equalities in (i) of Fubini's theorem hold for $|h|$ at the place of h. Thus h is $\mu_1 \otimes \mu_2$ integrable if and only if any of the iterated integral of $|h|$ is finite. Thus, simply compute, say,

$$\int_{\Omega_1} \left(\int_{\Omega_2} |h(\omega_1, \omega_2)| \, d\mu_2(\omega_2) \right) d\mu_1(\omega_2)$$

to see whether it is finite.

(d) In the discrete case, this is known in the context of *absolutely convergent series* (see any calculus text), namely the following. Let $\sum_{n \geq 1} x_n$ be an infinite series of numbers. If the sequence of partial sums $s_n = \sum_{i=1}^{n} x_i$ converges to $s < \infty$, then we say that the series $\sum_{n \geq 1} x_n$ is convergent and its sum is denoted by $\sum_{n \geq 1} x_n = s$. Of course a series of nonnegative numbers such as $\sum_{n \geq 1} |x_n|$ always has a sum $\leq \infty$. When $\sum_{n \geq 1} |x_n| < \infty$, we say that the series $\sum_{n \geq 1} x_n$ is *absolutely convergent*. (For that to happen, it is necessary and sufficient that the increasing sequence $\sum_{i=1}^{n} |x_n|$ is bounded from above). Note that if $\sum_{n \geq 1} |x_n| < \infty$, then $\sum_{n \geq 1} x_n$ is finite and is the same for any reordering of the sequence $\{x_n, n \geq 1\}$. For discrete random vectors, we are using *double series*. The double series

$$\sum_{\substack{m \geq 1 \\ n \geq 1}} x_{mn} \quad \text{is absolutely convergent when} \quad \sum_{\substack{m \geq 1 \\ n \geq 1}} |x_{mn}| < \infty,$$

and in this case, we have

$$\sum_{\substack{m \geq 1 \\ n \geq 1}} x_{mn} = \sum_{n \geq 1} \left(\sum_{m \geq 1} x_{mn} \right) = \sum_{m \geq 1} \left(\sum_{n \geq 1} x_{mn} \right) < \infty,$$

i.e., the finite value of the double series is computed via its iterated series and its value is independent of the order of summations. For this to happen, it is necessary and sufficient that $\sum_{m \geq 1} |x_{mn}| < \infty$ for each n *and*

$$\sum_{n \geq 1} \left(\sum_{m \geq 1} |x_{mn}| \right) < \infty.$$

Note that the limit sum of the double series given above is the limit of the convergent double sequence

$$\{S_{m,n}\}_{m,n=1}^{\infty}, \qquad S_{m,n} = \sum_{\substack{i \leq m \\ j \leq n}} x_{ij},$$

i.e., for any $\varepsilon > 0$, there exist N, M such that

$$|S_{m,n} - L| < \varepsilon \quad \text{for all} \quad m \geq M, \ n \geq N,$$

and we write

$$\lim_{\substack{m \to \infty \\ n \to \infty}} S_{m,n} = L.$$

A.4 Convergence of Random Elements

Let $\{X_n, n \geq 1\}$ be a sequence of random variables, defined on $(\Omega, \mathcal{A}, P)$. Since the X_n's are functions, $X_n : \Omega \to \mathbb{R}$, we first need to investigate the convergence of $\{X_n, n \geq 1\}$, as $n \to \infty$, as a sequence of functions. Convergence concepts related directly to the X_n's are grouped under the headline of *stochastic convergence*.

First, as functions, it is natural to look at

$$\lim_{n \to \infty} X_n(\omega) = X(\omega), \qquad \text{for all } \omega \in \Omega.$$

This is the *pointwise convergence* of random variables (as functions), or *convergence everywhere*, or X_n converges *surely* to X.

It turns out that this type of convergence of random variables is too strong in the context of probability and statistics. To see this, consider the experiment of tossing a biased coin repeatedly. Let Y_n denote the outcome of the nth toss.

$$P(Y_n = 1) = 1 - P(X_n = 0) = \theta.$$

Here $\Omega = \{0, 1\}^{\mathbb{N}}$ and $Y_n : \Omega \to \{0, 1\}$ with $Y_n(\omega)$ being the nth component of ω. Consider

$$X_n = \frac{1}{n} \sum_{k=1}^{n} Y_k.$$

For $\{X_n, n \geq 1\}$ to be a *consistent estimator* of θ, we require that X_n converges to θ. But X_n cannot converge to θ for *all* $\omega \in \Omega$. Indeed, for $\omega = (\omega_1, \omega_2, ...)$ containing a finite number of 1's, we have

$$\lim_{n \to \infty} \frac{1}{n} \sum_{k=1}^{n} Y_k(\omega) = 0.$$

Thus, we should weaken the sure convergence to make sense for consistency of estimators. As we will see in the next chapter, X_n will converge to θ in the following weaker concepts of convergence of random variables.

First, we weaken the above sure convergence by requiring only X_n to converge *almost surely*, i.e., X_n converges to X only on some subset A of Ω such that $P(A) = 1$.

DEFINITION *A sequence of random variables $\{X_n, n \geq 1\}$ is said to converge almost surely to the random variable X if*

$$P\left(\omega : \lim_{n \to \infty} X_n(\omega) \neq X(\omega)\right) = 0,$$

and is denoted as $X_n \xrightarrow{a.s.} X$.

Remark. Let

$$A = \{\omega : \lim_{n \to \infty} X_n(\omega) = X(\omega)\},$$

Then $P(A) = 1$. The a.s. convergence is also called the *convergence with probability one*.

DEFINITION *A sequence of random variables $\{X_n, n \geq 1\}$ converges to X in probability, denoted as $X_n \xrightarrow{P} X$, if for each $\varepsilon > 0$,*

$$\lim_{n \to \infty} P\left(|X_n - X| \geq \varepsilon\right) = 0.$$

or equivalently

$$\lim_{n \to \infty} P\left(|X_n - X| < \varepsilon\right) = 1.$$

A.5   Convergence in Distribution

Again, a typical situation in statistics is this. Let Y be a population with *unknown* distribution function F. Let $Y_1, Y_2, \ldots, Y_n$ be a random sample from Y. We are interested in using the statistic (i.e., an observable random variable) $X_n = (Y_1 + Y_2 + \ldots + Y_n)/n$ for statistical inference. For that to be possible, among other things, we need to know the (sampling) distribution of X_n:

$$F_n(x) = P(X_n \leq x), \qquad x \in \mathbb{R}.$$

Now, F_n can be written in terms of F, say,

$$F_n(x) = F^{*(n)}(nx),$$

where $F^{*(n)}$ denotes the n-fold *convolution* of F, i.e.,

$$F^{*(n)}(x) = (F * F * \ldots * F)(x) \ (n \text{ times}),$$

$(F * G)(x) = \int_{-\infty}^{+\infty} F(x - y) \, dG(y)$.

But F is unknown, and so is F_n. Can we approximate the distribution F_n of X_n when n is sufficiently large? This is the essentials of large sample statistics.

DEFINITION *The sequence of random variables $\{X_n, n \geq 1\}$ is said to converge in distribution to a random variable X, in symbol, $X_n \overset{D}{\longrightarrow} X$, if*

$$\lim_{n \to \infty} F_n(x) = F(x) \qquad \text{for} \quad x \in C(F),$$

where F_n's and F are distribution functions of X_n's and X, respectively, and $C(F)$ is the continuity set of F.

The concepts of convergence in distribution was developed for random elements taking values in Euclidean spaces (random vectors) in terms of distribution functions. This will be no longer possible for the case where random elements of interest take values in spaces for which, in one hand, there is no concept of distribution functions, and on the other hand, a counterpart of the Lebesgue-Stieltjes theorem (for $\mathbb{R}^d$) seems not possible. In such cases, we have to study directly the convergence of their probability laws.

In the following, after giving some typical situations arising from statistics, we set out to generalize the convergence in distribution to random elements taking values in *metric spaces*, by establishing various equivalences for the convergence in distribution of random vectors.

Standard statistical applications concern data that are vector-valued observations, where the probability theory, and induced statistical methods are well developed on Euclidean spaces. As Fréchet [35] has pointed out, there are many random elements in nature and technology.

PORTMANTEAU THEOREM *Let (U, ρ) be a metric space. Let Q, Q_n, $n \geq 1$, be probability measures on the Borel σ-field of U. Then the following are equivalent.*

(i) $\int_U g(x) \, dQ_n(x) \longrightarrow \int_U g(x) \, dQ(x)$ for all bounded, continuous, real-valued functions g, i.e., $g \in C_b(U)$.

(ii) $\int_U g(x) \, dQ_n(x) \longrightarrow \int_U g(x) \, dQ(x)$ for all bounded, uniformly continuous, real-valued functions g.

(iii) $\limsup_{n \to \infty} Q_n(A) \leq Q(A)$ for all closed sets A of U.

(iv) $\liminf_{n \to \infty} Q_n(A) \geq Q(A)$ for all open sets A of U.

(v) $Q_n(A) \longrightarrow Q(A)$ *for Q-continuity sets A of U.*

In view of this theorem, the convergence in distribution of random vectors, $X_n \xrightarrow{D} X$, is equivalent to the convergence of the sequence of probability measures $\{Q_n, n \geq 1\}$ to the probability measure Q on $\mathbb{R}^d$ (where Q_n's, Q are probability laws of X_n's, X, respectively) in the sense that

$$\int_{\mathbb{R}^d} g(x)\, dQ_n(x) \longrightarrow \int_{\mathbb{R}^d} g(x)\, dQ(x)$$

for all bounded, continuous, real-valued functions g, defined on $\mathbb{R}^d$. Thus, consistently, if $X, X_n, n \geq 1$ are random elements taking values in an arbitrary metric space U, with corresponding probability laws $Q, Q_n, n \geq 1$, being probability measures on U, we say that X_n converge in distribution to X, $X_n \xrightarrow{D} X$ if

$$\int_{\mathbb{R}^d} g(x)\, dQ_n(x) \longrightarrow \int_{\mathbb{R}^d} g(x)\, dQ(x) \quad \text{for all } g \in \mathcal{C}_b(U).$$

This convergence of probability measures is termed *weak convergence* of probability measures, and is denoted as $Q_n \xrightarrow{W} Q$.

Remarks. (a) If Q is a probability measure on U, then the map $f \in \mathcal{C}_b(U) \to \int_U f\, dQ$ is a positive, continuous linear form, since

$$\left| \int_U f\, dQ \right| \leq \int_U \|f\|\, dQ \leq \|f\| = \sup_{x \in U} |f(x)|.$$

Let

$$B = \{ f \in \mathcal{C}_b(U) : \|f\| \leq 1 \}, \qquad \|Q\| = \sup_{f \in B} \left| \int_U f\, dQ \right|.$$

We say that Q_n converges *strongly* to Q when $\|Q_n - Q\| \longrightarrow 0$ as $n \to \infty$. Let $u_n \in U$ such that $u_n \to u$ (in the metric ρ). Then we should expect that δ_{u_n} converges to δ_u. Consider the case where $a_n \neq a$ for $n \geq 1$. Now, for each n, let $f_n \in \mathcal{C}_b(U)$ such that $f_n : U \to [-1, 1]$ with $f_n(u_n) = 1$ and $f_n(u) = -1$, we see that $\|\delta_{u_n} - \delta_u\| = 2$, so that δ_{u_n} does not converge strongly to δ_u, whereas, clearly, $\delta_{u_n} \to \delta_u$ weakly. Thus, the *weak* convergence concept is more natural.

(b) *Weak topology.* The space of probability measures on a metric space U can be topologized in a compactable way with the weak convergence. Moreover, when U is separable, the weak topology is metrizable by *Prohorov's metric*. See Billingsley [10], pp. 236–239. Prohorov's metric can be defined in a simple way in Ethier and Kurtz [29], pp. 96 and 108–110.

(c) The equivalences in the Portmanteau theorem are useful in appropriate situations.

A.6 Radon-Nikodym Theorem

Various situations in statistics can be described as the process of seeking information (or knowledge) on some random variable X of interest based on the observations of some related random variable Y. For example, suppose that the random vector (X, Y) describes some stochastic system. Consider the problem of predicting the values of X (which are, say, not directly observable) when we can observe Y. If a is a predicted value for X, then the prediction error can be taken as $E(X - a)^2$. We have seen that, without information of Y, the expected value of X, $E(X)$, is the (unique) *best predictor* since $E(X)$ minimizes $E(X - a)^2$ over all a's. Recall that it is so because

$$E(X - a)^2 = E\left[(X - E(X)) + (E(X) - a)\right]^2 = V(X) + (E(X) - a)^2.$$

Now suppose that we are able to observe Y. Then we should use this information in predicting X. Roughly speaking, instead of the *unconditional expectation* $E(X)$, we should employ a *conditional expectation of X with respect to Y* in order to revise our predictor. Specifically, if the observed value of Y is y, then our update predictor should be $E(X|Y = y)$. Of course, like the unconditional case, we need to show that this leads to the best predictor in this new situation. Thus, at the design phase, we consider $E(X|Y)$ as our predictor.

As another example, in the search for "best" estimators of population parameters, we might want to revise an estimator S by conditioning it on another estimator T, i.e., considering $E(S|T)$.

In order to carry out the above statistical procedures, we need to define the concept of *conditional expectation* of a random variable with respect to another random variable.

The Discrete Case

Let X and Y be two random variables, defined on a probability space $(\Omega, \mathcal{A}, P)$, with Y being *discrete*. When $Y = y$ with $P(\{\omega : Y(\omega) = y\}) > 0$, we consider the probability measure $P(\cdot|B)$, where $B = \{Y = y\}$, on $\mathcal{A}$, i.e.,

$$P(A|B) = \frac{P(A \cap B)}{P(B)}, \qquad \text{for} \quad A \in \mathcal{A}.$$

Suppose that $\int_B X(\omega)\, dP(\omega)$ exists, then the expected value of X when $Y = y$ is the average of X with respect to $P(\cdot|Y = y)$, i.e.,

$$E(X|Y = y) = \int_\Omega X(\omega)\, dP(\omega|Y = y).$$

This is consistent with the fact that, the conditional probability is just the probability in a reduced or restricted space, conditional expectation is their expectation in that reduced space. Now

$$\int_\Omega X(\omega)\, dP(\omega|B) = \int_B X(\omega)\, dP(\omega|B) + \int_{B^c} X(\omega)\, dP(\omega|B)$$

$$= \frac{1}{P(B)} \int_B X(\omega)\, dP(\omega),$$

we have

$$E(X|Y=y) = \frac{1}{P(Y=y)} \int_{\{Y=y\}} X(\omega)\, dP(\omega).$$

Note that if X is P-integrable, i.e., $E(|X|) < \infty$, then $E(X|Y=y)$ is defined for all y such that $P(Y=y) > 0$. Indeed, for any $B_j = \{Y = y_j\}$, with $P(B_j) > 0$,

$$\int_\Omega I_{B_j}(\omega)|X(\omega)|\, dP(\omega) \le \int_\Omega |X(\omega)|\, dP(\omega).$$

Thus, in considering conditional expectation of X with respect to Y, we always assume that $E(|X|) < \infty$.

As stated before, at the design phase, we would like to consider a *random variable* denoted as $E(X|Y)$, before making the observations on the variable Y, so that when $Y(\omega) = y$, we should have

$$E(X|Y)(\omega) = E(X|Y=Y(\omega)) = \psi \circ Y(\omega),$$

where

$$\psi : \mathcal{R}(Y) \to \mathbb{R}, \qquad \psi(y) = E(X|Y=y),$$

and $\mathcal{R}(Y)$ is the range of Y. In other words, $E(X|Y=y)$ is a value of the random variable denoted as $E(X|Y)$, and $E(X|Y)$ will be called the *conditional expectation of X with respect to* (or given) Y.

First, for $E(X|Y=y) = \psi(y)$, a function of y, the variable $E(X|Y)$ should be $\sigma(Y)$-measurable, where

$$\sigma(Y) = \{Y^{-1}(D) : D \in \mathcal{B}(\mathbb{R})\}$$

is the sub-σ-field of $\mathcal{A}$, generated by Y. Do not confuse $\sigma(Y)$ with the standard deviation of Y here!

Now, in the above case where $B = \{Y = y\}$ with $P(B) > 0$. if we also assume that $P(B) < 1$, then $P(B^c) > 0$. The events $\{B, B^c\}$ form an $\mathcal{A}$-partition of Ω with positive probabilities, so that quantities like $E(X|B)$, $E(X|B^c)$ are well-defined. Note also that the σ-field generated by the partition $\{B, B^c\}$ is $\{\emptyset, \Omega, B, B^c\}$ which is the same as $\sigma(I_B)$. More generally, let Y be discrete, then Y induces a (countable) $\mathcal{A}$-partitions of Ω, namely, $B_j = \{\omega : Y(\omega) = y_j\}$, $j \ge 1$, recalling that

$$Y(\omega) = \sum_{j\ge 1} y_j I_{B_j}(\omega).$$

The σ-field of Y, $\sigma(Y)$, is the same as the σ-field generated by its induced partition $\{B_j, j \geq 1\}$, denoted as $\sigma(B_j, j \geq 1)$. We are going to define $E(X|Y)$ as follows.

$$E(X|Y)(\omega) = \begin{cases} E(X|B_j) & \text{if } \omega \in B_j, \quad P(B_j) > 0 \\ \text{arbitrary} & \text{if } P(B_j) = 0. \end{cases}$$

As such we have in fact a family of random variables, but these random variables are equal almost surely since they differ only on the union of the B_j's with $P(B_j) = 0$, which is of probability zero. Thus, in fact, by $E(X|Y)$, we mean an *equivalence class* of random variables. Any member of this class is called a *version* of $E(X|Y)$, and by *abuse of language*, we often take $E(X|Y)$ to be some version of it. For computations, suppose that $P(Y = y_j) > 0$ for all $j \geq 1$, we have

$$E(X|Y)(\omega) = \sum_{j \geq 1} E(X|Y = y_j) I_{\{Y = y_j\}}(\omega).$$

We realize immediately in this construction that $E(X|Y)$ is measurable with respect to the sub-σ-field (of $\mathcal{A}$) $\sigma(Y) = \sigma(B_j, j \geq 1)$, since each I_{B_j} is so. Sometimes, by *abuse of notation*, we also write $I_{B_j} \in \sigma(Y) = \sigma(B_j, j \geq 1)$ to denote the fact that the variable I_{B_j} is measurable with respect to $\sigma(Y) = \sigma(B_j, j \geq 1)$.

It is important to realize that the concept of σ-fields generated by random variables captures the concept of information provided by these random variables. Saying that $E(X|Y)$ is σ-measurable means that $E(X|Y)$ can be determined from Y, i.e., $E(X|Y)$ is a function $\psi(Y)$ of Y. Thus,

$$E(X|Y)(\omega) = E(X|Y = Y(\omega)) = \psi \circ Y(\omega),$$

where $\psi(y) = E(X|Y = y)$.

Thus, it is advantageous to view $E(X|Y)$ as $E(X|\sigma(Y))$, i.e., the conditional expectation of X with respect to a σ-field. This allows us to consider $E(X|Y_1, Y_2, \ldots)$ as $E(X|\sigma(Y_1, Y_2, \ldots))$, where $\sigma(Y_1, Y_2, \ldots)$ is the σ-field generated by the random variables $Y_1, Y_2, \ldots$.

We are led to consider $E(X|\mathcal{B})$ where $\mathcal{B}$ is a sub-σ-field of $\mathcal{A}$. For example, $\mathcal{B} = \sigma(Y)$, where Y is discrete here, which is the same as the σ-field generated by the countable partition $\{B_j, j \geq 1\}$, where $B_j = \{Y = y_j\}$. Note that if $\mathbf{X}$ is a d-dimensional random vector, $\mathbf{X} = (X_1, X_2, \ldots, X_d)$, then $E(\mathbf{X}|\mathcal{B})$ is the random vector

$$E(\mathbf{X}|\mathcal{B}) = (E(X_1|\mathcal{B}), E(X_2|\mathcal{B}), \ldots, E(X_d|\mathcal{B})).$$

In this case of discrete Y, i.e., $\mathcal{B}$ is $\sigma(B_j, j \geq 1)$, we recognize the following two basic properties of $E(X|\mathcal{B})$.

(i) $E(X|\mathcal{B})$ is $\mathcal{B}$-measurable.

(ii) For any $B \in \mathcal{B}$, we have

$$\int_B E(X|\mathcal{B})(\omega) \, dP(\omega) = \int_B X(\omega) \, dP(\omega).$$

Indeed, (i) follows from the construction of $E(X|\mathcal{B})$ above. As for (ii), observe that elements in $\sigma(B_j, j \geq 1)$ are of the form $\bigcup_{j \in J} B_i$, for $J \subseteq \{1, 2, \ldots\}$. Thus, let $B = \bigcup_{j \in J} B_j$,

$$\int_B E(X|\mathcal{B}) \, dP = \int_\Omega \sum_{j \in J} E(X|B_j) I_{B \cap B_j}(\omega) \, dP(\omega)$$

$$= \int_\Omega \sum_{j \in J} E(X|B_j) I_{B_j}(\omega) \, dP(\omega)$$

$$= \sum_{j \in J} E(X|B_j) P(B_j) = \sum_{j \in J} \int_{B_j} X(\omega) \, dP(\omega)$$

$$= \int_{\cup_{j \in J} B_j} X(\omega) \, dP(\omega) = \int_B X(\omega) \, dP(\omega).$$

The above two basic properties of $E(X|\mathcal{B})$, when $\mathcal{B}$ is generated by a countable partition, will be used to define $E(X|\mathcal{B})$ for arbitrary $\mathcal{B}$.

The General Case

In the view of the two basic properties of conditional expectation of X with respect to a sub-σ-field $\mathcal{B}$ of $\mathcal{A}$ in the previous section, we consider the following general definition.

DEFINITION Let X be an integrable random variable defined on $(\Omega, \mathcal{A}, P)$ and $\mathcal{B}$ be a sub-σ-field of $\mathcal{A}$. A version of the conditional expectation of X with respect to $\mathcal{B}$ is any random variable W such that

(i) W is $\mathcal{B}$-measurable, i.e., $W^{-1}(\mathcal{B}^{-1}(\mathbb{R})) \subseteq \mathcal{B}$.

(ii) For any $B \in \mathcal{B}$,

$$\int_B W(\omega) dP(\omega) = \int_B X(\omega) \, dP(\omega)$$

and we write W as $E(X|\mathcal{B})$.

Remarks. a) A general concept of conditional expectation is needed to make sense of a situation where the observable random variable Y is of continuous type, so that $P(Y = y) = 0$ for any $y \in \mathbb{R}$. And yet, it is intuitive that $P(X \in A | Y = y)$ or more generally, $E(X|Y = y)$ should make sense.

b) The above definition for conditional expectation in the general case only stands if random variables like $E(X|\mathcal{B})$ exist! For this crucial reason, we are going to lay down some necessary mathematics to establish the existence of $E(X|\mathcal{B})$.

Let $P_\mathcal{B}$ denote the restriction of P from $\mathcal{A}$ to $\mathcal{B}$. Consider the set function

$$\nu : \mathcal{B} \to \mathbb{R}, \qquad \nu(B) = \int_B X(\omega)\, dP_\mathcal{B}(\omega).$$

Since $E(|X|) < \infty$, $\nu(\cdot)$ is finite (or bounded). However, since X takes values in $\mathbb{R}$, and so does $\nu(\cdot)$. Also, $\nu(\emptyset) = 0$ and $\nu(\cdot)$ is σ-additive. As such, $\nu(\cdot)$ is a *signed measure*. Note that $P_\mathcal{B}$ is a nonnegative, finite measure. Moreover, if $P_\mathcal{B}(B) = 0$, then $\nu(B) = 0$, i.e., ν is *absolutely continuous* with respect to $P_\mathcal{B}$, and we write $\nu \ll P_\mathcal{B}$. In the above definition of ν, X is not necessarily $\mathcal{B}$-measurable. We seek a $\mathcal{B}$-measurable random variable Y (as a candidate for $E(X|\mathcal{B})$) so that

$$\nu(B) = \int_B Y(\omega)\, dP_\mathcal{B}(\omega), \qquad \text{for any} \quad B \in \mathcal{B}.$$

It turns out that the existence and uniqueness ($P_\mathcal{B}$-almost surely) of $E(X|\mathcal{B})$ are given in the simplest version of a result in measure theory known as the *Radon-Nikodym theorem*.

RADON-NIKODYM THEOREM *Let P be a probability measure on $(\Omega, \mathcal{B})$, and ν be a finite signed measure on $\mathcal{B}$, such that $\nu \ll P$. Then there exists a random variable $Y : \Omega \to \mathbb{R}$, (i.e., $\mathcal{B}$-measurable) such that*

$$\nu(B) = \int_B Y(\omega)\, dP(\omega), \qquad B \in \mathcal{B}.$$

Any two such random variables are equal P-a.s.

A.7   Large Deviations

The techniques of large deviations can be used to investigate certain aspects of *large sample theory of statistics*. As such, we provide here a *friendly introduction* of the topic of large deviations. Interested students should also consult a standard text such as Dembo and Zeitouni [20].

A.7.1   Some motivations

Consider the following problem in quality control. In a production line, a machine produces items with a proportion p of defectives. The population is a Bernoulli random variable X taking values in $\{0,1\}$ (good, defective) with mean $E(X) = p$.

If we wish to estimate p, we take a random sample $X_1, X_2, \ldots, X_n$ from X and form the sample mean $\bar{X}_n = S_n/n$ where $S_n = X_1 + X_2 + \ldots + X_n$. The estimator $\bar{X}_n$ of p is weakly consistent by the law of large numbers, or simply by Chebyshev's inequality, i.e., for any $\varepsilon > 0$,

$$P(|\bar{X}_n - p| \geq \varepsilon) \leq \frac{p(1-p)}{n\varepsilon^2} \longrightarrow 0, \quad \text{as } n \to \infty.$$

The rate at which $\bar{X}_n \to p$ can be made more explicit as follows.

Since

$$P(|\bar{X}_n - p| \geq \varepsilon) = P(\bar{X}_n \geq p + \varepsilon) + P(\bar{X}_n \leq p - \varepsilon)$$

we will investigate the asymptotics of $P(\bar{X}_n \geq a)$ for $a > p$, and $P(\bar{X}_n \leq b)$, for $b < p$. For $a > p$, we have

$$
\begin{aligned}
\{\bar{X}_n \geq a\} &\iff \{\bar{X}_n - a \geq 0\} \\
&\iff \{t(\bar{X}_n - a) \geq 0\}, \quad \text{for any } t > 0 \\
&\iff \{\exp[t(\bar{X}_n - a)] \geq 1\}, \quad \text{for any } t > 0.
\end{aligned}
$$

Let Y denote the nonnegative random variable $e^{t(\bar{X}_n - a)}$, for a fixed $t > 0$. Then

$$P(Y \geq 1) \leq \int_{\{Y \geq 1\}} Y\, dP \leq \int_{\{Y \geq 1\}} Y\, dP + \int_{\{Y < 1\}} Y\, dP = E(Y).$$

Thus

$$
\begin{aligned}
P(\bar{X}_n \geq a) &= P\left(e^{t(\bar{X}_n - a)} \geq 1\right) \\
&\leq E\left(e^{t(\bar{X}_n - a)}\right) = e^{-at}\left[\phi\left(\frac{t}{n}\right)\right]^n,
\end{aligned}
$$

where $\phi(t) = E(e^{tX})$ is the moment generating function of X or the *Laplace transform* of the distribution F of X, i.e.,

$$\phi(t) = \int_{\mathbb{R}} e^{tx}\, dF(x), \quad t \in \mathbb{R},$$

here $\phi(t) = 1 - p + pe^t$. Now,

$$
\begin{aligned}
e^{-at}\left[\phi\left(\frac{t}{n}\right)\right]^n &= \exp\left[-at + n\log\phi\left(\frac{t}{n}\right)\right] \\
&= \exp\left[-n\left(a\frac{t}{n} - \log\phi\left(\frac{t}{n}\right)\right)\right].
\end{aligned}
$$

Since $t > 0$ is arbitrary, we can replace t by nt leading to

$$P(\bar{X}_n \geq a) \leq \exp[-n(at - \log \phi(t))], \qquad t > 0.$$

Hence

$$P(\bar{X}_n \geq a) \leq \exp\left[-n \sup_{t>0}(at - \log \phi(t))\right].$$

Observe that the function

$$g_a(t) = at - \log \phi(t) = at - \log(1 - p + pe^t)$$

is concave with $g_a(0) = 0$ and $g_a'(0) = a - p > 0$. Thus g_a attains a positive maximum $I(a)$, say, on $t > 0$, and we have

$$P(\bar{X}_n \geq a) \leq e^{-nI(a)}. \qquad (A.2)$$

Similarly, for $b < p$,

$$P(\bar{X}_n \leq b) \leq \exp\left[-n \sup_{t<0}(bt - \log \phi(t))\right].$$

Now $g_b'(0) = b - p < 0$, and hence g_b attains a positive maximum $I(b)$ on $t < 0$. Thus,

$$P(\bar{X}_n \leq b) \leq e^{-nI(b)}. \qquad (A.3)$$

Hence, let $a = p + \varepsilon$ and $b = p - \varepsilon$, we have

$$P(|\bar{X}_n - p| \geq \varepsilon) \leq 2e^{-n[I(a) \wedge I(b)]},$$

and we say that $\bar{X}_n$ converges to p at an *exponential rate*, where $x \wedge y = \min\{x, y\}$. In general, if $P(\bar{X}_n \in A) \to 0$ as $n \to \infty$, then

$$\alpha(A) = -\lim_{n \to \infty} \frac{1}{n} \log P(\bar{X}_n \in A)$$

(if existed) is the exponential rate of convergence, i.e., for n large, $\bar{X}_n \in A$ with small probability of the order of $e^{-n\alpha(A)}$. For example, let $\bar{X}_n$ be the sample mean of a random sample of size n from the standard normal population. Consider $A = (-\infty, -a] \cup [a, \infty)$, for some $a > 0$. Then $P(\bar{X}_n \in A) = P(|\bar{X}_n| \geq a)$. For each n,

$$P(|\bar{X}_n| \geq a) = 1 - P(|\bar{X}_n| < a) = 1 - \frac{1}{\sqrt{2\pi}} \int_{-a\sqrt{n}}^{a\sqrt{n}} e^{-x^2/2} dx.$$

Thus

$$-\lim_{n \to \infty} \frac{1}{n} \log P(\bar{X}_n \in A) = \frac{a^2}{2} = \alpha(A).$$

The events $\{|\bar{X}_n - p| \geq \varepsilon\}$, $\varepsilon > 0$, are *very rare events* in the sense that their probabilities decay to zero exponentially fast. They represent *large deviations* of $\bar{X}_n$ from the mean $E(X) = p$, since $\bar{X}_n$ avoids some neighborhood of $E(X)$, whereas the events $\{|\bar{X}_n - p| < \varepsilon\}$, $\varepsilon > 0$, represent small deviations of $\bar{X}_n$ from p, i.e., $\bar{X}_n$ is in some neighborhood of p.

If we denote by P_n the probability law of $\bar{X}_n$, then (A.2) and (A.3) are rewritten as

$$a > p, \qquad P_n([a, \infty)) \le e^{-nI(a)} \qquad (A.4)$$

and

$$a < p, \qquad P_n((-\infty, a]) \le e^{-nI(a)}. \qquad (A.5)$$

In fact the above inequalities hold for $a \ge p$, and $a \le p$, respectively, since $\Lambda(p) = 0$, where $\Lambda(\cdot)$ is the function defined on $\mathbb{R}$ by

$$\Lambda(x) = \sup_{t \in \mathbb{R}} \left[xt - \log \phi(t) \right].$$

This function is called the *Cramer transform* of the distribution F of X. Indeed, since for any $x \in \mathbb{R}$, $g_x(0) = 0$, we have $\Lambda : \mathbb{R} \to [0, +\infty]$. Next, for each fixed t, Jensen's inequality implies

$$E(e^{tX}) \ge e^{tEX} = e^{tp} \quad \text{or} \quad tp \le \log \phi(t)$$

so that

$$tp - \log \phi(t) \le 0, \qquad \text{for any} \quad t \in \mathbb{R}$$

implying that $\Lambda(p) \le 0$. But $\Lambda(\cdot) \ge 0$, and hence $\Lambda(p) = 0$.

Now, note that in (A.4) and (A.5), the events (in $\mathcal{B}(\mathbb{R})$) $[a, \infty)$ and $(-\infty, a]$ are *closed sets* of $\mathbb{R}$. In fact, *every* closed set F of $\mathbb{R}$ satisfies a similar *upper bound*, namely,

$$\limsup_{n \to \infty} \frac{1}{n} \log P_n(F) \le - \inf_{x \in F} \Lambda(x). \qquad (A.6)$$

Also, every *open set* G of $\mathbb{R}$ satisfies the *lower bound*

$$\liminf_{n \to \infty} \frac{1}{n} \log P_n(G) \ge - \inf_{x \in G} \Lambda(x). \qquad (A.7)$$

Clearly (A.6) and (A.7) are equivalent to

$$- \inf_{x \in A^\circ} \Lambda(x) \le \liminf_{n \to \infty} \frac{1}{n} \log P_n(A)$$

$$\le \limsup_{n \to \infty} \frac{1}{n} \log P_n(A) \le - \inf_{x \in \bar{A}} \Lambda(x), \qquad (A.8)$$

for any $A \in \mathcal{B}(\mathbb{R})$. Thus, if $A \in \mathcal{B}(\mathbb{R})$ is such that, say,

$$\inf_{x \in A^\circ} \Lambda(x) = \inf_{x \in \bar{A}} \Lambda(x) = I(A)$$

then $P_n(A)$ will decay exponentially fast with rate $I(A)$, i.e.,

$$\lim_{n \to \infty} \frac{1}{n} \log P_n(A) = -I(A).$$

Remark. In our Bernoulli example,

$$\Lambda(x) = \begin{cases} x \log\left(\dfrac{x}{p}\right) + (1-x) \log\left(\dfrac{1-x}{1-p}\right) & \text{for } x \in [0,1] \\ \infty & \text{for } x \notin [0,1] \end{cases}$$

with the convention $0 \log 0 = 0$ (Use Stirling's formula: $n! \approx \sqrt{2\pi}\, n^{n+\frac{1}{2}} e^{-n}$).

The *rate function* Λ is *convex* since it is the upper envelope of a family of linear functions. It is *lower semicontinuous* (l.s.c.), i.e., the level-sets, for $\alpha > 0$, $\{x \in \mathbb{R} : \Lambda(x) \le \alpha\}$ are *closed*, but Λ is not continuous. However, this Λ has an *additional* property which is *stronger* than l.s.c., namely, having *compact* level-sets. This follows from the fact that $0 \in D^\circ(\phi)$, where $D(\phi) = \{x \in \mathbb{R} : \phi(x) < \infty\}$, recalling

$$\phi(t) = E e^{tX} = 1 - p + p e^t,$$

so that $D(\phi) = D^\circ(\phi) = \mathbb{R}$.

In general, Λ need not be of compact level-sets. Finally, note that the exponential rate of convergence of $P(\bar{X}_n \in A)$ depends not only on A, but also on the distribution of X.

Now, let us continue with our example in quality control. Suppose we are concerned with the quality of items produced by a machine, expressed as the proportion p of defectives. To detect whether the machine functions normally (i.e., $p \le p_0$ or $p > p_0$), we take a sample $X_1, \ldots, X_n$ from the production line and look at the number of defectives $S_n = X_1 + X_2 + \ldots + X_n$. Based on this observation, we will decide to let the machine keep running or call for repair. This decision problem is formulated as the problem of testing the null hypothesis $H_0 : p = p_0$ against the alternative hypothesis $H_1 : p = p_1$ ($p_0 < p_1$). The likelihood ratio test has a rejection region of the form $S_n \ge a$. The performance of the test is determined by the error probabilities of Type I and Type II

$$\alpha_n = P(S_n \ge a | H_0) \quad \text{and} \quad \beta_n = P(S_n < a | H_1).$$

The exponential rates of α_n, β_n are of interest, and can be studied within the framework of large deviations. Note also that a measure of (asymptotic) efficiency of a test as above can be defined by examining probabilities of large deviations.

The motivation for studying probabilities of large deviations is not restricted to i.i.d. samples. In fact, large deviations is a branch of probability theory that deals with *rare events*, and as such, large deviations techniques are useful for stochastic systems in which a qualitative theory for understanding rare events is desirable.

A.7.2 Formulation of large deviations principles

To investigate large deviations in many other situations, it is necessary to formulate the problem in its most general form. The above elementary exposition provides ideas for extensions.

From the above discussions, we see that the sequence of probability measures P_n on $\mathbb{R}$ converges weakly to the Dirac measure δ_p as $n \to \infty$ (as a consequence of the weak law of large numbers). Events $A \in \mathcal{B}(\mathbb{R})$ such that $p \notin \bar{A}$ are *rare* in the sense that $P_n(A) \to 0$ as $n \to \infty$. *Very rare events* are rare events such that $P_n(A) \to 0$ exponentially fast. Thus, in general, the problem is the study of the large deviations of a sequence of probability measures (or more generally, of a family P_ε, $\varepsilon > 0$) on some measurable space $(U, \mathcal{U})$.

On finite dimensional spaces like $U = \mathbb{R}^d$, the exponential rate of convergence is typically given by the *Cramer transform*, which is defined in terms of a reference measure on $\mathbb{R}^d$ (Lebesgue measure). Large deviations of sample paths of stochastic processes, such as Brownian motions, require infinite dimensional spaces, like $C[0, 1]$. In order to cover all cases, a general large deviations principle should be formulated *without reference to any reference measure*. In view of the properties of the Cramer transform and the result (A.8), the abstraction is this.

LARGE DEVIATIONS PRINCIPLE *Let U be a polish space (i.e., a complete, separable metric space), and $\mathcal{U}$ its Borel σ-field. A rate function is a function $I : U \to [0, \infty]$, which is lower semicontinuous (l.s.c.). A family $\{P_\varepsilon, \varepsilon > 0\}$ of probability measures on $\mathcal{U}$ is said to satisfy the large deviations principle with rate function I if, for all $A \in \mathcal{U}$,*

$$- \inf_{A^\circ} I \le \liminf_{\varepsilon \to 0} \varepsilon \log P_\varepsilon(A) \le \limsup_{\varepsilon \to 0} \varepsilon \log P_\varepsilon(A) \le - \inf_{\bar{A}} I. \qquad (A.9)$$

Remark. For a sequence of probability measures, we have $\varepsilon = 1/n$. If, in addition, the rate function I is assumed to have *compact level-sets*, then $\inf_{u \in U} I(u) = 0$ is attained at some u_0 since then I attains its infimum on a *closed* set.

As stated earlier, (A.9) is equivalent to

$$\limsup_{\varepsilon \to 0} \varepsilon \log P_\varepsilon(F) \le -I(F), \qquad \text{for} \quad F \text{ closed} \qquad (A.10)$$

and

$$\liminf_{\varepsilon \to 0} \varepsilon \log P_\varepsilon(G) \ge -I(G), \qquad \text{for} \quad G \text{ open}, \qquad (A.11)$$

where $I(A) = \inf_{u \in A} I(u)$.

Now, let us take a closer look at (A.10) and (A.11). The set function τ on $\mathcal{U}$, defined by $\tau(A) = e^{-I(A)}$ is an *idempotent probability*, i.e.,

$$\tau(A) = \sup_{u \in A} \tau(\{u\}), \qquad \tau(\{u\}) = \tau(u) = e^{-I(u)},$$

which is *upper semicontinuous* (u.s.c.). Next, $\{P_\varepsilon^\varepsilon, \; \varepsilon > 0\}$ is a family of *subprobability measures*. Rewrite (A.10) and (A.11) as

$$\limsup_{\varepsilon \to 0} [P_\varepsilon(F)]^\varepsilon \leq \tau(F) \tag{A.12}$$

and

$$\liminf_{\varepsilon \to 0} [P_\varepsilon(G)]^\varepsilon \geq \tau(G) \tag{A.13}$$

and take this as a *definition* for the convergence of the family of subprobability measures $\{P_\varepsilon^\varepsilon, \; \varepsilon > 0\}$ to the idempotent probability τ. Then, clearly, (A.12) and (A.13) remind us of the *weak convergence of probability measures*, in which, if $P_\varepsilon^\varepsilon$, τ are replaced by probability measures, then (A.12) and (A.13) are equivalent. Thus, the above concept of convergence of subprobabilities to idempotent probabilities (which is another way of stating the large deviations principle (LDP)) is a generalization of weak convergence of probability measures.

Note that u.s.c. "densities" of idempotent probability measures are *rate functions* for LDP. Such idempotent probability measures are capacity functionals of random closed sets on $\mathbb{R}^d$. By *subprobability measures*, we mean here set functions $\nu : \mathcal{B}(U) \to [0,1]$ such that

a) $\nu(\emptyset) = 0$.

b) $\nu(A) = \inf\{\nu(G) : A \subseteq G \text{ open}\}$, for any $A \in \mathcal{B}(U)$.

c) $\nu\left(\bigcup_{n=1}^{\infty} A_n\right) \leq \sum_{n=1}^{\infty} v(A_n)$.

d) For any open G, $\nu(G) = \lim_{\delta \to 0} v(G^{-\delta})$, where $G^{-\delta} = ((G^c)_\delta)^c$ with $A_\delta = \{u \in U : \rho(u, A) < \delta\})$.

Note that ν is monotone increasing in view of b). Also, the space $\mathcal{M}(U)$ of all subprobability measures contains all idempotent probability measures (also called *supmeasures*) and set functions of the form P^ε with $\varepsilon \in (0,1]$ and P a probability measure. See Dembo and Zeitouni [20]. The space $\mathcal{M}(U)$ is relevant to *capacity functionals* of *random closed sets* in locally compact spaces. Below are some examples of LDP.

The Finite Case

Large deviations for sample means in a Bernoulli population is an example of LDP in the finite case. Here is another one. Let $U = \{u_1, u_2, \ldots, u_k\}$ be a finite set with $|U| = k$. The space of all probability measures on U is identified with the simplex S_k of $\mathbb{R}^k$. A LDP for a sequence of probability measures $\{P_n, \, n \geq 1\}$ on $(S_k, \mathcal{B}(S_k))$ is given by Sanov as follows.

Let $X : (\Omega, \mathcal{A}, P) \to U$ be a random variable with $P_X = \tau \in S_k$. Let $\{X_n, \, n \geq 1\}$ be a sequence i.i.d. random variables distributed as X. Let $Y_n : \Omega \to S_k$ be defined as

$$\mathbf{Y}_n(\omega) = \begin{pmatrix} f(u_1, \omega) \\ \vdots \\ f(u_k, \omega) \end{pmatrix}$$

where $f(u, \omega) = $ fraction of u in $X_1(\omega), \dots, X_n(\omega)$. Let P_n be the probability law of Y_n on $\mathcal{B}(S_k)$. Then $\{P_n, \, n \geq 1\}$ satisfies the LDP with rate function $I : S_k \to [0, \infty]$ given by

$$I(\nu) = H(\nu|\tau) = \sum_{j=1}^{k} \nu(u_j) \log \frac{\nu(u_j)}{\tau(u_j)}.$$

Note that $H(\nu|\tau)$ is the *relative entropy* of ν with respect to τ. It can be checked that $I(\cdot)$ is indeed a rate function, and in fact $I(\cdot)$ has compact level-sets.

$\mathbb{R}^d$ **Case**

Cramér's large deviations principle for $\mathbb{R}$ can be extended to $\mathbb{R}^d$. The corresponding rate function is now called the *Fenchel-Legendre transform*. Specifically, let X be a random vector with values in $\mathbb{R}^d$, with distribution function F. The *Laplace transform* of dF is

$$\phi(t) = E\left(e^{\langle \mathbf{t}, \mathbf{X} \rangle}\right) = \int_{\mathbb{R}^d} \exp\left(\sum_{j=1}^{d} t_j x_j\right) dF(\mathbf{x}),$$

where $\mathbf{x} = (x_1, \dots, x_d)$, $\mathbf{t} = (t_1, \dots, t_d) \in \mathbb{R}^d$. The Fenchel-Legendre transform of dF is

$$\Lambda(\mathbf{t}) = \sup_{\mathbf{x} \in \mathbb{R}^d} \left[\langle \mathbf{t}, \mathbf{x} \rangle - \log \phi(t)\right].$$

For example, let S be a *random set* with values in $\mathcal{P}(U)$, U finite, and having capacity functional

$$T(A) = P(S \cap A \neq \emptyset), \qquad \text{for} \quad A \subseteq U.$$

Consider the associated *random* vector $\mathbf{X}$ with values in $\mathbb{R}^d$, with $d = |U|$, given by

$$X = \begin{pmatrix} I_{(S \cap A_1 \neq \emptyset)} \\ \vdots \\ I_{(S \cap A_d \neq \emptyset)} \end{pmatrix}$$

where the A_j's, $j = 1, \ldots, d = |U|$, are all the subsets of U. If we identify each map $h : \mathcal{P}(U) \to \mathbb{R}$ with a vector in $\mathbb{R}^d$, namely,

$$\begin{pmatrix} h(A_1) \\ \vdots \\ h(A_n) \end{pmatrix} = \begin{pmatrix} t_1 \\ \vdots \\ t_d \end{pmatrix} = \mathbf{t},$$

then we can consider the Laplace transform of $\mathbf{X}$ as a function defined on maps h

$$\psi_S(h) = \phi_X(t) = E\left(e^{\langle t, \mathbf{X} \rangle}\right)$$

$$= E\left[\exp\left(\sum_{j=1}^{d} h(A_j)I_{(A_j \cap S \neq \emptyset)}\right)\right]$$

$$= E\left[\exp\left(\sum_{A \cap S \neq \emptyset} h(A)\right)\right].$$

Also, by identifying the capacity functional T with the vector $(T(A), A \subseteq U)$ in $\mathbb{R}^d$, the Fenchel-Legendre transform of X is written as

$$\Lambda_S(T) = \sup_{h:2^U \to \mathbb{R}} \left[\sum_{A \subseteq U} h(A)T(A) - \log \psi_S(h)\right].$$

The Infinite Dimensional Case

Consider the Brownian motion $W(t)$, $t \in [0, 1]$, with paths in the space $C_0[0, 1]$ of continuous functions vanishing at zero. The space $C_0[0, 1]$ is topologized by the supremum norm as usual. In the study of random perturbations of dynamical systems one wishes to investigate the convergence of the solution of the perturbed system to the solution of the unperturbed system as the effect of the perturbation decreases. For example, consider the perturbed diffusion process expressed by the stochastic differential equation

$$dX(t, \varepsilon) = b(X(t, \varepsilon))dt + \sqrt{\varepsilon}\, dW_t,$$

for $t \in [0, 1]$, $x(0, \varepsilon) = 0$, $\varepsilon > 0$. The process $(\sqrt{\varepsilon}\, W_t, t \in [0, 1])$ is normal with covariance function $c_\varepsilon(t, s) = \varepsilon(t \wedge s)$. Let P_ε be the law governing this process, i.e., the probability measure (Wiener) on $C_0[0, 1]$ (with its Borel σ-field) determined by its covariance function. Since P_ε converges weakly, as $\varepsilon \to 0$, to δ_0, where 0 denotes the function in $C_0[0, 1]$, which is identically zero, one wishes to study the probabilities of large deviations of Brownian motion sample paths, i.e., LDP for the family $(P_\varepsilon, \varepsilon > 0)$ on $C_0[0, 1]$. An LDP exists with a rate function I given by

$$I : C_0[0,1] \rightarrow [0,\infty], \quad I(f) = \frac{1}{2} \int_0^1 (f'(t))^2 dt$$

for f absolutely continuous with a square integrable derivative $f'(t)$, otherwise $I(f) = \infty$.

A.7.3 Large deviations techniques

We have addressed the questions "What is the problem of large deviations?" and "Why do we need to study large deviations?" Now it is time to turn to the question "How to study large deviations?"

Let U be a polish space with its Borel σ-field $\mathcal{U}$. Probability measures on $\mathcal{U}$ are probability laws of random elements, defined on some $(\Omega, \mathcal{A}, P)$, with values in U, i.e., if $X_n : \Omega \rightarrow U$, then its probability law P_n is defined by

$$P_n(A) = P(X_n \in A), \quad \text{for} \quad A \in \mathcal{U}.$$

The *problem of large deviations* focuses on $\{X_n, \ n \geq 1\}$ or $\{P_n, \ n \geq 1\}$ for which $P_n(A) \rightarrow 0$, as $n \rightarrow \infty$, exponentially fast, with some rate I, for a class of Borel sets A in $\mathcal{U}$. Thus, the problem consists of establishing the *existence* of a LDP for $\{P_n, \ n \geq 1\}$, followed by the *determination* (identification) of the rate function I. Note that the rate function is unique. Below is a panorama of *tools* for achieving the above.

As in the case of weak convergence of probability measures, it is more convenient to deal with functions rather than with sets (via the Portmanteau theorem). The following *necessary condition* of LDP is crucial.

VARADHAN'S THEOREM *If* $\{X_n, \ n \geq 1\}$ *satisfies the LDP with a rate function* I *having compact level-sets, then for any bounded continuous function* h *on* U,

$$\lim_{n\to\infty} \frac{1}{n} \log E\left[e^{nh(X_n)}\right] = \sup_{u \in U}[h(u) - I(u)].$$

Remark. By taking $h \equiv 0$, we see that $\inf_{u \in U} I(u) = I(u_0) = 0$ for some $u_0 \in U$. Thus, in studying LDP for $\{P_n, \ n \geq 1\}$, we need to be able to evaluate, for all bounded continuous functions h, the asymptotics of

$$\frac{1}{n} \log \int_U e^{nh(u)} dP_n(u) \quad \text{as} \quad n \rightarrow \infty.$$

The existence of LDP for $\{P_n, \ n \geq 1\}$ is a consequence of the existence of this limit *and* some "compactness" condition.

The "compactness" condition for $\{P_n,\ n \geq 1\}$ is expressed as follows. The sequence $(P_n,\ n \geq 1)$ is said to be *exponentially tight* if for each $0 < \alpha < \infty$, there is a compact set K_α of U such that

$$\limsup_{n \to \infty} \frac{1}{n} \log P_n(K'_\alpha) \leq -\alpha.$$

Note that exponential tightness is *not* a necessary condition for LDP, i.e., $\{P_n,\ n \geq 1\}$ might not need to be exponentially tight in order to satisfy an LDP with a rate function having compact level-sets.

A form of the converse to Varadhan's theorem provides *a sufficient condition* for establishing LDP.

BRYC'S THEOREM If $\{P_n,\ n \geq 1\}$ is exponentially tight and

$$\lim_{n \to \infty} \frac{1}{n} \log \int_U e^{hh(u)} dP_n(u) = L(h)$$

exists for every $h \in \mathcal{C}_b(U)$, then $\{P_n,\ n \geq 1\}$ satisfies the LDP with the rate function I (having compact level-sets) given by

$$I(u) = \sup_{h \in \mathcal{C}_b(U)} [h(u) - L(h)].$$

Moreover,

$$L(h) = \sup_{u \in U} [h(u) - I(u)].$$

The following *contraction principle* is useful, for example, to bring down LDP on complicated spaces to simpler ones.

CONTRACTION PRINCIPLE If $\{P_n,\ n \geq 1\}$ satisfies the LDP with a rate function I having compact level-sets, and $f : U \to V$ (another polish space) continuous, then

(i) $J : V \to [0, \infty]$, defined by $J(v) = \inf\{I(u) : u \in f^{-1}(v)\}$ is a rate function on V, having compact level-sets,

(ii) $\{P_n f^{-1},\ n \geq 1\}$ satisfies the LDP on V with rate function J.

In the opposite direction, a method for establishing LDP on complicated spaces from simpler ones is *projective limits*. For *large deviations for projective limits*, see e.g., Dembo and Zeitouni [20].

References

[1] Aigner, M., *Combinatorial Theory*, Springer Verlag, New York, 1997.

[2] Akhmerov, R.R. et al., *Measures of Non-Compactness and Condensing Operators*, Birkhauser, Basel, Boston, 1992.

[3] Akian, M., Densities of idempotent measures and large deviations, *Trans. Amer. Math. Soc.*, 351, 4515–4543, 1999.

[4] Artstein, Z., Distributions of random sets and random selections, *Israel J. Math.*, 46, 313–324, 1983.

[5] Artstein, Z. and Vitale, R., A strong law of large numbers for random compact sets, *Ann Probab.*, 3, 879–882, 1975.

[6] Aumann, R.J., Intervals of set-valued functions, *J. Math. Analysis and Appl.*, 12, 1–12, 1965.

[7] Banas, J. and Goebel, K., *Measures of Non-Compactness in Banach Spaces*, Lecture Notes in Pure and Applied Math., Vol. 60, Marcel Dekker, New York, 1980.

[8] Bednarski, T., Binary experiments, minimax tests and 2-alternating capacities, *Ann. Statist.*, 10, 226–232, 1982.

[9] Beer, G., *Topologies on Closed and Closed Convex Sets*, Kluwer, Dordrecht, The Netherlands, 1993.

[10] Billingsley, P., *Convergence of Probability Measures*, J. Wiley, New York, 1968.

[11] Bollobas, B. and Varopoulos, N.Th., Representation of systems of measurable sets, *Math. Proc. Camb. Phil. Soc.*, 78, 323–325, 1975.

[12] Bosq, D., and Nguyen, H.T., *A Course in Stochastic Processes: Stochastic Models and Statistical Inference*, Kluwer Academic, Dordrecht, The Netherlands, 1996.

[13] Cassell, C.M., Sarndal, C.E., and Wretman, J.H., *Foundations of Inference in Survey Sampling*, J. Wiley, New York, 1977.

[14] Chateauneuf, A. and Jaffray, J.Y., Some characterization of lower probabilities and other monotone capacities through the use of Mobius inversion, *Math. Social Sciences*, 17, 263–283, 1989.

[15] Choquet, G., Theory of capacities, *Ann. Inst. Fourier*, V, 131–295, 1953–1954.

[16] Cressie, N., A central limit theorem for random sets, *Z. Wahrs. Geb.*, 49, 37–47, 1978.

[17] Cressie, N. and Laslett, G. M., Random set theory and problems of modeling, *SIAM Review*, 29, 557–574, 1987.

[18] Debreu, G., Integration of correspondences. In: *Proceedings of the Fifth Berkeley Symposium Math. Statistics and Probability*, Vol. 2, University of California Press, 351–372, 1967.

[19] Dellacharie, C., Quelques commentaires sur le prolongement de capacités, in *Lecture Notes in Mathematics*, Vol. 191, Springer-Verlag, Berlin, 77–81, 1971.

[20] Dembo, A. and Zeitouni, O., *Large Deviations Techniques and Applications*, Springer-Verlag, New York, 1998.

[21] Dempster, A.P., Upper and lower probabilities induced from a multivalued mapping, *Ann. Math. Statist.*, 38, 325–339, 1967.

[22] Diamond, P. and Kloeden, P., *Metric Spaces of Fuzzy Sets*, World Scientific, Singapore, 1994.

[23] Dubois, D. and Prade, H., *Possibility Theory*, Plenum, New York, 1988.

[24] Dudley, R.M., *Real Analysis and Probability*, Wadsworth & Brooks/Cole, Belmon, 1989.

[25] Dupuis, P. and Ellis, R.S., *A Weak Convergence Approach to the Theory of Large Deviations.* J. Wiley, New York, 1997.

[26] Durett, R., *Probability: Theory and Examples*, Duxbury Press, Belmont, 1996.

[27] Dutta, B. and Ray, D., A concept of egalitarianism under participation constraints, *Econometrika*, 57(2), 615–635, 1989.

[28] Engelking, R., *General Topology*, Polish Scientific Publ., Warszawa, 1977.

[29] Ethier, S.N. and Kurtz, T.G., *Markov Processes: Characterization and Convergence*, Wiley, New York, 1986.

[30] Everett, C.J. and Whaples, G., Representatives of sequences of sets, *Am. J. Math.*, 71, 287–293, 1949.

[31] Fagin, R. and Halpern, J.Y., Uncertainty, belief, and probability, *Computational Intelligence*, 7, 160–173, 1991.

[32] Falconer, K., *Techniques in Fractal Geometry*, J. Wiley, New York, 1997.

[33] Feng, De-Jun and Feng, D., On a statistical framework for estimation from random set observations, *J. Theoretical Prob.*, 17, 85–110, 2004.

[34] Feng, De-Jun and Nguyen, H.T., On Choquet weak convergence of capacity functionals, *Proceedings of the Fourth International Conference on Intelligent Technologies*, Chiang Mai, Thailand, 2003, pp. 473–477.

[35] Fréchet, M., Les éléments aléatoires de nature quelque dans un espace distancié, *Ann. Inst. Poincaré*, IV, 215–310, 1948.

[36] Gill, R.D., Van der Laan, M.J., and Robins, J.M., Coarsening at random: characterizations, conjectures, and counterexamples, in *Lecture Notes in Statistics*, Vol. 123, Springer-Verlag, 255–294, 1997.

[37] Girotto, B. and Holzer, S., Weak convergence of bounded, monotone set functions in an abstract setting, *Real Analysis Exchange*, 26(1), 157–176, 2000–2001.

[38] Goodman, I.R., Fuzzy sets as equivalence classes of random sets, in *Fuzzy Sets and Possibility Theory*, Yager, R. et al., Eds., Pergamon Press, New York, 1982.

[39] Goodman, I.R., Nguyen, H.T., and Rogers, G.S., On the scoring approach to admissibility of uncertainty measures in expert systems, *J. Math. Anal. and Appl.*, 159, 550–594, 1991.

[40] Goutsias, J., Morphological analysis of random sets: an Introduction, In: [41], 3–26.

[41] Goutsias, J., Mahler, R., and Nguyen, H.T., Eds., *Random Sets: Theory and Applications*, Springer-Verlag, New York, 1997.

[42] Grabisch, M., Nguyen, H.T., and Walker, E.A., *Fundamentals of Uncertainty Calculi with Applications to Fuzzy Inference*, Kluwer Academic, Dordrecht, The Netherlands, 1994.

[43] Graf, S., A Radon-Nikodym theorem for capacities, *J. Reine Ang. Math.*, 320, 192–214, 1982.

[44] Guiasu, S., *Information Theory with Applications*, McGraw-Hill, New York, 1977.

[45] Hajek, K., *Sampling from a Finite Population*, Marcel Dekker, New York, 1981.

[46] Hall, P., On representatives of subsets, *J. London Math. Soc.*, 10, 26–30, 1935.

[47] Halmos, P.R., *Measure Theory*, Springer-Verlag, Berlin-Heidelberg-New York, 1974.

[48] Halmos, P. and Vaughan, H.E., The marriage problem, *Amer. J. Math.*, 10, 214, 215, 1950.

[49] Halpern, J.Y., *Reasoning about Uncertainty*, MIT Press, Cambridge, Massachusetts, 2003.

[50] Hartigan, J.A., Estimation of a convex density contour in two dimensions, *J. Amer. Statist. Assoc.*, 82, 267–270, 1987.

[51] Heitjan, D.F. and Rubin, D.B., Ignorability and coarse data, *Ann. Statist.*, 19, 2244–2253, 1991.

[52] Hiriart-Urruty, J.B. and Lemaréchal, C., *Fundamentals of Convex Analysis*, Springer-Verlag, Berlin, 2001.

[53] Huber, P.J., The use of Choquet capacities in statistics, *Bull. Inst. Intern. Stat.*, XLV, book 4, 181–188, 1973.

[54] Huber, P.J. and Strassen, V., Minimax tests and the Neyman-Pearson lemma for capacities, *Ann. Statist.*, 1, 251–263, 1973.

[55] Jaffray, J.Y., On the maximum of conditional entropy for upper/lower probability generated by random sets. In: Goutsias, J., Mahler, R., and Nguyen, H.T., Eds., *Random Sets: Theory and Applications*, Springer-Verlag, New York, 107–127, 1997.

[56] Jaynes, E.T., Information theory and statistical mechanics, *Phys. Rev.*, 106, 620–630; 109, 171–182, 1957.

[57] Jaynes, E.T., Where do we stand on maximum entropy? In: Levine, R.D. and Tribus, M., Eds., *The Maximum Entropy Formalism*, MIT Press, Cambridge, Massachusetts, 1979, 15–118.

[58] Kampé de Fériet, J., Interpretation of membership functions of fuzzy sets in terms of plausibility and belief, in: Gupta, M. and Sanchez, E., Eds., *Fuzzy Information and Decision Processes*, North Holland, Amsterdam, 1982, 93–98.

[59] Keeney, R.L. and Raifa, H., *Decisions with Multiple Objectives: Preferences and Value Tradeoffs*, J. Wiley, New York, 1976.

[60] Kelley, J.L., *General Topology*, Van Nostrand, Princeton, 1955.

[61] Kendall, D.G., Foundations of a theory of random sets. In *Stochastic Geometry*, Harding, E.F. and Kendall, D.G., Eds., J. Wiley, New York, 322–376, 1974.

[62] Klein, E. and Thompson, A.C., *Theory of Correspondences*, J. Wiley, New York, 1984.

[63] Kuratowski, C., Sur les spaces complets, *Fund. Math.*, 15, 301–309, 1930.

[64] Li, B. and Wang, T., On consistency of estimators based on random set observations. In *Proceedings InTech'04*, University of Houston-Downtown, Houston, Texas, December 2004.

[65] Li, B. and Wang, T., Computational aspects of the CAR model and the Shapley value, *Information Sciences*, to appear.

[66] Li, S., Ogura, Y., and Kreinovich, V., *Limit Theorems and Applications of Set Valued and Fuzzy Valued Random Variables*, Kluwer Academic Publishers, Dordrecht, The Netherlands, 2002.

[67] Li, X., and Yong, J., *Control Theory of Distributed Parameter Systems and Applications*, Springer Lecture Notes in Control and Information Sciences, Vol. 159, Springer Verlag, Berlin, 1991.

[68] Lindley, D.V., Scoring rules and the inevitability of probability, *Intern. Statist. Rev.*, 50, 1–267, 1982.

[69] Little, R.J.A. and Rubin, D.B., *Statistical Analysis with Missing Data*, J. Wiley, New York, 1987.

[70] Mammen, E. and Tsybakov, A.B., Asymptotic minimax recovery of sets with smooth boundaries, *Ann. Statist.* 23(2), 502–524, 1995.

[71] Marinacci, M., Decomposition and representation of coalitional games, *Math. Oper. Res.*, 21, 1000–1015, 1996.

[72] Marinacci, M. and Montrucchio, L., Introduction to the Mathematics of Ambiguity. In: Gilboa, I., Ed., *Uncertainty in Economic Theory*, Routledge, New York, 46–107, 2004.

[73] Mathéron, G., *Random Sets and Integral Geometry*, J. Wiley, New York, 1975.

[74] Meyer, P.A., *Probabilités et Potentiel*, Hermann, Paris, 1966.

[75] Meyerowitz, A., Richman, F., and Walker, E.A., Calculating maximum entropy densities for belief functions, *International J. Uncert., Fuzziness, and Knowledge-Based Systems*, 2, 377–389, 1994.

[76] Michael, E., Topologies in spaces of subsets, *Trans. Amer. Math. Soc.*, 71, 152–182, 1951.

[77] Molchanov, I.S., *Limit Theorems for Unions of Random Sets*, Lecture Notes in Mathematics, Vol. 1561, Springer-Verlag, Berlin, 1993.

[78] Molchanov, I.S., *Statistics of the Boolean Model for Practitioners and Mathematicians*, J. Wiley, New York, 1997.

[79] Molchanov, I.S., *Theory of Random Sets*, Springer Verlag, Berlin, 2005.

[80] Nelsen, R.B., *An Introduction to Copulas*, Lecture Notes in Computer Science, Vol. 139, Springer-Verlag, 1999.

[81] Neveu, J., *Mathematical Foundations of the Calculus of Probability*, Holden-Day, San Francisco, California, 1965.

[82] Nguyen, H.T., Sur les mesures d'information de type Inf, *Springer-Verlag Lecture Notes in Mathematics*, 398, 62–74, 1974.

[83] Nguyen, H.T., On random sets with belief functions, *J. Math. Anal. and Appl.*, 65, 531–542, 1978.

[84] Nguyen, H.T., Survey sampling revisited and coarse data analysis, *Thai Statistical Jornal*, 2, 1–19, 2004.

[85] Nguyen, H.T. and Kreinovich, V., How to divide a territory? A new simple differential formalism for optimization of set-functions, *Intern. J. Intell. Systems*, 14, 223–251, 1999.

[86] Nguyen, H.T. and Nguyen, N., A negative version of Choquet theorem for Polish spaces, *East-West J. Math.*, 1, 61–71, 1998.

[87] Nguyen, H.T., Prasad, R., and Walker, E.A., *A First Course in Fuzzy and Neural Control*, CRC Press, Boca Raton, Florida, 2002.

[88] Nguyen, H.T. and Rogers, G.S., *Fundamentals of Mathematical Statistics*, Vols. I and II, Springer-Verlag, New York, 1989.

[89] Nguyen, H.T. and Walker, E.A., *A First Course in Fuzzy Logic*, 3rd ed., Chapman and Hall/CRC, Boca Raton, Florida, 2005.

[90] Nguyen, H.T. and Wang, T., *A First Course in Probability and Statistics*, Vols. I and II, Tsinghua University Press, Beijing, 2006.

[91] Nguyen, H.T. and Wu, B., *Fundamentals of Fuzzy Statistics*, Springer Verlag, Berlin, 2006.

[92] Nolan, D., The excess-mass ellipsoid, *J. Multivariate Anal.*, 39, 348–371, 1991.

[93] Norberg, T., Convergence and existence of random set distributions, *Ann. Prob.*, 12, 726–732, 1984.

[94] Norberg, T., Random capacities and their distributions, *Prob. Theory and Random Fields*, 73, 281–197, 1986.

[95] Norberg, T., Semicontinuous processes in multidimensional extreme value theory, *Stochastic Process. Appl.* 25(1), 27–55, 1987.

[96] Norberg, T., On the existence of ordered couplings of random sets with applications, *Israel J. Math.*, 77, 241–264, 1992.

[97] Owen, G., *Game Theory*, Saunders Co., Philadelphia, 1968.

[98] Parthasarathy, K.R., *Probability Measures on Metric Spaces*, Academic Press, New York, 1967.

[99] Pfanzagl, J., *Theory of Measurement*, Physica Verlag, Würzburg–Vienna, 1971.

[100] Polonik, W., Density estimation under qualitative assumptions in higher dimensions, *Journ. Multi. Analysis*, 55, 61–81, 1995.

[101] Polonik, W., Measuring mass concentration and estimating density contour clusters: an excess mass approach, *Ann. Statist.*, 23, 855–881, 1995.

[102] Puhalskii, A., *Large Deviations and Idempotent Probability*, Chapman and Hall/CRC Press, Boca Raton, 2001.

[103] Reiss, R.D., *A Course on Point Processes*, Springer-Verlag, New York, 1993.

[104] Ripley, B., Locally finite random sets: foundations for point process theory, *Ann Prob.*, 4, 983–994, 1976.

[105] Robbins, H.E., On the measure of a random set, *Ann. Math, Statist.*, 14, 70–74, 1944.

[106] Rota, G.C., Theory of Mobius functions, *Z. Wahrs. Geb.*, 2, 340–368, 1964.

[107] Saaty, T.L., *Fundamentals of Decision Making and Priority Theory, with the Analytic Hierarchy Process*, RWS Publications, 1994.

[108] Salinetti, G. and Wets, J.B., On the convergence of sequences of convex sets in finite dimension, *SIAM Review*, 21, 18–33, 1979.

[109] Salinetti, G. and Wets, J.B., On the convergence of closed-valued measurable multifunctions, *Transactions of the American Math. Society*, 266 (1), 1981.

[110] Salinetti, G. and Wets, J.B., On the convergence in distribution of measurable multifunctions (random sets), normal integrands, stochastic processes, and stochastic infima, *Math. Oper. Res.*, 11, 385–422, 1986.

[111] Sarndal, C.E., Swensson, B., and Wretman, J., *Model Assisted Survey Sampling*, Springer-Verlag, Berlin, Heidelberg, 1992.

[112] Schneidler, D., Integral representation without additivity, *Proceedings Amer. Math. Soc.*, 97, 255–261, 1986.

[113] Schreiber, T., Statistical inference from set-valued observations, *Prob. Math. Statist.*, 20, 223–235, 2000.

[114] Shafer, G., *A Mathematical Theory of Evidence*, Princeton University Press, Princeton, 1976.

[115] Shapley, L.S., Cores of convex games, *Intern. J. Game Theory*, 1, 11–26, 1971.

[116] Shilov, G.E. and Gurevich, B.L., *Integral, Measure and Derivative: A Unified Approach*, Dover, New York, 1977.

[117] Spiegel, E. and O'Donnell, C.J., *Incidence Algebras*, Marcel Dekker, New York, 1997.

[118] Strassen, V., The existence of probability measures with given marginals, *Ann. Math. Statist.*, 36, 423–439, 1965.

[119] Sugeno, M., Theory of fuzzy integrals and its applications, Ph.D. dissertation, Tokyo Institute of Technology, 1974.

[120] Tsybakov, A.B., On nonparametric estimation of density level sets, *Ann. Statist.*, 25, 948–969, 1997.

[121] Van der Vaart, A.W. and Wellner, J.A.. Preservation theorems for Glivanko-Cantelli and uniform Glivanko-Cantelli classes. In: Gine, E., Mason, D., and Werllner, J.A., Eds., *High Dimensional Probability II*, Birkhauser, Boston, 2000, pp. 113–132.

[122] Van der Laan, M.J. and Robins, J.M., *Unified Methods for Censored Longitudinal Data and Causality*, Springer-Verlag, New York, 2003.

[123] Vervaat, W., Narrow and vague convergence of set-functions, *Statist. and Prob. Letters*, 6, 295–298, 1988.

[124] Wagner, D.H., Survey of measurable selection theorems, *SIAM J. Control and Opt.*, 15, 859–903, 1977.

[125] Wagner, D.H., Survey of measurable selection theorems: an update, *Lecture Notes in Math.*, 749, 176–219, 1979.

[126] Walters, P., *An Introduction to Ergodic Theory*, Springer Verlag, New York, 1982.

[127] Wasserman, L., Prior envelopes based on belief functions, *Ann. Statist.*, 18, 454–464, 1990.

[128] Zadeh, L.A., Fuzzy sets as a basis for a theory of probability, *Fuzzy Sets and Systems*, 1, 3–28, 1978.

[129] Zadeh, L.A., Toward a perception-based theory of probabilistic reasoning with imprecise probabilities, *J. Statistic. Planning and Inference*, 105, 233–264, 2002.

Index